Eugenics

Recent Titles in Human Evolution, Behavior, and Intelligence

The Evolution of Love
Ada Lampert

The G Factor: The Science of Mental Ability
Arthur R. Jensen

Sex Linkage of Intelligence: The X-Factor
Robert Lehrke

Separation and Its Discontents: Toward an Evolutionary Theory of
  Anti-Semitism
Kevin MacDonald

The Biological Origins of Art
Nancy E. Aiken

The Culture of Critique: An Evolutionary Analysis of Jewish
  Involvement in Twentieth-Century Intellectual and Political
  Movements
Kevin MacDonald

Relating in Psychotherapy: The Application of a New Theory
John Birtchnell

The Evolution of the Psyche
D. H. Rosen and M. C. Luebbert, editors

Mind and Variability: Mental Darwinism, Memory, and Self
Patrick McNamara

The Darwinian Heritage and Sociobiology
Johan M.G. van der Dennen, David Smillie, and Daniel R. Wilson

The Culture of Sexism
Ignacio L. Götz

Evolution as Natural History: A Philosophical Analysis
Wim J. van der Steen

Eugenics
A Reassessment

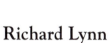

Richard Lynn

Human Evolution, Behavior, and Intelligence
Seymour W. Itzkoff, Series Editor

Westport, Connecticut
London

Library of Congress Cataloging-in-Publication Data

Lynn, Richard.
    Eugenics : a reassessment / Richard Lynn.
       p.   cm. — (Human evolution, behavior, and intelligence, ISSN 1063-2158)
    Includes bibliographical references and index.
    ISBN 0-275-95822-1 (alk. paper)
    1. Eugenics. I. Title. II. Series.

HQ751 .L9 2001
363.9'2—dc21   00-052459

British Library Cataloguing in Publication Data is available.

Library of Congress Catalog Card Number: 00-052459
ISBN: 0-275-95822-1
ISSN: 1063-2158

First published in 2001

Praeger Publishers, 88 Post Road West, Westport, CT 06881
An imprint of Greenwood Publishing Group, Inc.
www.praeger.com

Printed in the United States of America

∞™

The paper used in this book complies with the
Permanent Paper Standard issued by the National
Information Standards Organization (Z39.48-1984).

10 9 8 7 6 5 4 3 2 1

Contents

Preface: The General Theory of Eugenics   vii

Part I. Historical Introduction

1. Sir Francis Galton Lays the Foundations of Eugenics   3
2. The Rise and Fall of Eugenics   18

Part II. The Objectives of Eugenics

3. Historical Formulations   47
4. Genetic Diseases and Disorders   59
5. Mental Illness   72
6. Intelligence   78
7. Mental Retardation   97
8. Personality   108
9. Psychopathic Personality   116

Part III. The Implementation of Classical Eugenics

10. The Genetic Foundations of Eugenics   137
11. The Genetic Principles of Selection   150
12. Negative Eugenics: Provision of Information and Services   165
13. Negative Eugenics: Incentives, Coercion, and Compulsion   187
14. Licenses for Parenthood   205
15. Positive Eugenics   215
16. The Ethical Principles of Classical Eugenics   225

Part IV. The New Eugenics

17. Developments in Human Biotechnology 245

18. Ethical Issues in Human Biotechnology 258

19. The Future of Eugenics in Democratic Societies 274

20. The Future of Eugenics in Authoritarian States 292

21. The Evolution of the Eugenic World State 307

References 321

Index 355

~

Preface: The General Theory of Eugenics

During the course of the twentieth century a profound change took place in scientific and public attitudes to eugenics. In the first half of the century, virtually all biological scientists and most social scientists supported eugenics, and so also did many of the informed public. In the second half of the century, support for eugenics declined; and in the last three decades of the century, eugenics became almost universally rejected. In the history of science there is nothing particularly unusual in the rejection of a scientific theory. This has happened frequently as theories have come to be seen as incorrect and have been discarded. What is unusual is the rejection of a theory that is essentially correct. It is my objective in this book to establish that this is what occurred in the twentieth century with regard to eugenics.

There is such widespread lack of understanding about what eugenics is that it will be useful to begin with a summary statement of what can be called the general theory of eugenics. This consists of eight core propositions. These are:

1. Certain human qualities are valuable. The most important of these are health, intelligence, and what was described by eugenicists as "moral character," which consists of a well-developed moral sense, self-discipline, strong work motivation, and social concern.

2. These human qualities are valuable because they provide the foundation for a nation's intellectual and cultural achievements; its quality of life; and its economic, scientific, and military strength.

3. Health, intelligence, and moral character are to a substantial extent genetically determined. Hence it would be possible to improve these qualities genetically. This would produce an improvement of what can be described as the "genetic human capital" of the population. This is

the objective of eugenics. Eugenicists recognize that these qualities are also determined environmentally and support such attempts to improve these qualities. Nevertheless, environmental measures to improve these qualities are virtually universally supported and are not a distinctive part of eugenics.

4. During the second half of the nineteenth century and throughout the twentieth century, the populations of the Western democracies and most of the rest of the world have been deteriorating genetically with respect to the three qualities of health, intelligence, and moral character. This process is known as dysgenics and poses a threat to the quality of civilization and culture and to the economic, scientific, and military strength of the nation state. The first objective of eugenics is to arrest and to reverse this process.

5. It would be feasible to improve the genetic quality of the population with respect to its health, intelligence, and moral character. There are two broad kinds of program by which this could be accomplished. These can be designated "classical eugenics" and "the new eugenics." *Classical eugenics* consists of the application to humans of the methods used for many centuries by plant and animal breeders to produce plants and livestock of better quality by breeding from the better specimens. The application of such a selective breeding program to human populations would require policies for "positive eugenics," designed to increase the numbers of children of the healthy, the intelligent, and those with strong moral character; and for "negative eugenics," designed to reduce the numbers of children of the unhealthy and of those with low intelligence and weak moral character.

6. The *new eugenics* consists of the use of human biotechnology to achieve eugenic objectives. The techniques of human biotechnology comprise artificial insemination by donor (AID), prenatal diagnosis of genetic diseases and disorders, in vitro fertilization and preimplantation diagnosis, cloning, and genetic engineering by the implantation of new genes.

7. Eugenics serves the needs of individuals and of nation states. It serves the needs of individuals because people like to have children who are healthy and intelligent and of good moral character. It serves the needs of the nation state because a nation state whose population has good health, high intelligence, and good moral character is stronger and more likely to succeed in competition with other nation states.

8. Although there has been much discussion in the Western democracies about whether the biotechnologies of embryo selection, cloning, and the like are ethical and should be permitted, the prohibition of them will not be successful. No new technologies that serve human needs have ever been successfully suppressed. The important question about eugenics is

not if it should be allowed, but where it will be developed and how to counter the threat this will present to the Western democracies.

This book is concerned with the elaboration and establishment of these eight core propositions of eugenics. It may be helpful for the reader if I map out the framework in which this task is attempted. The book is divided into four parts. Part I gives a historical account of eugenics, its foundation by Sir Francis Galton in the second half of the nineteenth century and up to his death in 1911, the increasing acceptance of eugenics during the first half of the twentieth century, and the decline of support for eugenics in the second half of the century. Part II discusses the objectives of eugenics and whether these should be confined to the improvement of the genetic quality of the population in respect to the reduction of genetic diseases and disorders and the increase of its intelligence and the strengthening of its moral character. Part III is concerned with the policies for the achievement of these objectives by the use of the classical eugenics of selective reproduction. Part IV is concerned with the new eugenics of human biotechnology as it has been developed in the closing decades of the twentieth century, how it is likely to evolve in the future in democratic societies, and how it is likely to be used for the development of national strength by authoritarian states, leading ultimately to the establishment of a world state.

ACKNOWLEDGMENTS

I am greatly indebted to Harry Weyher for his encouragement in undertaking the task of writing this book, to the Pioneer Fund for support, and to Marian van Court for her critical comments on this work.

I

*Historical
Introduction*

1

⤳

Sir Francis Galton Lays the Foundations of Eugenics

1. The Objectives of Eugenics
2. Genetic Basis of Health, Intelligence, and Character
3. The Problem of Dysgenics
4. The Methods of Eugenics
5. Eugenic Policies for Western Societies
6. Postive Eugenics
7. Negative Eugenics
8. Immigration and Emigration
9. Improvement of the Environment
10. "Kantsaywhere": A Eugenic Utopia
11. A Critique of "Kantsaywhere"
12. Conclusions

⤳ ⤳ ⤳

Eugenics was first advanced by the Greek philosopher Plato in his book *The Republic*, written about 380 B.C. This book was a blueprint for a utopian state. The state would consist of three classes of rulers or "guardians," soldiers, and workers, each of which would be bred from the best individuals using the methodology of the selective breeding of livestock, which was well known in Athens in the fourth century B.C. From time to time, similar eugenic utopias have been proposed, such as that of the sixteenth-century Italian monk Tommaso Campanella (1613) in his *Civitas Solis* (City of the Sun).

In modern times, eugenics was founded by the English statistician, biologist, psychologist, and polymath Sir Francis Galton (1820–1911), who coined the word *eugenics* and set out its basic principles. Galton wrote about the genetics of human characteristics and the possibility of improving the genetic qualities of human populations first in an article published in 1865 and later in his book *Hereditary Genius* (1869). It was in his 1883 book, *Inquiries into Human Faculty*, that Galton proposed the term *eugenics* for these ideas, the word being constructed from the Greek to mean "good breeding." Galton (1883) wrote, "We greatly want a brief word to express the science of improving stock which takes cognizance of all influences that tend in however remote a degree to give the more suitable races or strains of blood a better chance of prevailing speedily over the less suitable, than they would otherwise have had. The word *eugenics* would sufficiently express the idea. . . . Eugenics was to be the study of agencies under social control that may improve or repair the racial qualities of future generations, either physically or mentally" (p. 17). It should be noted that in the nineteenth century the word *race* had the connotation of what in the twentieth century would be termed "population" or "subpopulation." Galton did not propose that one race, in the twentieth-century meaning of the word, should be assisted to prevail over another.

Galton wrote about eugenics in a series of books and articles over a period of 45 years, from his first essay of 1865 until the year of his death, 1911. During these years he considered virtually all the ramifications of the concept. He discussed the characteristics that eugenics would seek to improve, the genetic determination of these characteristics, the genetic deterioration taking place in modern populations that eugenics would seek to reverse, and the policies that might be implemented to promote eugenics. It is with Galton's ideas on these issues that an assessment of the concept of eugenics has to start.

1. THE OBJECTIVES OF EUGENICS

Galton proposed that the objective of eugenics should be the improvement of the genetic qualities of the population with respect to three characteristics: (1) health, (2) intelligence, and (3) what he called "moral character." He employed the concept of *health* broadly to include not only the absence of disease, but also the presence of energy, vigor, and what he sometimes called "physique." Galton (1909) believed there would be a widespread consensus on the desirability of health in this broad sense, writing of the discussions he had had with a number of people that "some qualities such as health and vigor are thought by all to be desirable and the opposite undesirable" (p. 66).

Galton argued in his *Hereditary Genius* that the *intelligence* of a population is a major component in its cultural, scientific, and economic achievements; that these are the defining characteristics of civilization; that civilization is better than barbarism; and that to maintain and promote an advanced civilization, the intelligence of the population needs to be improved and prevented from declining. Galton conceptualized intelligence as a single general ability that is capable of being channeled into a variety of fields of human endeavor. This has become the prevailing view in contemporary psychology.

Galton conceptualized moral character as a syndrome of personality qualities comprising a strong moral sense, energy and "zeal" for sustained work, integrity, trustworthiness, and a sense of social obligation. Those who lacked moral character he described in an early article as "men who are born with wild and irregular dispositions" (1865, p. 324). Galton believed that character was also an important quality for the maintenance of civilization and, therefore, that it also needed to be improved and prevented from declining.

Galton considered that the three qualities of health, intelligence, and character could be aggregated into a single broader characteristic that he called "worth." "By this," he wrote, "I mean civic worthiness, or the value to the State, of a person. . . . If I had to clarify persons according to worth, I should consider them under the three heads of physique (including good health), ability, and character" (1908a, p. 104). Thus Galton saw the objective of eugenics as the improvement of the population's worth in these three respects. It is difficult to improve on Galton's formulation of the qualities that it would be desirable to strengthen in human populations, and they will serve as the basic guidelines throughout the subsequent discussions.

2. GENETIC BASIS OF HEALTH, INTELLIGENCE, AND CHARACTER

Galton understood that the concept of eugenics was crucially dependent on health, intelligence, and character having some genetic basis or heritability. If they do not, these traits can not be improved genetically. Galton (1883) took it for granted that health has some genetic basis and asserted that energy, which he regarded as a component of good health, "is eminently transmissible by descent" (p. 19). As regards intelligence and character, many of his contemporaries, including Charles Darwin in the 1860s, believed that these were solely environmentally determined, and Galton realized that he had to argue the case that these traits are at least to some degree under genetic control. He first tackled this question at length in *Hereditary Genius*, where he put forward four arguments.

First, he constructed a number of family pedigrees of eminent men including lawyers, statesmen, scientists, writers, musicians, Cambridge scholars, and, to cover physical capacities, wrestlers. He showed that eminence in these fields tends to run in families and is more likely to be present among close relatives than among those who are more distant. Using various criteria for "eminence," he calculated that it occurred in 26 percent of the fathers of the eminent men, in 36 percent of their sons, in 7.5 percent of their grandfathers, and in 9.5 percent of their grandsons. These percentages are all high when compared with the presence of eminence in the general population, which he calculated at 0.025 percent, or 1 in 4,000 individuals. Hence Galton concluded that eminence is disproportionately represented in certain families. He argued that this, together with the greater frequency of eminence among close relatives as compared with more distant relatives, indicated genetic transmission.

Galton was aware of the possible objection that close relatives of eminent men would have environmental advantages that might explain these achievements. His second argument for evidence of a genetic basis for eminence was designed to counter this objection. The argument consisted of an examination of the lives of a number of the adopted sons of popes and showed that these did not achieve eminence to the same extent as the biological sons of eminent men. This work anticipated a number of adoption studies carried out in the twentieth century showing that adopted children resemble their adopted parents less for intelligence and for criminal propensities than they resemble their biological parents. Scarr and Weinberg (1978), for example, report correlations of .14 and .52, respectively, for intelligence; and a similar pattern has been found for crime by Hutchings and Mednick (1977).

Galton's third argument for a genetic basis for ability was that there are a number of exceptions to the general rule that high ability tends to run in families. He noted that quite frequently very gifted individuals have come from quite ordinary families. One of the individuals he cited was Jean D'Alembert, the brilliant French mathematician of the eighteenth century, who was a foundling reared in the family of a poor glazier in Paris. Galton (1869) argued that such cases showed that the environment had relatively little impact on achievement and asserted that "if a man is gifted with vast intellectual ability, eagerness to work, and power of working, I cannot comprehend how such a man should be repressed" (p. 79). If environmental effects are relatively unimportant, Galton argued, ability must be largely determined by genetic factors.

Galton distinguished between ability and achievement. His argument was that achievement is determined by ability (intelligence) and the character qualities he called "zeal" and "the capacity for hard labor." These, Galton wrote in Hereditary Genius, are "a gift of inheritance" (1869, p. 76); and 40 years later he reaffirmed that "character, including the aptitude for work, is heritable like every other faculty" (1908a, p. 291). He believed that any individual who inherits the three qualities of ability, zeal, and work capacity is likely to succeed. Thus Galton demonstrated that achievement has a high heritability, as shown by its transmission in elite families and by its spontaneous appearance in ordinary families, and he argued that this implies that the underlying components of achievement must also have high heritability.

Galton (1908a) presented a fourth argument for the genetic basis of human qualities. This centered on the resemblances of twins, which he thought could be studied to tease out the relative contributions of heredity and environment to human intelligence and personality. In describing this work later, he wrote, "It occurred to me that the after-history of those twins who had been closely alike as children, and were afterwards parted . . . would supply much of what was wanted" (p. 294). He found a number of twin pairs, including some who had been reared separately, and assessed their similarity. He found that twin pairs were closely similar in their abilities and character and concluded that "the evidence is overwhelming that the power of Nature was far stronger than that of Nurture" (p. 295). Galton's conclusion on this point has been confirmed by a number of twentieth century studies showing that identical twins reared in different families are closely similar in respect of intelligence and various measures of character. Bouchard (1993, p. 58) has summarized the studies for intelligence; and Tellegen et al. (1988) have summarized them for personality.

3. THE PROBLEM OF DYSGENICS

Galton believed that the population of Britain and other Western nations had begun to deteriorate genetically. He was one of the first to understand the phenomenon, which was later to become known as dysgenics. Galton gained this understanding from his reading of Charles Darwin's *Origin of Species*, published in 1859. He read the book shortly after it appeared and grasped the central point that the genetic quality of populations is maintained and enhanced through natural selection, the process by which nature ensures the "survival of the fittest," the phrase proposed by Herbert Spencer (1874), and eliminates weaker individuals by high mortality and low fertility. Galton realized that the cleansing function of natural selection, the elimination of the unfit, had begun to weaken in Britain and Western nations during the nineteenth century. He first made this point in 1865 when he wrote, "One of the effects of civilization is to diminish the rigor of the application of the law of natural selection. It preserves weakly lives that would have perished in more barbarous lands" (1865, p. 325).

Four years later, in his *Hereditary Genius*, Galton discussed a second way in which natural selection had broken down. This was the emergence of an inverse relationship between ability and fertility; that is to say, the more talented members of the population were having fewer children than the less talented. He argued that in the first stages of civilization, "the more able and enterprising men" tended to have large numbers of children, but that as civilization matured, these began to have fewer children than the less able and enterprising. He suggested that the principal reason for this was that the more able and enterprising tended to marry late or even not to marry at all because they perceived marriage and children as a distraction from the advancement of their careers. During the twentieth century, research confirmed Galton's thesis that the more intelligent and the better educated do tend to marry late or not at all and that they have fewer children than the less intelligent and the less well educated, and that this is true of women as well as men. I have summarized the considerable research on this phenomenon in my book *Dysgenics* (Lynn, 1996).

Galton suggested an additional explanation for the low fertility of the professional classes—that able and enterprising young men tended to seek out heiresses in order to rise in the social and occupational hierarchy. Unhappily, Galton argued, heiresses tended to come from relatively infertile families who only had one or two daughters. If the families had not had low fertility, they would have had several sons as well as daughters, none of whom would be heiresses. The marriage of able and enterprising men to heiresses tended to be childless because of the inherited low fertility of the wives, so the effect was to reduce the fertility of both partners to the marriage. This theory was later adopted by the geneticist R. A. Fisher (1929). However, it is questionable on two grounds: (1) whether fertility has any genetic basis and (2) whether the tendency of able and enterprising men to marry heiresses is present on a large enough scale to have any impact on the overall inverse relationship between ability and fertility.

Nevertheless, whatever the explanation of the inverse association between talent and fertility, Galton was right in believing that it existed and that this would lead to genetic deterioration. "There is," he wrote on the concluding pages of *Hereditary Genius*, "a steady check in an old civilization on the fertility of the abler classes: the improvident and unambitious are those who chiefly keep up the breed. So the race gradually deteriorates, becoming in each successive generation less fit for a high civilization" (1869, p. 414). Forty years later, in one of his last papers written shortly before his death, he reiterated that "it seems to be a tendency of high civilization to check fertility in the upper classes" (1909, p. 39).

Galton understood that the two processes through which natural selection works to keep populations genetically sound—the high mortality and the low fertility of the less fit—had broken down in the nineteenth century. The less fit were increasing in numbers through the reduction of their previous high mortality and through an increase in their fertility. This was producing a genetic deterioration in the populations of the economically developed nations and a concomitant deterioration in the quality of their civilizations. In his *Hereditary Genius*, Galton cited Spain as a historical example of a nation in which dysgenic processes, which he attributed to the celibacy of the priests and nuns, had reduced the genetic quality of the population and the strength of the nation as an economic, cultural, and military power. Because natural selection was failing to keep human populations fit, it would be necessary, Galton (1908a) wrote in his *Memories* shortly before his death, "to replace natural selection by other processes" (p. 323). The formulation and implementation of these processes was the objective of eugenics.

4. THE METHODS OF EUGENICS

Galton believed that the way to implement eugenics would be to adopt the selective breeding methods employed by animal and plant breeders. Galton knew that selective breeding had been used for centuries to obtain improved stocks. As noted earlier, Plato referred to it in *The Republic*, in which Socrates explains that selective breeding of domestic animals is known to produce improved varieties and proposes that the same methods should be used for humans in his utopian state. Curiously, Galton never referred to Plato's eugenic utopia, possibly because he thought it was an unattractive model or possibly because he was simply unaware of it; but he certainly knew that it was possible to breed improved strains of animals and plants. This was common knowledge among informed people in the nineteenth century. In the Middle Ages, people had bred new strains of strongly built horses able to carry knights with heavy armor. In the eighteenth century the English stockbreeder Robert Bakewell achieved considerable successes in breeding improved strains of cattle and sheep. Racehorses had been bred for their running speeds. Plants had also been improved by selective breeding. For instance, in 1806, Michael Keens, working in his garden near London, produced the first large, sweet-tasting strawberry of the kind known today and on which he bestowed the name of Keens' Imperial (Farndale, 1994).

Galton knew that the technique of breeding animals and plants for improved strains entailed selecting the best specimens and breeding them for a number of generations until the improved stock was obtained. He believed that this principle could be used to improve the genetic qualities of humans. He set this out in 1869 in the opening paragraph of the first chapter of *Hereditary Genius*: "As it is easy to obtain by careful selection a permanent breed of dogs or horses gifted with peculiar powers of running, or of doing anything else, so it would be quite practicable to produce a highly gifted race of men by judicious marriages during several consecutive generations" (p. 5).

Galton realized, however, that a eugenics program for humans, at least in Britain and other Western nations, would not be able to adopt precisely the methods of animal stock breeders. Typically, stock breeders select a small number of the best males and mate them with a larger number of the better females. Something approaching this technique has been used in many human societies in harem systems, in which a number of nubile and usually unwilling females have been coerced into harems to serve the sexual and procreative needs of powerful men, such as emperors in China, sultans in Turkey, and princes in India. Galton realized that though systems of this kind could be eugenically effective if the men possessing the harems had valuable qualities, their introduction would not be well received in Victorian England. He went out of his way to clarify his position on this subject, writing of the critics of eugenics, "The most common misrepresentations now are that its methods must be altogether those of compulsory unions, as in breeding animals. It is not so" (1908a, p. 311). Galton was always conscious that proposals not only must be effective genetically, but also had to be acceptable to public opinion.

5. EUGENIC POLICIES FOR WESTERN SOCIETIES

Galton understood that although the general principles of selective breeding for improved strains that had been employed for centuries by stock breeders could be used to improve humans, careful thought has to be given to the details of how these principles could be applied in Britain and other Western societies. He realized that policies had to be formulated that were politically feasible and humane. In a lecture delivered in 1905 and published in 1909, he set out his views on how eugenic policies could best be promoted. He argued for a four-stage strategy. First, eugenics would be established as an accepted academic discipline, which would include research demonstrating the genetic basis of certain diseases and of energy, intelligence, and character. Second, once this had been achieved, the case would be made to the public for the general principle of eugenics—that human quality is largely genetically determined and that it is capable of improvement. Third, detailed policies could then be formulated to achieve eugenic objectives. Finally, when all this had been accomplished, the population as a whole would become convinced of the desirability of eugenics and would approve of a range of eugenic policies (Galton, 1909).

Although Galton was cautious about setting out a detailed eugenic policy agenda, which he thought would be likely to alienate public opinion, he did make some policy proposals couched in rather general terms. He suggested that it would be useful to distinguish three broad classes of individuals in the population: the "desirables," the "passables," and the "undesirables." The "desirables" would be those who had an exceptional endowment of worth, consisting of health, intelligence, and character. The "passables" would be those who had average endowment of these qualities. The "undesirables" would be those in which these qualities were poor. He thought the best strategy for practical eugenics would be to attempt to increase the fertility of the desirables, and he proposed the term *positive eugenics* for policies designed to achieve this. At the same time, attempts should be made to discourage the fertility of the undesirables, which he designated *negative eugenics*. No action should be taken on the passables (1908a, p. 322).

Galton did not specify the proportions of the population falling into these three categories; but it seems that he thought of the desirables as being a small elite and the undesirables as also being quite small, perhaps each of these amounting to some 5 to 10 percent of the population. He seems to have thought that this general strategy would be acceptable to public opinion because the great majority of the population would be unaffected, and a general consensus might be secured on the desirability of curtailing the fertility of a small problem group consisting principally of the mentally retarded, habitual criminals, and psychopaths. At the same time, the passable majority might not object too strongly to, or perhaps even notice, measures designed to increase the fertility of the small elite.

6. POSITIVE EUGENICS

Positive eugenics was to consist of measures to encourage the procreation of the small elite of desirables. Galton made two proposals to advance this objective. The first was for the establishment of local eugenics associations, which would be staffed by eugenics enthusiasts to promote eugenic principles and policies in their localities. These associations would elect officers, hold meetings, organize lectures on eugenic topics, and attempt to raise public consciousness of the importance of eugenics. They would collect pedigrees of worthy families in their neighborhoods to identify eugenically desirable young couples and would give them financial assistance to have children. These young couples would be distinguished by what Galton called "civic worthiness." They would be drawn predominantly from the professional and the middle classes and from skilled artisans because, as Galton (1909) put it, "the brains of the nation lie in the higher of our classes" (p. 11). Nevertheless, these desirables could be present in all social classes, except the bottom, and would include "contented laborers" (p. 266). "The aim of eugenics is to represent each clan or sect by the best specimens" (p. 11).

Galton's second proposal for positive eugenics was that the desirable elite should be made conscious of their ethical duty to have children and thereby increase, or at least maintain, their numbers in future generations. To develop this consciousness, he proposed that families that had made valuable social contributions over several generations by virtue of their qualities of good health, abilities, and character should be identified, thus fostering their consciousness of being a genetic elite. Galton first floated this idea in 1883 when he wrote, "My object is to build up by extensive inquiry and publication of results, a sentiment of caste among those who are naturally gifted" (1883, p. 98). Some 20 years later he returned to this proposal in a lecture delivered to the Sociological Society at the London School of Economics in 1905. He announced on this occasion that he was funding a research fellowship at University College, London, to undertake the task of compiling pedigrees of elite families. The research was carried out by Edgar Schuster, and the first volume, which dealt with the pedigrees of eminent British scientists, was published a year later by Galton and Schuster (1906).

7. NEGATIVE EUGENICS

Galton's negative eugenics was to be directed at curtailing the fertility of the undesirables. This social group broadly approximates to what was to become known in the last quarter of the twentieth century as the "underclass" and is characterized by low intelligence and a serious deficiency of moral sense.

Galton had read the American sociologist Richard Dugdale's (1877) account of the degenerate Jukes family, which had produced seven generations of criminals, alcoholics, unemployables, and prostitutes. This account was the first detailed pedigree study of what was later to be called "the intergenerational cycle of transmitted deprivation." Such families, Galton (1883) believed, were not confined to the United States. England also was "overstocked and over-burdened by the listless and incapable" (p. 18). He believed that such families were "infamous" (p. 44) and that society would be better off without them. He considered that this class comprised "persons who are exceptionally and unquestionably unfit to contribute offspring to the nation" (1909, p. 66).

Galton realized that just how they could be prevented from doing this raised difficult problems. He doubted whether the undesirables could be induced to curtail their fertility by moral persuasion. Therefore, he believed that some kind of coercion would be required, writing that "stern compulsion ought to be exerted to prevent the free propagation of the stock of those who are seriously afflicted by lunacy, feeble-mindedness, habitual criminality, and pauperism" (1909, p. 311). He was confident that "our democracy will ultimately refuse consent to that liberty of propagating children which is now allowed to the undesirable classes" (1908a, p. 312), but he did not specify the details of how this would be achieved. In the twentieth century, this prediction was realized by the sterilization laws that were widely implemented in many Western nations.

8. IMMIGRATION AND EMIGRATION

Galton realized that immigration could have eugenic implications. He argued that immigrants with desirable qualities could strengthen what was later to be called the "gene pool" of the receiving country, whereas those with undesirable qualities would weaken it. Consequently, potential immigrants should be assessed for acceptance with this consideration in mind. In *Hereditary Genius*, Galton (1869) wrote that Britain should adopt "the policy of attracting eminently desirable refugees, but no others, and of encouraging their settlement and the naturalization of their children." He gave as an illustration the benefits Britain had gained from the immigration of the Huguenots, a group of skilled artisans expelled from France in the seventeenth century because their Protestant beliefs were offensive to the Catholic king. These refugees, he declared, "were able men and have profoundly influenced for good both our breed and our history." As an example of undesirable immigration, he instanced the French aristocrats who had taken refuge in Britain during the French Revolution. He thought that these "had but poor average stamina" and that it was fortunate that they "have scarcely left any traces behind them" (p. 413).

There were corresponding eugenics implications for emigration. Galton (1869) thought that a country would suffer a depletion of its genetic quality through the emigration of the talented, but it would also gain through the emigration of undesirables. He wrote, "England has certainly got rid of a great deal of refuse through emigration. She has been disembarrassed of a vast number of turbulent radicals and the like, men who are decidedly able but by no means eminent, and whose zeal, self-confidence, and irreverence far outbalance their other qualities" (p. 414).

Neither immigration nor emigration were particularly important issues in late Victorian and Edwardian Britain. The only significant immigration into Britain at this time was of Jews seeking refuge from persecution in Russia and Poland. Galton (1869) believed that, in general, they were of good genetic stock. He wrote that they "appear to be rich in families of high intellectual breeds" (p. 47), and he never voiced any criticism of their admission.

9. IMPROVEMENT OF THE ENVIRONMENT

Galton believed that heredity was more important than environment in determining human quality, but he recognized that environment also plays a part. Consequently, he argued that a comprehensive eugenics program would include attempting to improve the environment to allow all children to develop their genetic qualities to their maximum. The aims of eugenics, he thought, should include measures to ensure "the healthful rearing of children" (1908a, p. 323). He wrote that eugenics can be defined as "the science which deals with all influences that improve the inborn qualities of the race; and also with those that develop them to the utmost advantage" (1909, p. 45). He did not devote much attention to the ways in which this might be done, although he did advocate the more widespread provision of scholarships to enable gifted children from poor families to secure an education (1883).

Despite Galton's inclusion of environmental improvements in his concept of eugenics in some of his later writings, it is better to restrict the term to Galton's original (1883) definition of *eugenics* as a means for promoting the genetic improvement of the population. The concept becomes too wide if it is used to include, for instance, encouraging pregnant women to take nutritional supplements to have healthier babies, providing preschool head-start programs that are designed to raise children's intelligence, and the like. Some eugenicists have used the word *euphenics* for the environmental improvements in health and education that they believed should go hand in hand with attempts to produce genetic improvements in population quality, which should properly be called eugenics, and it is better to reserve this term for environmentalist programs.

10. "KANTSAYWHERE": A EUGENIC UTOPIA

Throughout most of his life, Galton considered it best to set out the principles of eugenics in general terms and was reluctant to propose a detailed policy program, which he thought would be likely to antagonize public opinion. Nevertheless, he departed from this principle in the last two years of his life, during which he wrote an account of a eugenic utopia. This was an

imaginary republic that he called Kantsaywhere. He completed the manuscript in 1910. He sent it to a publisher, who declined to publish it, and Galton
did not pursue the matter further. He eventually decided not to publish the
manuscript, and he destroyed some of it; but portions were found among his
papers after his death. Karl Pearson (1914) published the most important
passages in his *Life, Letters, and Labours of Francis Galton.*

Kantsaywhere is a small country of some ten thousand inhabitants. It is
governed by the Eugenic College, which elects a Council to carry out the
legislative and the executive functions of government. In this regard, the
constitution is similar to that of Great Britain, which is governed by a cabi-
net drawn from the House of Commons. However, whereas the members of
the British House of Commons are elected by the population in accordance
with the principles of democracy, the members of the Eugenic College in
Kantsaywhere are recruited by examination. The way the system works is that
each year approximately 20 new members are recruited to the Eugenic College to replace the number who die or retire each year, thereby keeping the
size of the College constant.

Each year the citizens are permitted to apply for membership of the Eugenic College. The applicants are subjected to an initial screening procedure,
from which about 80 are selected for a more intensive selection process. This
consists of an examination in four parts: (a) physical and anthropometric,
involving tests of the strength of the arm in pulling a bow, strength of grip,
reaction time, visual acuity, acuteness of hearing, and the like; (b) aesthetics
and literary, consisting of singing and essay writing; (c) health and medical
history; and (d) genealogical, consisting of an assessment of the candidate's
ancestors' accomplishments. Each test is weighted equally, and the marks
obtained in the four tests are summed, producing a rank order of candidates.
Each year, top performers are admitted to the Eugenic College, the precise
number depending on the number of vacancies in the particular year and
generally being about 20. Thus, recruitment to the Eugenic College is similar
to the recruitment of mandarins in Imperial China for a period of some two
thousand years, and to the elite of the British Civil Service in the second half
of the nineteenth century, on which no doubt Galton modeled his system.
The selection procedure ensures that the ruling elite in Kantsaywhere is a
self-perpetuating oligarchy.

One of the functions of the Eugenic College is to preserve and enhance
the genetic quality of the population. It does this by issuing licenses for parenthood. Couples wishing to have children are required to apply for a license
to do so, and to obtain this they have to take an examination. The details of
this examination are not described, but it can be reasonably assumed that it
resembles the examination for admission to the Eugenic College. The marks
obtained by the couples are summed, and on the basis of the total they are
classified into five grades containing approximately equal numbers. Those in
grade one, the top grade, are permitted to have as many children as they wish;

those in grade two are allowed to have three children; those in grade three may have two children; those in grade four are allowed one; while those in grade five, the bottom grade, are not allowed to have any children. Provision is made for the possibility that some couples might have more children than they are allowed. These would be punished by fines, confinement in segregated labor colonies, or deportation.

11. A CRITIQUE OF "KANTSAYWHERE"

Galton's "Kantsaywhere" is a valuable thinking exercise for concentrating the mind on what a serious eugenic society might be like. It raises three issues in particular that need to be pondered. First, it contains strong negative eugenic provisions in the form of curtailing the right to have children of a substantial proportion of the population considered to be of poor genetic quality; second, these measures would need counterbalancing with provisions for positive eugenics to induce the genetic elite to have more children than they would otherwise choose; and third, it raises the question of whether a eugenic state of this kind and severity could be attained in a democracy.

The negative eugenics of Kantsaywhere involve the curtailment of the right to have children of about 40 percent of the population—the 20 percent or so in grade five who fail to secure a license to have any children, and the further 20 percent or so in grade four who are permitted to have one child. Opinion surveys carried out in the second half of the twentieth century have shown that in the economically developed world, the great majority of people would like to have two children (e.g., Vining, 1982); so individuals in grades four and five would have their reproductive rights severely restricted. This restriction raises the problem of how it could be enforced. Galton proposed that couples who exceeded the permitted number of children would be punished by fines, long-term imprisonment, or deportation. It is doubtful whether the punishments would work effectively to deter childbearing among many of these couples because the couples in question would be those with low intelligence and conscientiousness, precisely the people for whom the prospect of punishment has little deterrent effect. If they were fined, many of them would probably not be able to pay and would have more children. Permanent incarceration in prisons or deportation would certainly prevent further transgressions, but these penalties would not be practical alternatives for what would probably be large numbers of offenders. A further problem with the scheme lies in securing the consent of the population in a democratic society. There can be little doubt that this would not be possible and that such a scheme could only be introduced in an authoritarian state.

The positive eugenics of "Kantsaywhere" raises equally difficult problems. Because 40 percent of the population is limited to one child or not allowed to have any children, the size of the population could only be maintained by grade one couples having four children and grade two couples having their permitted quota of three. Galton did not discuss this problem in the part of the manuscript that survived, and perhaps he thought that there would not be a problem here because most couples at the time he was writing had at least three or four children. However, by the end of the twentieth century, most of the genetic elite, who can be broadly equated with the professional class, had at most two children, and many of them were opting to remain childless. There would be considerable problems in inducing them to have the numbers of children required to keep the size of the population stable. These problems might be overcome by a combination of moral pressure and financial incentives, but there can be no doubt that there would be substantial practical difficulties in implementing these components of Galton's plan.

The third major issue raised by "Kantsaywhere" is that it is not a democracy but an oligarchy run by a genetic elite. It is not difficult to see why Galton envisaged Kantsaywhere as an oligarchy. He must have come to the conclusion that his eugenic state would have to exert so much coercion on its citizens in the form of restricting the childbearing of some and increasing the childbearing of others that it would not be viable as a democracy. If an elected government were to introduce these measures, the people would reject them at the next election. Galton must have realized this, and this must be why he gave Kantsaywhere a conservative oligarchic constitution in which power is held by the Eugenic Council, which would consist largely of the old, rather like that in the former Soviet Union and present-day China. Oligarchies of this kind are resistant to change, except after prolonged and cautious consideration. Galton must have concluded that a constitution of this kind would be necessary to maintain his eugenic state.

As one reflects on the oligarchic constitution of Kantsaywhere and the kinds of eugenic policies the oligarchy would have to enforce to secure its eugenic objectives, it becomes easy to understand why Galton was ambivalent about publishing his eugenic utopia and why, elsewhere in his writing, he preferred to argue for eugenics in general terms, rather than to go into details of precisely what policies would be required in an eugenic society. To have spelled out the details could have been to risk alienating public opinion to such an extent that the whole idea of eugenics would be likely to be rejected. The major lesson of "Kantsaywhere" is that it is extremely difficult to formulate eugenic programs using the classical eugenics of selective reproduction that would have a major impact and that would be politically feasible in democratic societies. This does not mean, however, that it would be impossible to introduce them in authoritarian states.

12. CONCLUSIONS

Francis Galton set out all the essential components of the case for eugenics. He understood that in preindustrial human societies populations were kept genetically sound through the operation of natural selection, which ensured "the survival of the fittest" by the two processes of high mortality and low fertility. He realized that natural selection against ill health, low intelligence, and weak moral character had slackened and even gone into reverse in the nineteenth century and that the consequence of this was a genetic deterioration of the population. He believed that these qualities are significantly under genetic control and assembled pedigree studies to show that this is the case. He thought that civilization depended on the intelligence and the sound character qualities of the population and that the deterioration of these would impair its quality. He believed that once natural selection had ceased to operate, society would have to devise methods of consciously contrived selection to correct the genetic deterioration that had begun to take place. To this consciously contrived selection, he gave the name eugenics. He set out the broad principles of how genetic deterioration could be arrested and reversed, which consisted of measures to encourage those possessing health, intelligence, and character to have more children and those deficient in these qualities to have fewer.

Galton (1908a) believed that eugenics would promote the interests of both individuals and society as a whole: "eugenics covers for both" (p. 322). Nevertheless, the principal objective of eugenics is to strengthen societies rather than to advance the well-being of individuals because, as he wrote at the conclusion of his *Memories*, "individuals appear to me as detachments from the infinite ocean of Being" (1908a, p. 323).

2

❧

The Rise and Fall of Eugenics

1. Eugenics in Britain
2. Eugenics in the United States
3. Eugenics Spreads Worldwide—Germany
4. Research on the Scientific Basis of Eugenics
5. Promotion of Birth Control
6. Sterilization
7. Immigration Control
8. The Decline of Eugenics
9. The Last Eugenicists
10. Conclusions

❧❧❧

Francis Galton's ideas on eugenics won considerable acceptance in the late nineteenth century and in the early and middle decades of the twentieth century. The majority of biologists, geneticists, and social scientists and many informed laypeople, accepted Galton's arguments that the quality of civilization and national strength depended on the genetic quality of the population, that natural selection was no longer operating to keep the quality of the population sound in contemporary populations, and that eugenic policies were needed to counteract this deterioration. There was also considerable progress in gaining public acceptance of this analysis, in consolidating the research base of the general theory of eugenics, in formulating policies to overcome the dysgenic problem, and in securing the implementation of these policies. By the middle decades of the twentieth century, eugenics had become widely accepted throughout the whole of the economically developed world, with the exception of the Soviet Union where genetics was proscribed for ideological reasons.

1. EUGENICS IN BRITAIN

Galton's ideas on eugenics began to win acceptance in Britain in the closing decades of the nineteenth century. Charles Darwin was one of those who
came to agree with Galton. Darwin was originally skeptical about whether
inheritance played any part in the determination of intelligence in humans,
but he changed his mind in 1869 after reading Galton's *Hereditary Genius*. In
1870 Darwin wrote to Galton, "You have made a convert of an opponent in
one sense, for I have always maintained that men did not differ much in
intellect, only in zeal and hard work" (Galton, 1908a, p. 290). In his *Descent
of Man*, Darwin (1871) expressed his agreement with Galton's view that natural
selection against the less fit had become relaxed in the nineteenth century as
a result of which "the weak members of civilized societies propagate their kind.
No one will doubt that this must be highly injurious to the race of man" (p.
501). He asserted that ways needed to be found to prevent the reproduction
of those "in any marked degree infirm in body or mind" (p. 918). Alfred Russell
Wallace (1890) recorded that when he discussed these problems with Darwin
in the 1880s, Darwin spoke pessimistically about the large number of children being produced by "the scum" and was "very gloomy on the future of
humanity, on the ground that in our modern civilization, natural selection
had no place and the fittest did not survive" (p. 93).

Another early supporter of Galton's views was Herbert Spencer. He also
recognized that in nineteenth-century Britain natural selection had largely
ceased to operate against the less fit. He laid the blame for this partly on
charities for aiding the indigent, and he asserted that "institutions which foster
good-for-nothings commit an unquestionable injury because they put a stop
to that natural process of elimination by which society continually purifies
itself" (Spencer, 1874, p. 286). In the last decade of the nineteenth century,
eugenics was taken up by the Fabian Socialist Sydney Webb. He carried out
research in the 1890s that showed that "the improvident" had larger than
average numbers of children. He wrote of the high fertility of the "degenerate
hordes of a demoralized 'residuum' unfit for social life" (Webb, S., 1896, p.
28).

In the early decades of the twentieth century, eugenics began to win widespread acceptance in Britain among biological and social scientists and the
informed public. In 1907 the Eugenic Education Society was founded, abbre-
viating its name to the Eugenics Society in 1926. Early British eugenicists
included Caleb Saleeby (1910), a physician who wrote one of the first books
on eugenics and who coined the term *dysgenic* for the high fertility of the
socially undesirable. It was to prove a useful word as the evidence that genetic deterioration was taking place accumulated. The leading exponent of
eugenics in Britain in the early 1900s was Karl Pearson (1857–1936), professor of applied mathematics at University College in London, who is mainly
remembered for working out the statistical method of the Pearson correlation
coefficient. In his book *National Life*, Pearson (1901) restated the case that

natural selection had largely ceased to operate in Western populations: "While modern social conditions are removing the crude physical checks which the unrestrained struggle for existence places on the overfertility of the unfit, they may at the same time be leading to a lessened relative fertility in those physically and mentally fitter stocks, from which the bulk of our leaders in all fields of activity have hitherto been drawn" (p. 101). He wrote that the problem of genetic deterioration could only be solved by eugenic intervention: "The only remedy, if one be possible at all, is to alter the relative fertility of the good and the bad stocks in the community" (Pearson, 1903). Nine years later he reaffirmed that "the less fit were the more fertile" and consequently "the process of deterioration is in progress" (Pearson, 1912, p. 32). This deterioration was "a grave national danger" (Pearson, 1914, pp. 3, 413).

From the 1920s to the 1960s most of the leading biological and social scientists in Britain subscribed in varying degrees to eugenics. Among the biologists and geneticists, the foremost of these were Sir Ronald Fisher, Sir Julian Huxley, Sir Peter Medawar, J. B. S. Haldane, and Francis Crick. Ronald Fisher (1890–1962) was professor of genetics at Cambridge and was the leading mathematical geneticist of the early and middle decades of the twentieth century. He integrated single-gene Mendelian genetics with multiple-gene (polygenetic) inheritance by demonstrating mathematically that polygenetic processes could be explained as the joint action of a number of genes acting in accordance with Mendelian principles. Fisher (1929) discussed the breakdown of natural selection and the onset of genetic deterioration in his book *The Genetical Theory of Natural Selection*. He set out evidence for an inverse socioeconomic status gradient of fertility, such that the highest social classes had the fewest children and the numbers of children increased steadily with declining socioeconomic status. Fisher argued that centuries of social mobility had led to a concentration of the genes for intelligence and moral character in the higher social classes and hence that their low fertility must entail genetic deterioration. He suggested that dysgenic fertility appears in all mature civilizations, giving as examples ancient Greece and Rome and Islam, and argued that dysgenic fertility was responsible for their ultimate collapse. Fisher (1929) went on to make some recommendations on how Britain might avert this fate. "The most obvious requirement is that reproduction should be somewhat more active among its more successful than among its less successful members" (p. 276).

Fisher proposed two means by which this could be achieved. First, he believed that the higher social classes limited their fertility because children are expensive and that this disincentive to having children could be alleviated by child allowances paid by the state from taxes imposed on the childless. This might raise the problem that those who responded to these incentives by having large numbers of children might be predominantly the lower social classes. To overcome this difficulty, Fisher proposed that child allowances should be paid as a proportion of the earnings of the fathers. Thus the high earning genetic elite would receive higher child allowances than the average earners and much higher allowances than the genetically impoverished low earners. Second, Fisher (1929) proposed that moral pressures might be brought to bear on those with desirable genetic qualities to have more children and that steps should be taken to instill "the knowledge that parenthood by worthy citizens constitutes an important public service" (p. 283).

Julian Huxley (1887–1975) was a biologist and geneticist who was successively chairman of the genetics department of Rice Institute in Texas, professor of physiology at Kings College, London, and director of the United Nations Educational, Scientific and Cultural Organization (UNESCO). Huxley's (1942) major work was in the field of evolutionary biology, in which his *Evolution, the Modern Synthesis* integrated genetics with the theory of evolution. Huxley was a leading member of the British Eugenics Society, of which he was president from 1959–62. The British Eugenics Society stages an annual conference, the highlight of which is the Galton lecture delivered by an eminent scholar. Huxley gave the Galton lecture on two occasions, in 1936 and in 1962. In the first of these he reaffirmed that natural selection had become greatly relaxed in contemporary civilizations, noting that "the elimination of natural selection is largely, though of course by no means wholly, rendered inoperative by medicine, charity, and the social services" and that dysgenic fertility was leading to "the tendency to degradation of the germ plasm," the result of which will be that "humanity will gradually destroy itself from within, will decay in its very core and essence, if this slow but insidious relentless process is not checked." To counteract this deterioration, "we must set in motion counterforces making for faster reproduction of superior stocks" (Huxley, J. S., 1936, p. 30). Five years later Huxley (1941) expanded these views in *Man Stands Alone*, in which he reaffirmed that the inverse association between socioeconomic status and fertility was having a dysgenic effect because the professional class is "a reservoir of superior germ plasm of high average level, notably in regard to intelligence" (p. 70). He envisaged that eugenics would in due course become universally accepted and that "once the full implications of evolutionary biology are grasped, eugenics will inevitably become part of the religion of the future" (p. 22).

J. B. S. Haldane (1892–1964) was professor of genetics at University College, London, and one of the leading mathematical geneticists of the middle decades of the twentieth century. He was a member of the British Eugenics Society and in 1924 published a book *Daedalus*, which set out in favorable terms a eugenic future. In 1936 he expressed his approval of Hermann Muller's scheme for a sperm bank containing the semen of the genetic elite, among which Haldane numbered himself and to which he volunteered to make a personal contribution (Clark, R., 1968).

Peter Medawar (1915–1992) was professor of zoology at University College, London. In 1960 he wrote *The Future of Man*, in which he discussed the issue of whether intelligence in Britain was declining. He argued that intelligence is significantly determined by heredity and that there was extensive evidence that children from large families had lower intelligence than those from small families. From these facts he concluded that "there is a fair case for the belief that intelligence is declining" (Medawar, 1960, p. 81). This decline would need eugenic measures to correct, but he wrote, "I advocate a humane solution of the problems of eugenics" (p. 99). In the same year, Medawar was awarded the Nobel Prize for his work on immunological intolerance.

Francis Crick is a geneticist who in 1953, in collaboration with James Watson, discovered the double-helix structure of deoxyribonucleic acid (DNA), for which they were jointly awarded the Nobel Prize. In 1963, Crick attended a conference organized by the Ciba Foundation on the theme "Man and His Future." One of the sessions at the conference was concerned with dysgenic fertility and genetic deterioration, and Crick was a discussant. He began by saying that he accepted that genetic deterioration was occurring and that he thought "we would all agree that on a long-term basis we have to do something" (Crick, 1963, p. 274). He suggested that a possible solution would be to levy a tax on children, payable by their parents, which would deter the reproduction of the poor more than that of the rich. This suggestion was premised on the assumption that the rich were in general better endowed with the genetically desirable qualities of intelligence and character than the poor were. He also suggested that it was time to challenge the belief that everyone has a right to have children; he suggested that some people are not fit to be parents, and that a system of licensing procreation might be introduced so that "if the parents were genetically unfavorable, they might be allowed to have only one child, or possibly two under certain special circumstances" (p. 274).

Many of the leading British psychologists of the early and middle decades of the twentieth century supported eugenics. The supporters included Charles Spearman, Sir Cyril Burt, Sir Godfrey Thomson, and Raymond Cattell. Charles Spearman (1863–1945) was professor of psychology at University College, London, and is remembered for his formulation of the construct of g, the general factor present in the performance of all cognitive tests. Spearman was a member of the British Eugenics Society, and in 1912 he wrote a paper in which he proposed that only those individuals who scored reasonably highly on g should be permitted to vote and to have children (Hart & Spearman, 1912).

Cyril Burt (1883–1971) was also professor of psychology at University College, London. He carried out one of the first studies on the heritability of intelligence, in which he collected a set of identical twins who were separated shortly after birth and were reared in different families. He found they were closely similar for intelligence, the correlation between the twin pairs being .77, indicating a high heritability (Burt, 1966). Later in the century, critics like Kamin (1974) were to assert that Burt's work was fraudulent, but

Bouchard (1993) has shown that subsequent studies of the correlation of separated identical twins for intelligence yield a figure of .72, very close to Burt's figure. Burt also did some of the early work on the calculation of the rate of decline of intelligence in Britain. He collected data on the tendency of the intelligence of children to fall with increasing family size, from which he inferred that parents with low IQs were having large numbers of children, and from which he calculated that the mean IQ in Britain was declining at a rate of approximately two IQ points a generation (Burt, 1952).

Sir Godfrey Thomson (1881–1955) was a professor at the University of Edinburgh who also worked on the decline of intelligence in Britain, collected data on the inverse association between intelligence and numbers of siblings, and reached the same conclusion as Burt on the rate of deterioration (Thomson, 1946).

Raymond Cattell (1904–1997) also collected data on the association between the intelligence of children and the numbers of their siblings and also found that it was significantly negative. He estimated that the deterioration of the national intelligence was 3.2 IQ points per generation (Cattell, 1937). It was Cattell who spelled out most fully the consequences of deteriorating intelligence, which he predicted would be a decline in the quality of cultural life and an increase in mental retardation, unemployment, and crime.

Leading British geneticists and psychologists were not the only ones who supported eugenics in the middle decades of the twentieth century. Many prominent academics in other specialties and many laypeople supported eugenics. They included Bertrand Russell, the philosopher whose *Principia Mathematica* set out the logical foundations of mathematics and who wrote, "There can be no doubt that the civilization produced by the white races has this singular characteristic, that in proportion as men and women absorb it, they become sterile. The most civilized are the most sterile, the least civilized are the most fertile, and between the two there is a continuous gradation. At present the most intelligent sections of the Western nations are dying out" (Russell, 1930, p. 125).

Other prominent individuals who supported eugenics included Marie Stopes, the indefatigable birth control campaigner who established the first family planning clinics in London in 1921 in order to spread the knowledge and practice of contraception to the working classes; George Bernard Shaw, whose play *Man and Superman* concerned a eugenic union between a British country squire and a "highly civilized Jewess," which resulted in a superior son; Beatrice Webb (1948), the Fabian Socialist, who considered that Shaw's play raised "the most important of all questions, this breeding of the right sort of man" (p. 256); Sir Winston Churchill, Britain's prime minister during World War II; Maynard Keynes, the economist whose prescription of deficit financing cured for half a century the problem of mass unemployment in Western nations; Sir William Beveridge, the principal architect of the welfare state; William Inge (1927), the dean of St. Paul's Cathedral in London, whose

Outspoken Essays asserted the Christian case for eugenics; Leonard Darwin (1926), the youngest son of the great biologist and the author of *The Need for Eugenic Reform*, which set out a series of eugenic policy proposals; and H. G. Wells (1905), whose *A Modern Utopia* described a utopian state based on eugenic and socialist principles. In Well's utopian society the citizens are not permitted to have children until they have worked for a certain number of years and are free of debts. Genetic undesirables are identified as "idiots and lunatics, perverse and incompetent persons, people of weak character who become drunkards, drug takers, and the like, and persons tainted with certain foul and transmissible diseases. All of these people spoil the world for others. They may become parents, and with most of them there is manifestly nothing to be done but to seclude them from the great body of the population" (Wells, 1905, p. 83). This would be done by deporting them to an island that would be guarded to prevent escapes.

2. EUGENICS IN THE UNITED STATES

Eugenics was taken up in the United States in the first decade of the twentieth century. In 1906, the American Breeders' Association, renamed the American Genetics Association in 1913, set up a Committee on Eugenics to promote work on the concept; and in 1910 the Eugenics Record Office was established at Cold Spring Harbor on Long Island, New York, as a center for eugenic research and publication. The American Eugenics Society was formed in 1923. Many prominent American biological and social scientists subscribed to eugenics in the early and middle decades of the twentieth century. Biologists included Charles Davenport, Harry Laughlin, Hermann Muller, Linus Pauling, and Joshua Lederberg.

Charles Davenport was a geneticist and was the first director of the Eugenics Record Office, where he carried out early research on the action of dominant and recessive genes in humans. His book *Heredity in Relation to Eugenics* (Davenport, 1911) was the first major statement of the case for eugenics to be made in the United States. In 1910 Davenport invited Harry Laughlin to become superintendent of the Eugenics Record Office at Cold Spring Harbor. Laughlin was more of a practical than a theoretical eugenicist and was responsible for drafting and promoting a number of the sterilization laws introduced in a number of U.S. states from the time of World War I onward.

Hermann Muller (1890–1967) was one of the foremost geneticists of the middle decades of the twentieth century. His principal work was on the effect of X rays on increasing the number of genetic mutations, for which he was awarded the Nobel Prize. Muller believed the rate of mutations was increasing in industrial societies as a result of a variety of pollutants, such as the use of pesticides on crops. He argued that as most mutations are harmful, this was increasing the number of adverse genes and was having a dysgenic effect. He was also concerned about dysgenic fertility in respect to intelligence and character qualities, and he proposed the establishment of an elite semen bank, which women would be encouraged to use to produce genetically superior children (Muller, 1935). In 1939 Muller drew up "The Geneticists' Manifesto," which addressed the issue of how the world's population might be improved genetically. The "Manifesto" stated that "some kind of conscious guidance of selection is called for" (Muller, 1939, p. 64). The "Manifesto" was signed by a number of the leading geneticists of the time, including J. B. S. Haldane and Julian Huxley. In the early 1960s Muller (1963) restated his fears about genetic deterioration and reaffirmed the necessity of taking corrective eugenic measures.

Another American biologist who subscribed to eugenics was Linus Pauling (1901–1994). In 1959 he wrote that the "human germ plasm, which determines the nature of the human race, is deteriorating. . . . Defective genes are not being eliminated from the pool of human germ plasm so rapidly as in the past, because we have made medical progress and have developed feelings of compassion such as to make it possible for us to permit the individuals who carry the bad genes to have more progeny than in the past" (Pauling, 1959, p. 684). Some years later Pauling (1968) suggested that carriers of the sickle cell anemia gene should be branded on the forehead so that they could identify other carriers and be careful to avoid having children with them. Possibly this suggestion was made to illustrate the point of the desirability of eugenic matings and was not intended to be taken seriously. Pauling's major research achievement was his discovery of the biochemical abnormality of the hemoglobin molecule in the gene responsible for sickle cell anemia, for which he was awarded the Nobel Prize in 1954.

Another eminent American geneticist who supported eugenics was Joshua Lederberg, head of the department of genetics at Stanford University from 1959 to 1978. In the early 1960s Lederberg (1963) wrote on the problem of genetic deterioration in modern societies and what might be done to correct it: "Most geneticists are deeply concerned over the status and prospects of the human genotype. Human talents are widely disparate; much of the disparity has a genetic basis. The facts of human reproduction are all gloomy—the stratification of fecundity by economic status, the new environmental insults to our genes, the sheltering by humanitarian medicine of once lethal defects" (p. 264). He went on to discuss possible eugenic solutions to the problem of genetic deterioration and suggested that in the future genetic engineering might be the best way ahead, rather than the traditional eugenic proposals for altering fertility. In 1958 Lederberg was awarded the Nobel Prize for his work on genetic engineering in bacteria by the introduction of new genes.

Many prominent American psychologists and other social scientists were also committed eugenicists in the early and middle decades of the twentieth century. They included Edward Thorndike, Lewis Terman, William McDougall, and Frederick Osborn. It was Thorndike who formulated the "law of effect," the first and most fundamental law of psychology, which states that actions that are followed by rewards tend to be repeated and those followed by punishments tend to be discontinued. In regard to eugenics, Thorndike (1913) wrote that "selective breeding can alter man's capacity to learn, to keep sane, to cherish justice or be happy. There is no more certain and economical a way to improve man's environment as to improve his nature" (p. 134).

Lewis Terman was professor of psychology at Stanford University and a leading psychologist of the time. He came to know of the work of Alfred Binet in France on the construction of the first intelligence test. Terman had the test translated into English and standardized in the United States, where it was published as the Stanford-Binet test. Terman's early studies showed that the intelligence of children varied in accordance with the socioeconomic status of their parents, and he asserted that "the children of successful and cultivated parents test higher than children from wretched and ignorant homes for the simple reason that their heredity is better" (Terman, 1922, p. 670). It was Terman who set up a study of approximately 1,500 highly intelligent California children. These children were followed up over their life span and provided an important source of data on their subsequent educational and occupational attainments, health and lifestyle, and marriage and on the intelligence of their children. Virtually all of the 1,500 children did well in life, showing that intelligence is an excellent predictor of occupational achievement and success (Terman & Oden, 1959).

William McDougall was one of the most influential psychologists of the early decades of the twentieth century. He was born in England, worked first at the University of Oxford, and then emigrated to the United States to take up a professorship at Harvard. McDougall believed in the importance of genetics as a determinant of human behavior and proposed the existence of a dozen or so *instincts*, such as those of love, sex and aggression, which he defined as genetically programmed "propensities" to behave in particular ways. This theory became discredited with the rise of environmentalism in the middle decades of the twentieth century; but in the 1970s it was rediscovered by Edward Wilson and given the name *sociobiology*. In his book *Is America Safe for Democracy?* McDougall (1921) argued that the genetic quality of the population was important for civilization and that the rules on immigration should take into account the quality of immigrants as a criterion for acceptance. In a later book, McDougall (1939) argued that the fall of civilizations could be explained by genetic deterioration caused by the low fertility of the elites and that "looking at the course of history, we may see in the tendency of the upper strata to fail to reproduce themselves an explanation of the cyclic course of civilization" (p. 260).

Frederick Osborn (1890–1975) was another influential American eugenicist of the middle decades of the twentieth century. Osborn's major work was his *Preface to Eugenics* (1951), in which he presented further evidence that fertility in the United States and other Western nations was dysgenic in regard to intelligence and character qualities. He expressed doubts about whether any corrective policies would be feasible in democratic societies. Except for the sterilization of the mentally retarded, which he supported, he proposed that eugenicists should concentrate on attempting to improve the environment for the poorer classes to enable them to realize their genetic potential more effectively. He believed that if this could be done, dysgenic fertility would soon be reversed as a result of improved living standards and better education and of information about birth control, so that contraception would come to be practiced efficiently throughout all sections of society. Once this happened, the better genetic stocks would have more children. He gave the name "the eugenic hypothesis" for what he hoped would be the emergence of a spontaneous solution to the problem of dysgenic fertility.

Support for eugenics in the United States was not confined to academic specialists in population biology, genetics, and psychology. As was the case in Britain, many prominent nonspecialists in the United States supported eugenics. They included Theodore Roosevelt, U.S. president from 1901 to 1909; Charles Wilson, president of Harvard; Irving Fisher, president of Yale, who was also president of the Eugenics Research Association in the 1920s; Margaret Sanger, the feminist and birth control campaigner who established the first family planning clinics in New York City; and many leading judges who were in favor of sterilization on eugenic grounds, such as Oliver Wendell Holmes (Holmes delivered the verdict of the Supreme Court supporting the sterilization of the mentally retarded teenage mother Carrie Buck in 1927 in *Buck v. Bell*). Several of the major U.S. foundations supported eugenic research, including the Carnegie Institution, which funded Davenport's eugenic studies at Cold Spring Harbor, and the Rockefeller Foundation, which gave grants in the 1930s for eugenic research to the Galton Laboratory at University College in London and to the Cornell Medical School in New York.

3. EUGENICS SPREADS WORLDWIDE—GERMANY

In the early decades of the twentieth century, eugenics gained support throughout the economically developed world. Societies for the promotion of eugenics were established throughout Europe; in Australia, Canada, and Japan; and in several countries of Latin America. Prominent eugenicists in Continental Europe included the French physician Alexis Carrell (1873–1944). Carrell won the Nobel Prize for Medicine in 1912 for his research on human tissue and organ transplants; his 1935 book, *L'Homme, cet inconnu*, advocated the execution on eugenic grounds of murderers and those convicted of serious criminal offences, such as child kidnappers, and argued that Western societies contained genetic aristocracies of the highly gifted whose fertility should be encouraged.

A useful exposition of the development of eugenics in these various countries has been given by Broberg and Roll-Hansen (1996), and no further account will be given here, with the exception of Germany. A eugenics society was established in Germany in 1905, and a number of leading geneticists, biologists, and social scientists were members. For the first quarter century or so of its existence, it functioned like the eugenics societies of other countries as a forum of debate. In 1923 a professorship of "race hygiene" was established at the University of Munich, to which Fritz Lenz was appointed. At this time the word *race* was used for what would later be called "population" and did not have the meaning it acquired in the second half of the twentieth century. In 1927 the Kaiser Wilhelm Institute for Anthropology, Human Heredity, and Eugenics was established in Berlin. The first director was Eugen Fischer. In 1942, Otmar von Verschuer succeeded to the directorship of the Institute. In later years he was often attacked because Joseph Mengele, who had worked with him as a graduate student, later became a physician at the concentration camp at Auschwitz, where he carried out experiments on the inmates.

In 1933 the Nazis came to power and began to put eugenics into practice. Their Eugenic Sterilization Law required physicians to report to Hereditary Health Courts any patients with mental retardation, psychosis, and serious genetic physical defects. These courts, which consisted of two physicians and a lawyer, decided whether the patients should be sterilized. A positive decision could be appealed to a higher court, and if the decision was upheld the sterilization was carried out. The procedures were investigated by Marie Kopp (1936), a U.S. sociologist, who reported that she was "convinced that the law is administered in entire fairness." It has been estimated that between 1933 and 1939, 300 to 400 Germans were sterilized, about half of whom were mentally retarded and the other half were mentally ill and physically disabled (Bok, 1983; Muller-Hill, 1988). The program was reduced in 1939 partly because of public opposition and partly because so many of the physicians carrying out the work were drafted into military service. The Nazi program of sterilization was heavily criticized in the later decades of the twentieth century, but it did not differ in principle from similar programs that had been in place in a number of the states in the United States since World War I, and also in a number of countries in Europe. As a proportion of the population, more sterilizations were carried out in Sweden than in Nazi Germany.

It has often been asserted that the Nazi euthanasia program consisting of the killing of the mentally retarded and the mentally and the incurably ill in hospitals was motivated by eugenics. This is a misconception. The objective of this program was to save the expense of maintaining these patients. As Hanauske-Abel (1996), an historian of the period, has written, "The program was geared towards economic performance in the health care market, improvement of institutional and national revenues, and cost-efficient utilization of limited resources" (p. 1458). The program was also designed to release resources for military purposes. As another historian of the period has written: " Hitler's wartime authorization of an adult euthanasia program was conceived as an economy measure, a means of creating emergency bedspace, and hostels for ethnic German repatriates from Russia and Eastern Europe" (Burleigh, 1994, p. 223).

In 1941 the Nazis set up the concentration camps for the killing of the Jews. It has often been asserted that the killing of the Jews was motivated by eugenics and indeed that eugenics inevitably leads in due course to the killing of "undesirables." For instance, Kevles (1992) writes that "the eugenics movement prompted the sterilization of several hundred thousand people and helped lead, of course, to the death camps" (p. 101). Contrary to this frequent assertion, the Nazis did not kill the Jews on eugenic grounds. It is true that Hitler and the Nazis and many of the German academic and scientific eugenicists of the Nazi era were anti-Semitic, but they did not assert that the Jews were intellectually inferior. D. J. Galton and C. J. Galton (1998) have correctly written on this issue that "Hitler did not justify his social policies on the basis of Darwinism or eugenics" (p.101). Hitler's anti-Semitism was based not on eugenic grounds but on two related beliefs: first, that many Jews were communists intent on overthrowing the German state and securing world domination by a Jewish-dominated communist elite; and, second, that the Jews and the Germans, or "Aryans," were the two most able races and that there would be a struggle between them for world domination (Gordon, 1984; MacDonald, 1994). In his *Mein Kampf*, Hitler (1943) described the Jews as "the mightiest counterpart to the Aryan" (p. 300), and he wrote that he asked himself "whether inscrutable Destiny . . . did not desire the final victory of this little nation" (p. 64). It was to attempt to secure victory in this struggle that Hitler determined to exterminate the Jews. The high abilities of the Jews were well recognized in Germany in the 1930s. As the German Jewish novelist Jacob Wasserman wrote, "Nearly all the people with whom I came into intellectual or cultural contact were Jews. I soon recognized that all public life was dominated by Jews. The banks, the press, the theater, literature, social functions, all were in the hands of the Jews" (Gay, 1988, p. 21). Much of the anti-Semitism in pre–World War II Germany was fueled by resentment and envy over Jewish achievements and power.

Some of the German scientific eugenicists in the 1930s held views about the undesirability of the Jews that were misconceived. The first of these was that Jews diluted the purity of the Aryan race. Some believed that racial hybrids tended to degenerate, and many shared the view that racial intermarriage customarily produced weakness, degeneracy, and illness, rather than hybrid vigor, and that race mixing produced the decline of civilizations (Winston, 1997, p. 183). There is, however, no evidence that race mixing is dysgenic. The second theme in the anti-Semitism of some of the German scientific eugenicists was that the Jews had evolved genetic qualities that made them good as middlemen in such occupations as money lenders and traders but that they were not good at production. They viewed the Jewish qualities as "specialized for a parasitic existence" (Tucker, 1994, p. 125). The idea that money lenders and middlemen are parasitical and do not make a positive contribution to a nation's economy is, of course, economically illiterate, but it was nevertheless held by a certain number of German biologists and geneticists in the 1930s.

Just how far the beliefs of German scientific eugenicists contributed to the Holocaust is difficult to determine. Hitler was certainly virulently anti-Semitic but not on eugenic grounds and would very likely have set in train "the final solution" even if his ideas had not been supported by some of the German scientific eugenicists. Although it was frequently asserted in the later decades of the twentieth century that the Holocaust was motivated by eugenics, it is doubtful whether eugenics had anything to do with the extermination of the Jews.

4. RESEARCH ON THE SCIENTIFIC BASIS OF EUGENICS

In the closing decades of the nineteenth century, when Galton advanced the concept and the objectives of eugenics, and in the early decades of the twentieth century, when eugenics societies were founded throughout the developed world, the scientific basis of eugenics was quite weak. Eugenics consists of a number of interlocking propositions, the principal of which are that many diseases have genetic causes; that intelligence and character qualities are significantly determined by heredity; that health, intelligence, and character could be improved by selective reproduction; and that modern populations are experiencing genetic deterioration of intelligence and character because of dysgenic fertility.

During the twentieth century, research was carried out on all these issues. They were all found to be correct, and this considerably strengthened the scientific basis of eugenics. The laws of heredity and the mode of operation of genes were worked out in the 1860s by Gregor Mendel, but his work was not recognized by his contemporaries. It was not until the first decade of the twentieth century that Mendel's work was discovered and understood. This enabled eugenicists to think more precisely about the genetic processes involved in eugenics. They understood that there are single genes responsible for many diseases and for some forms of mental retardation and that there are many genes determining intelligence and character. They realized that the objective of eugenics should be to increase the number of desirable genes and to reduce the number of the undesirable ones.

As the twentieth century progressed, the genes responsible for increasing numbers of genetic disorders were discovered, including those for certain forms of mental retardation and mental illness. In the second decade of the twentieth century, the American eugenicist Charles Davenport (1916) discovered that Huntington's disease, the crippling physical and mental disorder of early middle age, is determined by a single dominant gene and hence could be eliminated entirely from the population in one generation (except for a small number of mutations) if no one with the condition had children. In the early 1930s, the Norwegian physician Ivar Folling (1934) discovered that the disorder of phenylketonuria (PKU), which causes profound mental retardation, is inherited through a recessive gene and is consequently much more difficult to eliminate.

The genes responsible for intelligence and character remain largely undiscovered, but it came to be understood that these characteristics are almost certainly determined by a number of them. In the 1930s, it was shown that both intelligence and character have a substantial genetic basis. In England, Herrman and Hogben (1932) showed that identical twins were much more similar for intelligence than fraternal (nonidentical) twins were. In the United States, Newman, Freeman, and Holzinger (1937) showed that identical twins reared in different families were closely similar for intelligence, indicating that intelligence has a strong genetic basis. In Denmark, Lange (1929) found that identical twins were much more similar in respect of having criminal records than nonidenticals were, and, therefore, that there is a strong genetic basis for criminality, one of the foremost expressions of weak moral character. Numerous studies were to confirm these results as the twentieth century unfolded, until by the end of the century, no serious scholar disputed them.

The development of the intelligence test by Alfred Binet in France in 1905 and the adoption of these tests in the United States and Britain made it possible to examine quantitatively the relationship between intelligence and socioeconomic status. It was soon found that the social classes differ substantially in intelligence, such that the professional class and their children averaged some 25 IQ points higher than the unskilled working class and their children (Terman, 1922; Cattell, 1937; Johnson, 1948). Studies of adopted children showed that the intelligence of children was strongly related to the socioeconomic status of their biological fathers, whom they had never seen (Jones & Carr-Saunders, 1927; Lawrence, 1931). These studies showed that socioeconomic status differences in intelligence have some genetic basis.

Early in the twentieth century, demographic studies were carried out using census data on the relationship between social class and fertility. The censuses of 1900 and 1910 in the United States and of 1911 in Britain showed an inverse social class gradient for fertility. The professional classes were having only around half the number of children of the lowest social class (Fisher, 1929). Taken in conjunction with the studies showing that there are genetically based differences in intelligence between the social classes, the conclusion that the populations of the economically developed countries were deteriorating genetically became inescapable. This realization was further confirmed by numerous studies showing that both intelligence and crime are related to numbers of siblings, such that both criminals and those with low intelligence have a strong tendency to come from large families. This implies that parents with low intelligence and criminal propensities have higher fertility. Later in the century direct evidence was to appear showing that the unintelligent and criminals have larger than average numbers of children. (I have reviewed all this evidence in my book *Dysgenics*.) By the end of the twentieth century, the research evidence showed beyond dispute that the eugenicists were right in their belief that genetic deterioration is taking place in modern populations throughout the economically developed world.

The eugenicists' contention that intelligence and character are socially valuable was also substantiated by numerous studies. Both were found to be important determinants of educational, occupational, and intellectual achievement, while low levels of both were found to be determinants of the social problems of long-term unemployment, crime, and welfare dependency (e.g., Herrnstein & Murray, 1994; Lynn, 1996; Jensen, 1998). Thus, by the end of the twentieth century, all the essential propositions of eugenics had been confirmed by research.

5. PROMOTION OF BIRTH CONTROL

The eugenicists were successful in achieving the implementation of three principal policies. These were, first, the promotion of knowledge of birth control; second, the sterilization of those with genetic diseases and disorders, low intelligence, and weak moral character; and third, the control of immigration.

As far as the promotion of knowledge about birth control was concerned, in the early decades of the twentieth century the eugenicists believed that following the development of the modern latex condom in the 1870s and the cervical diaphragm in the 1880s, these reliable forms of contraception were used disproportionately by the better educated, more intelligent, more prudent, and more responsible and those with stronger moral character, and that the result of this was that these people were having fewer children than the poorly educated, unintelligent, imprudent, and less responsible and those with weaker moral character. They believed that this was the root cause of dysgenic fertility. They perceived that the most straightforward way of overcoming this problem was to promote knowledge about contraception to the less intelligent, the less well informed, and the less responsible and to provide family planning clinics in which these people could obtain contraceptives and contraceptive information.

There were problems in promoting knowledge about birth control and providing birth control facilities in the late nineteenth and early twentieth centuries because anyone who attempted this provoked opposition from those who thought these matters should not be made public and because there was a danger of prosecution under obscenity laws. The severity of these obstacles varied from country to country. One of the first countries in which these difficulties were overcome was the Netherlands, where the first family planning clinic was opened in 1882 (Soloway, 1990). However, in the United States and Britain, the problems of disseminating information and providing facilities for birth control were more formidable.

In the United States the leading campaigner for the legalization and provision of birth control was Margaret Sanger, who set up the first family planning clinic in New York City during World War I. The clinic was declared illegal, and Margaret Sanger was imprisoned for contravention of the obscenity laws. Nevertheless, after further campaigning, the legal impediments to the provision of family planning advice were overcome. In the interwar years, the obscenity laws prohibiting family planning clinics were repealed in most of the U.S. states. By the 1940s, contraception had been legalized in all states except Connecticut and Massachusetts, where it remained a criminal offence for a physician to give out information on birth control under any circumstances, even to married women, until the 1950s.

In Britain the leading campaigner for the dissemination of knowledge about birth control in the early decades of the twentieth century was Marie Stopes. Like Margaret Sanger, Stopes was a keen eugenicist. She summarized her objective as "more children from the fit, less from the unfit—that is the chief issue of birth control" (Hall, 1977, p. 82). She also had a feminist agenda to promote the release of women from the burden of having more children than they were able to look after. Marie Stopes set up the first two birth control clinics in Britain in London in 1921. It was largely through her tireless work that similar campaigns to promote the knowledge and the use of birth control were undertaken by eugenicists in a number of other countries and that an initially reluctant British medical profession came to accept the desirability of providing birth control services. In 1930 the Royal Institute of Public Health introduced courses on contraception for physicians and medical students, and the Ministry of Health issued a memorandum authorizing local authorities to establish birth control clinics. By 1930 the long struggle for the acceptance and the provision of birth control in Britain had been won in principle, although another generation was to pass before birth control clinics were widely established throughout the country.

In Germany, eugenicists successfully lobbied for the establishment of state-financed genetic and marriage counseling centers to provide advice on contraception. As a result of all these campaigns, knowledge and use of contraceptives spread throughout the populations of the economically developed nations, and the magnitude of the socioeconomic-status fertility differentials declined as the century progressed, although they were not completely eliminated. The promotion of the knowledge and the use of birth control among lower socioeconomic status groups made a significant contribution to the reduction of dysgenic fertility and was a major policy achievement for eugenics.

6. STERILIZATION

The eugenicists' second major policy achievement was securing the legalization and implementation of the sterilization of the mentally retarded, the insane, and criminals. The eugenicists believed that mental retardation, insanity, and crime are social evils, are partly under genetic control, and tend to be transmitted from parents to children. It followed that the incidence of mental retardation, insanity, and crime could be reduced if those with these conditions could be prevented from having children, and the most straightforward way of securing this was to sterilize them. The eugenicists' objective of securing the sterilization of these groups achieved its first success in the United States in 1907, when the state legislature of Indiana passed a law "to prevent the procreation of confirmed criminals, idiots, imbeciles, and rapists" by sterilizing them. The law specified that potential cases for sterilization should be considered by a board of physicians and other professionals who would decide whether sterilization was appropriate.

By 1913 this law had been adopted by 12 states; and by 1931, by 30 states, and these laws were implemented on a substantial scale. Orders for sterilization were sometimes challenged in the courts. In 1927 a landmark decision on the sterilization of the mentally retarded in the United States was made in the Supreme Court in *Buck v. Bell*. The case concerned the legality of sterilizing a young woman named Carrie Buck on the grounds that she was mentally retarded. The decision of the Virginia court to permit the sterilization was upheld by the Supreme Court. It has been estimated by Ludmerer (1972) that by 1935 about 20,000 sterilizations had been carried out and that by 1970 this figure had risen to about 60,000, of whom about half were mentally retarded and half were psychiatric patients and criminals. From the early 1970s some sterilizations continued to be performed in the United States, but the number was greatly reduced as a result of legal challenges, changes in sentiment, and pressures from civil liberties groups.

Sterilization laws similar to those in the United States were introduced in 1928 in the Canadian province of Alberta, in Denmark, and in Switzerland; in 1933 in Germany, in 1934 in Norway and Sweden, and in 1935 in Finland. In 1997 it emerged that approximately 60,000 people had been sterilized in Sweden between 1934 and 1976. This was about double the number as a proportion of the population that had been sterilized in Nazi Germany between 1933 and 1939. Eugenic sterilization was introduced in Japan in 1948, allowing physicians to sterilize those with mental or physical handicaps or certain hereditary diseases without their consent. It is estimated that 16,520 Japanese women were involuntarily sterilized between 1949 and 1995. The Japanese Sterilization Law was revoked in 1996.

Britain was one of the few countries among the economically developed nations that did not have a sterilization program until the early 1990s, when a judicial decision ruled that mentally retarded girls and young women could be sterilized, subject to the consent of their parents or guardians and of their physicians. For most of the twentieth century, the eugenic objective of preventing the mentally retarded and mentally ill from having children was secured by the segregation of men and women in institutions. In 1913, the British Parliament passed The Mental Deficiency Act, which provided for the compulsory custody of the mentally retarded in institutions and the segregation of male and female inmates to prevent them from having children. This provision was explicitly based on the eugenicists' arguments that mental retardation is largely hereditary, that many of the children of retardates are themselves retarded, that these are a social problem, and that it would be desirable to prevent the mentally retarded from having children. It may be thought that the British solution of segregating the sexes was more humane than sterilization; but it can equally well be regarded as harsher because those who are sterilized can continue to enjoy personal and sexual relationships, whereas these are denied by segregation.

7. IMMIGRATION CONTROL

Eugenicists in the United States secured a third success in the control of immigration in the 1924 Immigration Act. From the 1880s onward there had been a large influx of immigrants into the United States from southern and eastern Europe, and eugenicists of the time were concerned that these people were of inferior genetic stock. Foremost among these eugenicists was Harry Laughlin (1912), who published a study showing that disproportionate numbers of southern and eastern European immigrants were in institutions for the mentally retarded and insane, from which he inferred that these populations were generally of poor genetic quality. In 1920 Laughlin was appointed "Expert Eugenical Agent" to the House of Representatives' Committee on Immigration and Naturalization, and in 1922 he reported to the Committee that the evidence showed that "recent immigrants were biologically inferior and that they jeopardized the blood of the nation" (Kevles, 1985, p. 103).

In the early 1920s eugenicists were not the only people in the United States concerned about large-scale immigration from southern and eastern Europe. For a number of years the annual numbers of immigrants had been in excess of one million. There was a widespread feeling that this was too many for the United States to be able to assimilate and that the numbers needed to be curtailed. Accordingly, in 1924 Congress passed the Immigration Restriction Act. The major provision of the act was that annual immigration from each European nation should be limited to 2 percent of U.S. residents born in that country. Since the great majority of Americans were from northwest Europe, the effect of this was to greatly reduce immigration from eastern and southern Europe. Kenneth Ludmerer, one of the leading historians of U.S. eugenics, has called the passing of the 1924 Immigration Act the greatest triumph of the eugenics movement. However, how much weight the eugenics argument carried in the minds of the Congressmen who passed the act has been disputed. Kamin (1974), Kevles (1985), and Gould (1981) maintain that eugenic considerations played a major part in the quota restrictions imposed by the act, but Herrnstein and Murray (1994) doubt this, pointing out that no reference to the intelligence of immigrants appears in the Congressional records of the time. However, politicians do not always like to put on paper their motives for passing legislation; and after the elapse of three quarters of a century, it is impossible to assess precisely the degree to which eugenic arguments contributed to the national quota restrictions imposed by the 1924 Immigration Act.

Although eugenic considerations may have played a part in the passing of this act, the belief of some U.S. eugenicists of the time that the peoples of eastern and southern Europe were genetically inferior to those of northwestern Europe was undoubtedly mistaken. Some studies had shown that immigrants from eastern and southern Europe scored lower in intelligence tests than those from northwest Europe. Foremost among these was Brigham's (1923) analysis of the test results of military conscripts in World War I, which concluded that U.S.-born Americans of northern European descent had a mental age of 13.3 years; foreign-born, non-English-speaking northern Europeans had a mental age of 13.4 years; foreign-born central Europeans, 11.7 years; foreign-born Mediterranean Europeans, 11.5 years; foreign-born Jews from eastern Europe, 11.5 years; and blacks, 10.7 years. These results were taken to indicate that immigrants from eastern and southern Europe had intelligence levels intermediate between northwestern European whites and blacks. The explanation for these results is probably that the tests were biased against immigrants from eastern and southern Europe, and this was not sufficiently understood at the time. There is substantial evidence from later work that indicates that immigrants from eastern and southern Europe do not have lower levels of intelligence than those from northwestern Europe. In fact Russian immigrants, who were largely Jewish, have scored above other whites with average IQs of around 115 (MacDonald, 1994). Furthermore, the average earnings of Americans of southern and eastern European origin have been about the same as those of northwestern European origin and sometimes higher. The earnings of all major ethnic groups in the United States have been analyzed from the 1980 census by Lieberson and Waters (1988). They present figures as percentages above or below overall U.S. average earnings. For men, the top group is the Russians, earning 28 percent above the average; next come the Hungarians, at 18 percent above average; the Poles are 14 percent and the Italians 4 percent above average; the Germans and the English stand at 11 percent and 7 percent above respectively. Earnings are a good proxy for intelligence and character, and hence these data show that immigrants from southern and eastern Europe have not been inferior to those from northern and western Europe.

While the restriction of immigration from eastern and southern Europe imposed by the 1924 Act cannot be justified on eugenic grounds, the fact that the act also imposed tight restrictions on immigration of Hispanics and blacks from the Caribbean and Africa was a eugenic achievement because there is considerable evidence that the intelligence levels of these peoples are lower than those of whites (Levin, 1997; Lynn, 1997; Jensen, 1998).

8. THE DECLINE OF EUGENICS

In the 1960s, support for eugenics began to decline. All over the world eugenics societies put themselves into voluntary liquidation. A critical year was 1969, in which the American Eugenics Society ended publication of its journal *Eugenics Quarterly* and replaced it with *Social Biology*, and the British Eugenics Society ended publication of the *Eugenics Review* and replaced it with the *Journal of Biological Science*. Neither of these two new journals devoted much of its content to eugenics. In 1972 the American Eugenics Society changed its name to The Society for the Study of Social Biology and dissociated itself from eugenics. Two years later the president of the new society, Frederick Osborn (1974), wrote of this change, "The society was groping for a wholly new definition of purpose. It was no longer thinking in terms of 'superior' individuals, 'superior' family stocks, or even of social conditions that would bring about a 'better' distribution of births. It was thinking in terms of diversity, in terms of the genetic attributes appropriate to different kinds of physical and social environments" (p. 126). This amounted to a total repudiation of eugenics. The British Eugenics Society lasted another two decades before it, too, distanced itself from eugenics in 1988 by changing its name to the Galton Institute. Eugenics was also repudiated in Continental Europe. In France, the Nobel prize winner, physician, and eugenicist Alexis Carrell had been honored by naming the medical faculty after him at the University of Lyons, and streets were named after him in Bordeau, Strasbourg, and several other cities. In the 1970s and 1980s Carrell's name was removed from the Lyons Medical Faculty and from the streets.

From the 1960s, eugenics began to be repudiated as regards both its policies and its intellectual basis. With respect to the policies, sterilization of the mentally retarded and criminals came to be considered an unacceptable denial of the human right to have children. Immigration control in favor of good quality immigrants was also attacked and, in the United States, largely removed by the Immigration Act of 1965, which abolished the preferences previously accorded to those from northwest Europe. With respect to eugenics's intellectual basis, it came to be increasingly asserted that the principal propositions of eugenics were wrong, that eugenics is a "pseudoscience" (Paul, 1995), that the eugenicists did not understand genetics, that all genes are equally valuable, that intelligence and personality traits have no genetic basis, and that eugenic policies would not work because they are based on false genetics. I have discussed these misconceived criticisms of eugenics in my book *Dysgenics* and will not consider them again here. These criticisms have an emotive rather than a rational basis and are expressions of a value system hostile to eugenics. The essence of this value system opposed to eugenics lies in the increasing precedence accorded to individual rights over social rights in the later decades of the twentieth century. In all societies, a balance has to be struck between individual rights and social rights. Individual rights consist of the personal liberties of individuals, whereas social rights consist of the rights of society to curtail the liberties of individuals in the interests of the society. In the early and middle decades of the twentieth century, social rights were recognized in the Western democracies, and the acceptance of the desirability of eugenic measures was an expression of this. For instance, in the early years of the twentieth century, the eugenicists Popenoe and Johnson (1918, p. 170) wrote that "so few people would now contend that two feeble-minded persons have any 'right' to marry and perpetuate their kind that it is hardly worth while to argue the point"; and Laughlin (1912, p. 110) wrote that "society must look upon germ-plasm as belonging to society and not solely to the individual who carries it." But by the closing decades of the twentieth century, as Diane Paul (1995, p. 71) has rightly said, "Almost no one today would profess such a belief. Indeed, the dominant view is now the opposite: that the nature of reproductive decisions should be no concern of the state."

The reason for this shift in opinion was that in the second half of the twentieth century, individual rights were extended over social rights over a broad front, such as in the legalization of abortion (women's right to choose), of pornography, and of the depiction of violence in films and on television. One of the most striking expressions of the priority accorded to individual rights over social rights in the second half of the twentieth century was the freedom allowed to those with HIV (human immunodeficiency virus) and AIDS. In previous historical times, people suffering from infectious diseases were compulsorily isolated from the healthy population. Examples were the permanent confinement of those with leprosy in leper colonies and the detention in quarantine of those with cholera, bubonic plague and the like, until they either recovered or died. Yet in the late twentieth century, people with HIV and AIDS were allowed complete liberty in the Western democracies, including the liberty of infecting others, and were allowed to travel freely and to enter the countries without any checks on whether they had HIV or AIDS. Some of those with these conditions have inflicted high social and individual costs in spreading the infection. They have been allowed to do so because of the priority accorded to individual rights over social rights.

In the second half of the twentieth century, a component of this general trend was an increasing acceptance of the right of those with genetic diseases and disorders, those with mental retardation, and criminals to have children, despite the social costs imposed by the genetic transmission of these pathologies; and this right came to be regarded as more legitimate than the social right of society to curtail the reproductive liberties of these groups. The fact that social rights ultimately involve the welfare of actual human beings was overlooked. Eugenics is premised on the assertion of social rights and in particular the right of the state to curtail reproductive liberties in the interests of preserving and promoting the genetic quality of the population. It was this change in values toward according greater precedence to individual rights at the expense of social rights that was the fundamental reason for the rejection of eugenics in the Western democracies in the closing decades of the twentieth century.

9. THE LAST EUGENICISTS

From 1970 until the end of the century, eugenics had only four articulate supporters in the Western democracies; Robert Graham, William Shockley, Raymond Cattell, and Roger Pearson. Robert Graham's (1970) book, *The Future of Man*, set out the evidence on dysgenic fertility for intelligence in Western societies and suggested a number of ways in which this might be reversed. He was also encouraged by the geneticist Hermann Muller to make a practical contribution to eugenic goals by setting up a sperm bank for the storage of semen donated by Nobel prize winners and other intellectually distinguished men for use by married women whose husbands were infertile and by others. By the mid-1980s he was able to report that more than 40 women had taken advantage of this facility and produced children whose IQs were well above average (Graham, 1987), and by 1997 this number had risen to 207 (Hotz, 1997).

William Shockley was a physicist who had won the Nobel prize for his invention of the transistor. He recorded that towards the end of his life he became interested in the problem of genetic deterioration of modern populations after reading a newspaper account of a mentally retarded woman with an IQ of 55 who had produced 17 children, most of whom were themselves mentally retarded or backward and a number of whom were criminals. Shockley examined the issues in a series of papers collated by Pearson (1992). Shockley concluded that the evidence showed that intelligence is largely under genetic control, that in the United States the intelligent were having fewer children than the unintelligent, that this was entailing genetic deterioration, and that steps needed to be taken to correct this. He also opened up the racial dimension of this problem by arguing that whites have higher intelligence but lower fertility than blacks and that dysgenic fertility was greater among U.S. blacks than among whites, increasing the IQ gap and leading to a deterioration of the social and economic position of the black population. Shockley pondered on how dysgenic fertility might be overcome and suggested in what he called "a thinking exercise" that everyone with an IQ of less than 100 should be offered a payment to be sterilized. He repeatedly urged the American National Academy of Sciences to sponsor research on the genetics of intelligence and on race differences, but the Academy declined to respond to these requests.

A new presentation of the case for eugenics was made by Raymond Cattell (1972) in his book *Beyondism*, a second and extensively modified edition of which appeared in 1987. The word *beyondism* is more or less a synonym for eugenics and was constructed to give emphasis to the ethical dimension of eugenics. Cattell argued for the development of a sense of moral obligation to future generations to enhance the evolutionary process and to produce an improved race of humanity "beyond" the present. The two editions consist partly of a restatement of the traditional arguments for eugenics and also break new ground in setting out a proposal for the advancement of eugenic principles through competition between a number of eugenic states. Cattell's work received little attention. His first book was not widely reviewed and was not even mentioned by Kevles (1985) in his historical account of the eugenics movement. A review of the second book by Jahoda (1989) condemned the whole program and expressed the hope that it would fail.

The remaining eugenicist of note is Roger Pearson, an anthropologist and director of the Institute for the Study of Man in Washington, D.C. In his book *Heredity and Humanity: Race, Eugenics and Modern Science*, Pearson (1996) presents a history of eugenics from classical times and argues that the eugenicists were right in their understanding of genetic deterioration in modern populations. He attributes the decline of eugenics and the loss of this understanding in the later decades of the twentieth century to an ideologically motivated denial of the role of heredity in the determination of human differences by a number of politically leftist academics, among whom he names Franz Boaz, Stephen Jay Gould, Leon Kamin, Richard Lewontin, and Stephen Rose. Pearson argues that the principal dysgenic effects acting over centuries in Western societies have been the requirement of celibacy for priests and nuns in the Roman Catholic Church, which prevented many of the most able from having children, and the adverse impact of a succession of wars, particularly those of the twentieth century in which greater mortality was suffered by the more able. He ends his book on a pessimistic note in which he predicts that if eugenic methods for the correction of dysgenic fertility and genetic deterioration are not found, the human species faces a real possibility of extinction.

Although, with these exceptions, eugenics became overwhelmingly rejected in the Western democracies in the closing decades of the twentieth century, eugenics has been more favorably regarded in Singapore. In Singapore, Lee Kuan Yew, the prime minister from 1959 to 1990, was a committed eugenicist who viewed with dismay the low fertility of women college graduates in Singapore. In an attempt to correct this, he introduced a variety of measures, including tax rebates for children, and a state-sponsored dating agency for graduates to assist them in finding suitable husbands and wives. In a speech delivered in 1983 he asserted, "If we continue to reproduce ourselves in this lopsided way, we will be unable to maintain our present standards. Levels of competence will decline, our economy will falter, the administration will suffer, and the society will decline." This and other excerpts from the speech are reported in Gould (1985), who ridiculed the argument on the grounds that education attainment has nothing to do with genetic quality. However, this is incorrect, because there are numerous studies showing that educational level has a substantial heritability and is strongly determined by intelligence and work motivation, both of which also have substantial heritability, and which I have summarized in detail in *Dysgenics*. There is no question that Lee Kuan Yew was right in regarding low fertility of women college graduates in Singapore as a symptom of dysgenics and genetic deterioration.

Eugenics was also favorably regarded in China in the 1990s. A Eugenic Law of 1994 made it compulsory for pregnant women to undergo prenatal diagnosis for the presence of genetic and congenital disorders in the fetus and to have abortions where these disorders have been diagnosed. Chinese physicians and geneticists are much more sympathetic to eugenics than are those in Western democracies. For instance, in a survey of attitudes to eugenic practices carried out from 1994 to 1996, 82 percent of Chinese physicians and geneticists supported the mandatory sterilization of, for example, a single blind woman on public welfare who has already had three blind children by three different men (all absent from the household), as compared with around 5 percent of physicians and geneticists in the Western democracies (Wertz, 1998). Between 92 percent and 98 percent of Chinese physicians and geneticists in this survey supported the termination of pregnancies where the father was diagnosed as having a serious genetic or congenital disorder and said they would give slanted advice to the women concerned to persuade them to consent to this, whereas only about 5 percent of physicians and geneticists in the Western democracies said they would do this. The reason for this difference between China and the West is that greater priority is given to social rights in China, whereas greater priority is given to individual rights in the Western democracies.

10. CONCLUSIONS

Galton's ideas on eugenics had considerable success during the first half of the twentieth century. His objectives were to convince the scientific community and the informed public that natural selection had ceased to function in Britain and other Western democracies; that as a consequence contemporary populations were deteriorating genetically in respect of health, intelligence, and moral character; that this posed a serious threat to the quality of civilization, and that eugenic policies were required to counteract these dysgenic trends.

These objectives were largely achieved during the 50 years or so following Galton's death in 1911. Eugenic societies for the promotion of eugenic research and programs were established throughout the economically developed world. Virtually all the leading geneticists and biologists accepted his analysis, including Ronald Fisher, the greatest mathematical geneticist of the century, and the Nobel laureate geneticists Alexis Carrel, Hermann Muller, Linus Pauling, Peter Medawar, Francis Crick, and Joshua Lederberg. Many of the leading social scientists and those from the wider intellectual community also supported eugenics.

During the course of the twentieth century, the scientific basis of eugenics was strengthened by research that clarified the nature of genetic transmission; that revealed the presence of gene action; that showed that health, intelligence, and moral character have significant heritabilities; and that dysgenic fertility has been present throughout the economically developed nations since the middle decades of the nineteenth century. The policy objectives of eugenics also secured some successes, of which the most important were the diffusion of knowledge and the practice of birth control throughout much of society, which had the effect of reducing the magnitude of dysgenic fertility; the reduction of the fertility of the mentally retarded and criminals by sterilization or segregation; and the restriction of immigration in favor of those with useful qualities and qualifications.

The achievements of the eugenics movement did have some limitations, however. The programs for the spread of the knowledge and the use of birth control did not succeed in eliminating dysgenic fertility, and large numbers of unplanned births continued to occur disproportionately among the least educated and the least intelligent and those with the weakest moral character. The sterilization programs had only a small eugenic impact. In the United States, the 60,000 or so sterilizations amounted to less than 0.1 percent of the mentally retarded and psychopathic, so the effect on the prevalence of mental retardation and psychopathic personality must have been negligible. Even in Sweden, the 60,000 or so sterilizations amounted to only about 1 percent of the population and will have had little significant impact on the gene pool. Furthermore, programs to promote positive eugenics by increasing the fertility of the most valuable members of society, those who were the most highly educated and the most intelligent and those with the strongest moral character, proved impossible to find; and the fertility of these groups remained disappointingly low. Nevertheless, despite these limitations, until the 1960s, the eugenics movement commanded wide assent and made substantial progress.

Few would have predicted that in the last three decades of the twentieth century, eugenics would come to be almost universally rejected and condemned. Yet this is precisely what occurred. Throughout the Western democracies the eugenics societies closed themselves down; sterilization of the mentally retarded and criminals largely ceased; immigration was permitted without regard to human quality; and only a tiny handful of social scientists continued to support eugenics. This reversal in attitudes towards eugenics was due principally to the increasing priority accorded to individual rights over social rights, in particular the right of those with genetic disorders and mental retardation and criminals to have an unlimited number of children and to transmit their disabilities and pathologies to future generations at the expense of social rights, one of which is the right of society to protect itself against the social costs incurred when these groups have children. In Singapore and China, however, social rights have been accorded greater precedence; and the political leaders, physicians, and geneticists remained sympathetic to eugenics.

II

~

The Objectives of
Eugenics

3

Historical Formulations

1. Francis Galton
2. Hermann Muller and "The Geneticists' Manifesto"
3. Carlos Blacker
4. Luigi Cavalli-Sforza and Walter Bodmer
5. Walter Bodmer and Robert McKie
6. The Promotion of National Economic, Scientific, Cultural, and Military Strength
7. The Promotion of Happiness
8. Nationalist and Universalist Eugenics
9. The Reversal of Dysgenics
10. Conclusions

An exposition and reassessment of eugenics needs to start with a consideration of the objectives of eugenics, with what eugenics is intended to achieve and why the attainment of eugenic objectives would be desirable. In this chapter we examine the major historical formulations of this issue, beginning with a closer look at Galton's writings on the subject and continuing with the views of subsequent geneticists and eugenicists. The chapter concludes with a discussion of the general principles upon which eugenic objectives and programs should be based.

1. FRANCIS GALTON

It was noted in Chapter 1 that Galton formulated the objectives of eugenics as the *improvement of health*, which he used broadly to include physical stamina, energy, and physique, as well as the absence of disease; *the improvement of intelligence*, which he regarded as a single entity that has subsequently become known as general intelligence or Spearman's g; and the *strengthening of moral character*, by which he meant honesty, integrity, self-discipline, the capacity for sustained work effort, and a sense of social obligation and commitment. Galton never spelled out in detail why he regarded the strengthening of these three characteristics as the objectives of eugenics. He seems to have regarded it as self-evident that these characteristics are valuable and that it was unnecessary to set out the case for the desirability of improving them. We can nevertheless infer Galton's views. In the last paragraph of his autobiography, his last major work written three or four years before his death, Galton (1908a, p. 323) wrote that eugenics had two aims: "Its first object is to check the birth-rate of the unfit . . . ; the second object is the improvement of the race by furthering the productivity of the fit. " Here, as already noted, Galton used the word *race* in its nineteenth-century sense to designate the population of the nation state and not in the broader twentieth-century sense. Galton seems to have believed that the reason why it would be desirable to improve the genetic quality of a nation's population is that this determines the quality of its civilization and the economic and military strength of the nation. Thus in *Hereditary Genius*, Galton (1869) proposed that the population of classical Athens had the highest intelligence of any human population and that this was responsible for the high level of civilization. He also contended that when the intelligence and the moral character of a society deteriorate through dysgenic fertility, the quality of its civilization declines. He cited the decline of Spain in the seventeenth century as an instance in which the deterioration of intelligence, which he attributed to the extensive celibate priesthood, had been responsible for national decline in the quality of civilization and of economic and military power.

It is apparent that Galton thought of eugenics as promoting the good of societies or populations rather than that of individuals. "Individuals," he wrote in his last book, "appear to me as partial detachments from the infinite ocean of Being, and their world as a stage on which Evolution takes place, principally hitherto by Natural Selection, which achieves the good of the whole with scant regard to that of the individual. Eugenics will replace Natural Selection by other processes to secure the improvement of the genetic quality of the population" (1908a, p. 323). Thus eugenics, in Galton's view, is primarily concerned with promoting the good of the population, not that of the individual. This idea that the well-being of the population is more important than that of individuals fell increasingly into disfavor in the second half of the twentieth century and is one of the major reasons that eugenics became almost universally rejected.

2. HERMANN MULLER AND "THE GENETICISTS' MANIFESTO"

The first important reformulation of the objectives of eugenics occurred in 1939 when Hermann Muller drew up a document called "The Geneticists'

Manifesto," which was signed by a number of leading geneticists of the period, including J. B. S. Haldane, S. C. Harland, L. Hogben, J. Huxley, and J. Neadham. The manifesto posed the question, "How could the world's population be improved genetically?" It began by stating that in the modern world selection for genetically desirable traits had become greatly relaxed and that as a result "some kind of conscious guidance of selection is called for." The qualities needing this conscious guidance were listed as health, intelligence, and "those temperamental qualities which favor fellow-feeling and social behavior" (Muller, 1939, p. 64). These temperamental qualities are broadly similar to Galton's concept of moral character, although Galton's concept was wider because it included self-discipline and the capacity for sustained hard work in addition to a sense of social obligation.

Nevertheless, "The Geneticists' Manifesto" should be regarded as essentially a restatement of Galton's position that natural selection for health, intelligence, and moral character was failing to operate effectively in contemporary society and that these are the characteristics that need to be improved by eugenic intervention. Many of the leading geneticists of the 1930s added their names to the manifesto, showing how widespread was the consensus on the objectives of eugenics at this time.

3. CARLOS BLACKER

Carlos Blacker was a British physician, a prominent eugenicist, and general secretary of the British Eugenics Society in the middle decades of the twentieth century. In *Eugenics: Galton and After*, Blacker (1952) proposed that the objectives for eugenics should be as follows: "Firstly, physical courage, especially in warfare—physical courage is eugenically valuable. . . . War-like qualities still commend themselves as eugenic virtues" (p. 286). The second quality on Blacker's list for eugenic improvement was intelligence because "a strong case can be made out for the social value of intelligence" and "most parents would prefer to have a son or daughter with an intelligence quotient of a hundred and thirty than of seventy" (pp. 286–87). Nevertheless, Blacker advanced three reservations. The first of these was that perhaps society needs unintelligent people to carry out undemanding jobs; the second was that high intelligence alone is not sufficient for the production of work of high quality but needs to be complemented with strong motivation; and the third was that high intelligence can be directed into socially undesirable activities, such as crime and the making of destructive weapons. Thirdly, Blacker concluded rather perfunctorily that further qualities for eugenic improvement are "serenity or contentment" and "if there is such a thing, an instinct for co-operation"; "health, physical beauty and fecundity"; and "freedom from genetic taints" (p. 289).

Blacker proposed a eugenic agenda that is to some degree a restatement of Galton's, but he adds several points. He restates Galton on the importance of health in his phrase, "freedom from genetic taints"; he reiterates the importance of intelligence, with some reservations, and of moral character, insofar as physical courage and "an instinct for co-operation" can be regarded as components of moral character. Blacker evidently regarded the promotion of the social and the individual well-being as objectives of eugenics, although he did not explicitly make this distinction. The objectives of the enhancement of physical courage, intelligence, health, fecundity, and the "instinct for co-operation" are evidently designed to promote social well-being. However, the promotion of "serenity and contentment" and "physical beauty" would not seem to serve any useful social purpose and seem designed to promote the well-being of individuals.

Insofar as Blacker's eugenic agenda departs from that of Galton, it is open to objection on four grounds. First, it is doubtful whether the enhancement of physical courage should have any place among the objectives of eugenics. We may grant that physical courage is a valuable attribute for those in occupations that require it, such as the military, the police, firefighters, and the like. Probably, however, there are sufficient numbers of people with the necessary physical courage at present, and it is doubtful whether it would be desirable to attempt to strengthen this characteristic. Foot soldiers certainly need physical courage and warlike qualities, but these are probably also sufficiently strong today and not in need of eugenic strengthening. Arguably, people are too warlike, and it would be more desirable to reduce the strength of this trait in the direction of making people more peaceable. It should be remembered that Blacker served in World War I and was writing shortly after the end of World War II and that at this period physical courage was no doubt an important component of military success. By the beginning of the twenty-first century, wars could be won by firing missiles from a distance, and the physical courage of soldiers is far less important than it was in previous historical times. For these reasons, it is impossible to endorse Blacker's characteristics of physical courage and warlike qualities as having any place in a contemporary eugenic agenda.

Second, with regard to the improvement of intelligence, there is confusion in Blacker's reservation that some of the more intelligent people produced by eugenic measures might devote their intelligence to the production of destructive weapons, and that this is an undesirable outcome. Blacker's first objective for eugenics was the breeding of a population with enhanced physical courage and warlike qualities, presumably in order to fight wars more effectively. Yet a further component of fighting wars more effectively is the ability to produce destructive weapons, such as, in the contemporary world, nuclear weapons and long-range missiles. Contrary to Blacker's view, one of the major arguments for eugenics is precisely that a more intelligent population would be able to produce more effective weapons and hence to succeed in the competitive struggles between nations that frequently erupt into warfare.

Third, it is doubtful whether the enhancement of "serenity and contentment" would make any contribution to the quality of civilization or to the strength of the nation state, the two primary objectives of eugenics as formulated by Galton. Creative geniuses, productive scientists, innovative entrepreneurs, successful corporate managers, and political and military leaders are not strikingly characterized by "serenity and contentment." To the contrary, the work of Post (1994) has shown that many of these people have experienced periods of depression and that a population bred for increased serenity and contentment would probably suffer a loss of creativity and productive energy.

Fourth, the promotion of fecundity need not play any part in a eugenic program. Eugenics is concerned with the improvement of the genetic quality of the population, and it is not necessary that everyone should have children. In a genetically improved population, a failure of some to have children could be offset by an increase in the fertility of others. Our conclusion has to be that Blacker's reformulation of the agenda of eugenics is sound insofar as it restates the objectives of Galton; but insofar as it departs from Galton, it is unimpressive.

4. LUIGI CAVALLI-SFORZA AND WALTER BODMER

Luigi Cavalli-Sforza and Sir Walter Bodmer (1971) are two population geneticists who considered the objectives of eugenics in their textbook *The Genetics of Human Populations*. In their discussion of this issue, they wrote, "Most people would agree on the desirability of some traits like intelligence, social responsibility, artistic talent, generosity and beauty" (p. 770). Some of these traits are the same as those originally proposed by Galton, and some are additions. Like Galton, Cavalli-Sforza and Bodmer include intelligence; their "social responsibility" and "generosity" are broadly similar to although rather narrower than Galton's "moral character." Their artistic talent is an addition, and it is difficult to understand why it should be singled out among a number of special abilities or aptitudes that might include musical, literary, scientific and technological talents. The list of special aptitudes might be extended further to include professional, business, entrepreneurial, administrative, and military talents on the grounds that all of these make important contributions to the well-being of society.

A curious feature of the list of characteristics proposed by Cavalli-Sforza and Bodmer for eugenic improvement is that it does not include health. A reduction of genes for genetic diseases and disorders is the most widely acceptable of eugenic objectives. Even the most strident opponents of eugenics have rarely disputed that health is in general preferable to disease and that it would be desirable if genetic diseases could be reduced. It is difficult to understand why Cavalli-Sforza and Bodmer omitted the desirability of eugenic measures to reduce genetic diseases and disorders among their eugenic objectives.

The final characteristic suggested by Cavalli-Sforza and Bodmer for eugenic improvement is beauty. As noted in the previous section, this proposal was also made some 20 years earlier by Blacker (1952). It may have instant appeal, but it poses problems. It cannot be claimed that an increase in the physical beauty of the population would make any contribution to the quality of civilization as expressed in the production of high-quality achievements in science, the arts, and public life or to the strengthening of the economic, scientific, or military base of society. It can be inferred that Cavalli-Sforza and Bodmer have a broader concept of the objectives of eugenics, a concept that includes the quality of life of the population, and that they believe that a eugenically contrived increase of beauty would contribute to this. In defense of this position, it has to be conceded that many of our fellow citizens are not beautiful. However, it is doubtful whether this causes significant psychic distress or whether the quality of our lives would be appreciably increased if the general level of beauty were to be increased. It may be that we appreciate beauty largely because it is quite rare and that if everyone were beautiful we should appreciate it less or even not at all. A further problem is that beauty is a more subjective quality than health, intelligence, and moral character. Many people find their own kind more beautiful than others. For instance, among African Americans it is widely considered that "black is beautiful," but many whites and Asians regard members of their own race as more beautiful than those of other races. For instance, in Ecuador in 1995, whites were outraged when a panel of judges chose a young black woman named Monica Chala as Miss Ecuador, apparently on the grounds that the contest for Miss Universe later that year was to take place in South Africa and that the black South African judges would regard a black Miss Ecuador as more beautiful than a white (Rahier, 1998).

Because of the doubtfulness of the social gains likely to accrue from an increase in beauty and the difficulties of reaching an agreed consensus on what beauty is, we should reject the proposal that the improvement of beauty should be one of the objectives of eugenics.

5. WALTER BODMER AND ROBERT MCKIE

Bodmer returned to the question of the objectives of eugenics in 1994 in *The Book of Man*, written jointly with Robert McKie. They discuss the eugenic potential of recent advances in medical genetics, human biotechnology, and genetic engineering, including the genetic assessment of embryos, gene therapy, and the like, which open up new possibilities for eugenics. Bodmer and McKie (1994) welcome some of these advances as the beginning of a new "golden era" in which genetic analysis and engineering will make it possible to reduce the incidence of genetic diseases and disorders. They welcome also the likelihood that the use of these techniques will make it possible to increase people's heights, abolish baldness, and improve the eyesight of the myopic. They predict that it will also become possible for couples to select children for intelligence and athletic ability by growing a number of embryos in vitro, assessing them for their genetic potential for intelligence and athletic ability, and then selecting for implantation those whose genes have the best potential for these traits. However, they disapprove of these potential developments. They write that they would be unacceptable because "the very notion of the sanctity of human individuality would be grossly offended" (p. 246).

Although Bodmer and McKie are correct in anticipating that these new techniques for eugenics will become available, it is impossible to accept their judgments on which applications of these are desirable and which are undesirable. No significant advantage either for society or for the individuals concerned would be gained from a general increase in height or reduction in baldness, to which Bodmer and McKie look forward, and myopia is so easily corrected by glasses or contact lenses that eugenic intervention to reduce it would be pointless. Bodmer does not explain why he has changed his mind on the desirability of eugenic intervention for the improvement of intelligence, which he favored in 1971. It is a weakness of Bodmer and McKie's discussion that they offer no general principles of the objectives of eugenics, which are required as the starting point for a consideration of what eugenic interventions would be desirable.

6. THE PROMOTION OF NATIONAL ECONOMIC, SCIENTIFIC, CULTURAL, AND MILITARY STRENGTH

We have now seen enough of the formulations of the objectives of eugenics by leading eugenicists and geneticists to realize that these objectives have been far from satisfactory to the extent that they depart from Galton's original three objectives of improving health, intelligence, and moral character. The problem is that none of these formulations set out the general principles of eugenic objectives from which the desirability of the promotion of particular objectives can be derived. This is the issue that we now need to address.

It is proposed that we should follow Galton in formulating the primary objective of eugenics as the improvement of the genetic quality of the population with respect to its health, intelligence, and moral character. A secondary and subsidiary objective is the improvement of happiness. In considering the primary objective, we need to specify the population that it is the objective of eugenics to improve. This may be the population of the nation state, which can usefully be designated "nationalist eugenics," or the population of the entire human species, which can usefully be designated "universalist eugenics."

Several of the classical eugenicists have sought to promote nationalist eugenics. Their objective has been to improve the genetic quality of the population as a means of increasing the economic, scientific, cultural, and military strength of their own nation state. This was the objective of Plato, the first eugenicist, who wrote his account of the eugenic state, *The Republic*, after his own city of Athens had been defeated in war by Sparta in 404 B.C. Plato wanted to prevent this from happening again. When he formulated his blueprint for an ideal state in *The Republic*, he stipulated that the rulers, soldiers, and workers would be bred for quality and efficiency, with the primary object of ensuring that the state would be powerful economically and militarily and would be able to defeat its enemies in warfare. Galton did not explicitly make the enhancement of national strength the ultimate objective of eugenics, but this is implicit in his writings insofar as he wrote that eugenics and dysgenics play a major role in the rise and fall of civilizations and in the outcome of conflicts between nations. Among Galton's successors, the nationalist objective of eugenics is most evident in the writings of Carlos Blacker, when he specified the first objective of eugenics as the increase of physical courage and warlike qualities. Among those who have put eugenics into practice, Hitler's objective was to improve the genetic quality of the Germans, and Lee Kuan Yew's objective was to improve the genetic quality of the people of Singapore. Nationalist eugenics should be seen, therefore, as one of a number of means for the enhancement of the economic, scientific, cultural, and military strength of the nation state. The term *cultural* in this formulation includes the promotion of a high level of civilization and quality of life as expressed in the efficiency with which goods and services are produced and delivered and in a low level of crime and antisocial behavior. Political leaders normally attempt to promote the national economic, scientific, cultural, and military strength by a variety of means, such as policies designed to increase economic growth, encouragement and subsidies for science and culture, and expenditure on military research and development for the production of more effective weaponry and maintenance of their armed forces. A eugenic program is an additional means available to political leaders concerned with maintaining and increasing their national power by the improvement of what can be called the "human genetic capital" of the population.

7. THE PROMOTION OF HAPPINESS

There is a further question of whether the promotion of happiness should be regarded as one of the objectives of eugenics. The formulation of the objectives of eugenics proposed in the previous section as the enhancement of the economic, scientific, cultural, and military strength of the nation state does not include the enhancement of the happiness of the citizens. Nevertheless, some of the writers whose ideas have been summarized earlier in this chapter have regarded the promotion of happiness as one of the objectives of eugenics. This was evidently the thinking of Blacker, Cavalli-Sforza, Bodmer, and McKie when they suggested that eugenics should be used to improve beauty, to eliminate baldness, and the like. These measures would make no contribution to the economic, scientific, cultural, or military strength of the nation state, but they might arguably increase the happiness of those afflicted by the misfortunes of ugliness and baldness and presumably have been proposed as among the objectives of eugenics on these grounds.

The promotion of happiness should be regarded as a subsidiary objective of eugenics or, in certain instances, as a desirable by-product of eugenic programs. The major area in which this is the case is that of genetic diseases and disorders, which are undesirable because they weaken the nation state, because they exact great costs, and because they cause unhappiness to the individuals concerned and to their families. Measures to reduce these genetic disorders would fulfill the primary eugenic objective of strengthening the nation state and at the same time would achieve the minor eugenic objective of reducing unhappiness. In certain other cases, eugenic measures that strengthen the nation state may simultaneously cause unhappiness to individuals. The most obvious example is the sterilization of those with genetically undesirable qualities, such as the mentally retarded and criminals. In these instances the benefits of a eugenic program for the nation state should override the loss of happiness to the small number of individuals concerned.

8. NATIONALIST AND UNIVERSALIST EUGENICS

It is useful to distinguish between "nationalist" and "universalist" eugenics. *Nationalist eugenics* is concerned with promoting the genetic quality of the population of the eugenicist's own nation state. Thus, Plato was concerned with promoting the genetic quality of the population of Athens, Blacker with that of Britain, Hitler with that of Germany, and Lee Kuan Yee with that of Singapore. In contrast to nationalist eugenics, *universalist eugenics* has the objective of improving the genetic qualities of the entire human species. The leading exponent of universalist eugenics has been Cattell (1972), who advocated a world system of what he called "cooperative competition," in which each nation would adopt its own unique eugenic program. Nations would compete to develop strong economics and successful cultures, and a process of group selection would determine which were the most successful. This competition would involve both biological and social selection. The nations that were biologically and socially more successful in producing superior civilizations would either replace the less successful nations or force them to adopt the eugenic policies and social structure of the more successful. At the biological level, some nations would produce biologically improved populations, particularly with respect to intelligence but also with regard to personality, and populations with lower aggressive and sexual drives, which Cattell regarded as too high for modern civilizations. Cattell envisioned that some nations might evolve genetically enhanced populations that became so genetically different from others than they would form new species of Homo sapiens that were no longer able to interbreed and produce fertile offspring with other Homo species. This outcome is possible if some nations were to use genetic engineering to introduce new genes for greater intelligence, greater longevity, and improved personality qualities.

At the level of social systems and organizations, the less successful nations would be expected to adopt the social structures of the more successful. This would be a form of social Darwinism, such as occurred in the course of the twentieth century in the competition between the socialist social and economic systems of the Soviet Union and China and the market economy systems of the Western democracies. The outcome of this competition was the acknowledged superiority of the market economies, and this forced the Soviet Union and China to abandon socialism and to adopt the economic system of the free market. In the twenty-first century, the major competition between social systems is likely to be between the democracies of the West and the authoritarian oligarchy of China. The outcome of this competition is unpredictable and is a good illustration of the rationale of Cattell's model of a world structure of competing nations, each pursuing different policies to improve its own economic, scientific, cultural, and military strength. Cattell envisioned that competition between nations pursuing different eugenic and social programs would be regulated by a world supervisory body, akin to the United Nations, that would ensure that the competition did not escalate into warfare.

In practice, nationalist and universalist eugenics both involve continuing competition between nation states that adopt eugenic programs. The difference lies in the degree of detachment of the eugenicist from the outcome of these conflicts. The nationalist eugenicist wishes to see his or her own nation win and endeavors to promote eugenics in his or her own nation to secure that end. The universalist eugenicist is indifferent to which nation wins, taking the view that as long as one nation adopts eugenics it will be so successful in developing its economic, scientific, cultural, and military strength that either it will force its rivals to adopt their own eugenic programs in order to compete or it will take control of the world and implement a program of global eugenics. It does not matter greatly to the universal eugenicist which nation does this because over the long term the results will be the adoption of eugenics throughout much of the world. To consider a parallel, it did not matter over the long term that the industrial revolution was pioneered in England in the eighteenth and nineteenth centuries. The advantage that England enjoyed was only temporary because it forced other nations to industrialize and within a few decades the other nations had caught up.

In this book I adopt the universalist vantage point of Cattell, although I doubt whether his proposal for a world supervisory body to regulate competition between competing eugenic states would work in practice. Cattell was nonpartisan about which nations would introduce the eugenic policies that would enable them to succeed in future competition with other nations. I adopt the same view. Like Cattell, my concern is to promote the genetic enhancement of the whole human species and the further development of culture and civilization that will follow from this.

9. THE REVERSAL OF DYSGENICS

The immediate objective of eugenics should be the reversal of the dys-genic processes that have been present in the economically developed na-tions since the middle of the nineteenth century and in most of the economi-cally developing world in the twentieth century. These dysgenic processes have been taking place with respect to health, intelligence, and moral char-acter. These processes were first identified by Francis Galton (1869) in En-gland and by Benedict Morel (1857) in France and were widely understood by geneticists, biologists, and social scientists in the early and middle decades of the twentieth century. The evidence for these processes and the reasons for them are set out in *Dysgenics: Genetic Deterioration in Modern Populations* (Lynn, 1996). To summarize briefly, the genetic deterioration in health has taken place as a result of medical advances that have preserved the lives of many of those with genetic disorders and have enabled them to have chil-dren. This has perpetuated the genes for these disorders, which were previ-ously eliminated from the population by the early deaths of those carrying them. The genetic deterioration of intelligence and moral character has been caused largely by the more efficient use of contraception in limiting fertility by the more intelligent and those with stronger moral character, who include the better educated and those in the higher socioeconomic classes, and by the inefficient use of contraception by the less intelligent and those with weak moral character, which is responsible for their higher fertility. The presence of dysgenic fertility in the United States has been confirmed in subsequent studies by Loehlin (1997) and myself (Lynn, 1998; 1999b).

Dysgenic fertility is not the only dysgenic process in the Western democ-racies. A second dysgenic factor that has been identified in the United States by Herrnstein and Murray (1994) is the immigration of large numbers of Hispanics and Africans, whose mean IQs Herrnstein and Murray estimate at 91 and 84, respectively, as compared to a mean IQ of 100 for white Ameri-cans. The increasing proportions of Hispanics and Africans in the United States will inevitably reduce the average intelligence level of the population. Europe has also been experiencing fairly large scale immigration of third world peoples with low intelligence, principally from Africa and the Caribbean, and this has been having the same dysgenic impact on the intelligence of the population that Herrnstein and Murray have identified for the United States.

Although dysgenic fertility and immigration are cause for concern in the Western democracies, dysgenic fertility is far more serious in a number of the economically developing countries. This is particularly the case in Latin America, where the most highly educated and intelligent women are having two or three children while the least educated are having six to eight children. Dysgenic fertility has begun to appear in sub-Saharan Africa and is likely to become more pronounced as the demographic transition to lower fertility gets under way and appears first among the most intelligent and the best educated. Dysgenic fertility has become a worldwide problem. The dysgenic processes of differential fertility and immigration are both likely to prove exceedingly difficult to correct. Indeed, Pearson (1996) envisions that it may prove impossible to halt these processes and that the result may be the eventual extinction of the human species. An alternative scenario is that although these dysgenic processes may well prove impossible to counter in the liberal Western democracies, some more authoritarian states are likely to find ways of correcting them. This is a further argument for Cattell's model of a world system of independent nation states each trying to solve the problem of dysgenics and to formulate politically feasible and genetically effective programs of eugenics.

10. CONCLUSIONS

Galton's proposal that the objectives of eugenics should be the improvement of the health, the intelligence, and the moral character of the population is still the most satisfactory and, for the most part, has been accepted by later geneticists and eugenicists. Some of these people have added further qualities and characteristics for eugenic improvement, such as physical courage and warlike qualities (Blacker), artistic talent and beauty (Cavalli-Sforza and Bodmer), the elimination of baldness and myopia (Bodmer and McKie), and the like; but no persuasive case for these additional objectives has been made. They have relied on common sense and on what can be called "gut reactions" to argue that certain eugenic objectives are desirable and others undesirable. This is not satisfactory. A proper discussion of the objectives of eugenics requires a consideration of the general principles on which these objectives should be based.

It is proposed that the principal objective of eugenics is to improve the genetic quality of the population. Nationalist eugenics seeks to improve the genetic human capital of the population with respect to health, intelligence, and moral character as a means of enhancing the economic, scientific, cultural, and military strength of the nation state with the objective of improving its competitive position in relation to other nation states. Most eugenicists, starting with Plato, have had nationalist objectives consisting of the improvement of the genetic qualities of their own national populations to enable them to compete more effectively with their competitors and enemies. Universalist eugenics seeks to secure the genetic improvement of all human populations and the entire human species. The promotion of an increase in happiness can be regarded as a subsidiary objective of eugenics that will be achieved principally by the reduction of genetic diseases and disorders.

4

Genetic Diseases and Disorders

1. Types of Genetic Diseases and Disorders
2. Dominant-Gene Disorders
3. Recessive-Gene Disorders
4. X-Linked Disorders
5. Multifactorial Disorders
6. Chromosome Disorders
7. Are Genetic Disorders Valuable?
8. Experiences of Parents of Children with Genetic Diseases and Disorders
9. Dysgenic and Eugenic Trends for Genetic Disorders
10. A Eugenic Assessment of Genetic Disorders
11. Conclusions

In this chapter we examine the eugenic case for reducing genetic diseases and disorders. Of all the objectives of eugenics, this is the most likely to win general agreement because virtually everyone accepts that good health is better than disease. There are still a few geneticists who dispute this and maintain that all genes are equally valuable. This contention needs to be considered, but it does not command wide assent. The overwhelming majority of people would prefer health to disease for themselves and their families, for the population of their countries, and for the whole world.

1. TYPES OF GENETIC DISEASES AND DISORDERS

There are about seven thousand known genetic diseases and disorders, which have a great variety of symptoms, degree of severity, and age of onset. These diseases and disorders fall into three major categories: (1) single-gene diseases and disorders, caused by one malfunctioning gene, which can be either dominant, recessive, or X-linked; (2) multifactorial diseases and disorders, caused by the joint action of several genes and environmental factors; and (3) chromosome disorders, caused by a defect in one of the chromosomes.

The best study of the prevalence of the different types of genetic diseases and disorders was carried out by Baird, Anderson, Newcombe, and Lowry (1988) in Canada in the mid-1980s. They examined a population-based registry of more than one million consecutive live births and compiled the prevalence rate of genetic diseases and disorders diagnosed before the age of 25 years. Their results are summarized in Table 4.1. It will be seen that, by far, the greatest number of genetic disorders are multifactorial, comprising about 90 percent of the total and with a prevalence of 47 per 1,000 live births.

A later and exceptionally thorough study of the prevalence of genetic disorders in Canada was carried out by Johnson and Rouleau (1997). They report a total prevalence of 51.95 cases per 1,000 live births for infants in their first year of life. This figure is very close to the 53.7 percent obtained by Baird and her associates, although there is a difference in the samples—Baird's sample consists of cases in which the disorder appears by the age of 25, and Johnson and Rouleau's sample consists of cases in which it appears by the time the child is one year old. Nevertheless, the two studies give closely consistent results for the prevalence of genetic diseases and disorders.

2. DOMINANT-GENE DISORDERS

Dominant-gene disorders tend to run in families. The affected parent transmits (on average) half of the defective genes to his or her offspring. *Huntington's*

Table 4.1
Prevalence Rates of Genetic Disorders per 1,000 Live Births in Canada

Disorder	Prevalence
Single-gene disorders:	
Dominant	1.4
Recessive	1.7
X-linked	0.5
Multifactorial disorders	47.0
Chromosome disorders	1.9
Unknown genetic causes	1.2
TOTAL	53.7

chorea, an incurable degenerative disease of the central nervous system, is one of the best known of the dominant-gene disorders. It begins in early middle age and consists of progressive deterioration of the physical and mental faculties. Affected individuals suffer from worsening memory loss and frequently develop delusions. They lose control of their muscles and fall into convulsions and bizarre dancing movements. Eventually, this becomes violent flailing, which is so severe they cannot eat or be fed, resulting in death.

One of the more curious of the dominant-gene disorders is *Gilles de la Tourette's syndrome*. Symptoms include motor tics such as frequent grimaces, jerking of the hand and arms, sucking, sniffing, throat clearing, and the like. About a third of those affected have uncontrollable urges to shout obscenities in inappropriate places, such as in classrooms or lecture theatres or at social gatherings. Most of them have a poor attention span, which adversely affects their progress in school and performance at work. The birth incidence is about 1 in 2,500 boys and about 1 in 7,500 girls. The disorder is not life threatening, but those affected find it difficult to find employment. They also have difficulty in attracting mates, but if they succeed in doing so, they can and do have children, an average of half of which inherit the disorder.

Marfan's syndrome is another of the rare dominant-gene disorders, which manifests itself as exceptional tallness, chest deformities, and visual and heart defects, typically causing death in early middle age. Another is *Leopard's syndrome*, consisting of leopardlike brown spots on the face and body, retardation of physical growth, wide-set eyes, abnormalities of the sex organs, and, in about a quarter of cases, deafness. In both these disorders intelligence is normal. However, in a number of other dominant-gene disorders, mental retardation is part of the syndrome of disabilities. For example, in *Alpert's syndrome*, there is a premature fusion of the bones of the skull, which prevents normal brain growth. Symptoms include deafness, deformities of the hands, and abnormal facial features, such as prominent forehead, lower jaw, and wide-set eyes. About half of the cases have mild mental retardation. Another disorder in which there are both physical and mental impairments is *Soto's syndrome*, which causes disproportionately large hands, arms, and feet; clumsiness; and unusual facial features. About 60 to 70 percent of sufferers are mildly mentally retarded.

Achondroplasia, or dwarfism, is one of the least serious of the dominant-gene disorders. The torso or trunk is of normal size, but the limbs are short. The incidence of dwarfism is about 1 in every 25,000 births. Dwarfs are fertile, and the transmission of the condition follows the principle of Mendelian dominant inheritance—in matings between a dwarf and a normal-sized person, an average of half the children inherit the condition; in matings between dwarfs, who are both carriers of the recessive normal gene, an average of 75 percent inherit dwarfism. Dwarfs have normal intelligence and a normal life span. Nevertheless, they tend to suffer from several physical disabilities, including respiratory problems, abnormalities of the spine, and proneness to ear infection causing deafness. They also encounter a number of difficulties in everyday living, such as climbing stairs, reaching books from library shelves, and getting clothes that fit properly. In addition, many dwarfs experience psychological distress from a consciousness of being different, as a result of which they find it hard to form romantic attachments and find mates.

3. RECESSIVE-GENE DISORDERS

When two carriers of a recessive gene have children, an average of one quarter of the children will inherit two copies of the recessive gene, and thus will inherit the disease. There are several thousand recessive-gene disorders. One of the first to be identified was *phenylketonuria*; its recessive mode of inheritance was discovered by the Norwegian physician Ivor Folling in 1934. Phenylketonuria consists of an enzyme defect and causes profound mental retardation unless it is corrected by a protein-reduced diet. The restricted diet is difficult to follow and is not wholly effective, and even those who do follow it have IQs a little below average. The single recessive gene is present in about 2 percent of Caucasians in Europe and North America, and about 1 in 16,000 births inherits the double recessive.

Cystic fibrosis is the most common recessive-gene disorder in European and North American populations. About 4 percent of the population are carriers, and the birth incidence is about 1 in 2,500. The disorder consists of a fault in the protein CFTR, which is responsible for the passage of salt ions across the membranes of cells in the body. A sticky mucus accumulates in the lungs, pancreas, liver, and sweat glands. The lungs are the most seriously affected, causing frequent and severe respiratory infections that can be cleared by antibiotics but that eventually cause permanent damage to the lungs. Sufferers from this disease typically die in childhood, adolescence, or early adulthood, and few survive beyond the age of 30.

Among Africans from sub-Saharan Africa, *sickle cell anemia* is the most common recessive disorder. The principal symptoms are anemia, bone pain, and damage of the heart, lungs, and kidneys. In the United States, the incidence of the disease is about 1 in 500. The disease is debilitating and sometimes life threatening. The disease is unusual in that a single copy of the gene provides some resistance to malaria and is therefore advantageous in environments where malaria is common.

Tay-Sachs disease is the most common recessive-gene disorder among Ashkenazi Jews of Eastern European descent. The double recessive is inherited by about 1 in 3,600 babies born to Ashkenazi Jews. Affected babies appear normal at birth, but the disease impairs the nervous system in infancy and the babies suffer mental retardation, blindness, and loss of muscle control and usually die before they reach the age of four.

Most of the recessive-gene disorders are rare and unknown to the general public. An example is *Sjorgen-Larssen syndrome*. The first symptom to appear is a red fish-scale-like skin rash for which no treatment is effective. Later, children develop spasticity, as a result of which about three-quarters are confined to wheelchairs. Almost all have some mental handicap ranging from severe to mild, and about half experience degeneration of parts of the retina and consequent visual disability for which no treatment is available.

4. X-LINKED DISORDERS

X-linked disorders are caused by rare recessive genes on the X chromosome. Females are carriers of the defective gene, but the disorders generally only appear in males. The reason for this is that males have only one X chromosome. Females have two X chromosomes and almost always have one normal, dominant gene, which prevents the expression of the disorder. The best known of the X-linked disorders are color blindness and hemophilia. There are several different kinds of *color blindness*. The commonest form causes difficulty in distinguishing green from red. Among Europeans, about 8 percent of males and 1 percent of females have either green or red color blindness. The prevalence of color blindness is lower among Asians and lower still among blacks.

Hemophilia is a disorder in which the blood does not form clots and the sufferer bleeds continuously from even the smallest scratch. It has become well known because of its presence in the royal families of Europe. Hemophilia is thought to have been the result of a mutant gene for the disease arising in Prince Albert, the husband of Britain's Queen Victoria. Until the middle of the twentieth century, boys who inherited hemophilia generally died in childhood as a result of bleeding to death. In the second half of the century, however, it became possible to control bleeding by infusions of factor VIII, a blood-clotting agent.

The commonest of the severe X-linked recessive disorders is *fragile X syndrome*. It is unusual insofar as it occurs quite frequently in females as well as in males, although the symptoms are less disabling in females. It is present in about 1 in 1,250 males and 1 in 2,000 females. All affected males are severely mentally retarded with average IQs around 40, whereas females are mildly retarded and have IQs typically in the range of 50 to 70 (Dykens et al., 1989).

Another relatively common X-linked disorder is *Duchenne's muscular dystrophy*. The principal symptom of the disease is a progressive weakening of the muscles, which begins around the age of three. Affected boys walk with a waddle and find it difficult to climb stairs. By about the age of 12, they are unable to walk at all, and they normally die in adolescence, usually from a chest infection or heart failure. Typically they have below average intelligence. There is no effective treatment. The disease occurs in about 1 in 3,000 boys.

There are many rare X-linked diseases. An example is *Lesch-Nyhan syndrome*. Sufferers have uncontrollable fits of self-mutilation and frequently chew off their own fingers. Another rare and little known disorder is *Lowe's syndrome*, a metabolic disorder affecting the eyes, kidneys, and brain, often, although not invariably, causing mild or severe mental retardation. Most boys who inherit the disorder die before age 10.

5.   MULTIFACTORIAL DISORDERS

Multifactorial disorders are caused by the joint action of several genes and adverse environmental conditions, such as poor nutrition. For the most part, genes for these disorders have not been identified. However, it is known that there is some genetic basis for them because studies have shown that identical twins are more similar in respect to the diseases and disorders than are nonidenticals. As shown in Table 4.1, the multifactorial genetic diseases and disorders are much more common than the single-gene disorders are, comprising about 90 percent of the totality of genetic diseases and disorders and collectively amounting to about 4.7 percent of live births. Some of the commonest of the multifactorial genetic diseases and disorders are spina bifida, diabetes, multiple sclerosis, Alzheimer's disease, hypertension, heart disease, cancer, epilepsy, schizophrenia, manic depressive psychosis, and gallstones.

Most of these diseases and disorders are already well known, so a brief summary of the symptoms and prevalence of the first four will suffice. *Spina bifida* is a disorder in which part of one or more vertebrae fails to develop completely, leaving part of the spinal cord exposed at birth and causing leg weakness or complete paralysis, urinary incontinence, epilepsy, and mental retardation. There is a birth incidence of about 3 per 10,000. *Diabetes* is a disorder in which insufficient insulin is produced by the pancreas. Levels of glucose in the blood become abnormally high, causing excessive thirst and passing of urine, weight loss, hunger, fatigue, muscle weakness, and blurred vision. About 1 percent of Caucasian populations inherit the disorder, which is frequently not expressed until middle or old age. *Multiple sclerosis* is the commonest genetic disease of the nervous system not present at birth, and it usually first appears in young adults. It has a prevalence of about 1 per 1,000. Its symptoms include spasticity (inability to control movement), paralysis, slurred speech, unsteady gait, blurred or double vision, numbness, weakness, pain, and depression. Many sufferers become progressively more disabled and are bedridden and incontinent by their late thirties or forties. *Alzheimer's disease* is a progressive condition in which the brain's nerve cells degenerate, causing loss of memory and, eventually, total disorientation. It is now the single most common cause of intellectual and personality deterioration in old age and is present in about 30 percent of those over the age of 85.

6.   CHROMOSOME DISORDERS

Chromosome disorders are caused by a defect occurring in one of the chromosomes. They are not normally inherited through genetic transmission but arise spontaneously through a process akin to genetic mutation. About a hundred different chromosome disorders have been identified. By far the most common is *Down's syndrome*. Normally each person has two sets of chromosomes; but in Down's syndrome, a third twenty-first chromosome spontaneously appears. In the majority of cases, defective egg formation in the mother is responsible. The disorder occurs in about 0.13 percent of babies, but it is about 10 times more common when mothers are over the age of 35.

Babies born with Down's syndrome suffer a number of impairments. Their IQs are typically in the range of 30 to 70 and average around 50. In adults, an IQ of 50 approximately represents the intellectual capacity of the average eight-year-old, who is able to read simple texts and do simple arithmetic but is not capable of functioning independently. In addition, about a quarter of them have congenital heart defects, and many are susceptible to diseases of various kinds. About half of the affected individuals are partially or completely deaf, have defective vision, and are obese; and about 30 percent of them have psychiatric disturbances. Until the 1960s, most died by the age of 20, but from around 1970 onwards, medical advances have enabled most of them to live to early middle age, by which time they have a high incidence of premature senile dementia, characterized by loss of memory and self-care skills, wandering, and incontinence (Prasher, 1996). In middle age virtually all of them develop Alzheimer's disease, the symptoms of which include the inability to walk, incontinence, epilepsy, and memory loss. These patients become entirely dependent on nursing care (Mann, 1993; Visser et al., 1997).

Klinefelter's syndrome is another relatively common chromosome disorder. It results from an additional X chromosome and occurs in about 1 male child per 500. Symptoms include retarded sexual development and mild mental retardation. A similar chromosome disorder, called simply *XYY syndrome*, has attracted attention because affected males tend to be aggressive and antisocial, as well as being retarded in language abilities (Patton, Beirne-Smith, & Payne, 1990). *Turner's syndrome* is the most common chromosome disorder in females, in which affected girls have only one X chromosome. They usually have below average intelligence and very poor spatial abilities. The incidence is about 1 per 2,500 births.

The great majority of those with chromosome disorders are either infertile or severely subfertile. Males with Klinefelter's syndrome and females with Turner's syndrome are infertile. Females with Down's syndrome are subfertile, and as of 1990 only 31 female pregnancies had been reported. Ten of the babies inherited the disorder, 18 were normal, and 3 were aborted (Rani, Jyothi, Reddy, & Reddy, 1990).

7. ARE GENETIC DISORDERS VALUABLE?

To the great majority of people, genetic disorders are undesirable afflictions, like any other kind of disease. Most people would very much prefer that neither they nor their children should suffer from them. This is demonstrated by the fact that when it became possible to diagnose genetic disorders prenatally around 1970, the great majority of women chose to terminate pregnancies when a genetic disease or disorder was discovered. Also, the great majority of physicians and medical practitioners agree that this is sensible; they carry out the genetic testing, advise their patients of the option of pregnancy terminations, and carry out terminations when they are requested. Despite this general consensus shared by the public and the medical profession, a number of geneticists, biologists, and others have maintained that there is "no such thing as a bad gene," and that the genes for disease and disorders are as valuable and desirable as those for health.

An early exponent of this paradoxical view was Theodosius Dobzhansky, professor of genetics at Columbia University. In his 1962 Silliman Lecture delivered at Yale, Dobzhansky (1962) informed his audience that there are no such things as good or bad genes because "a gene harmful in one environment may be neutral or useful in another; what is good in the Arctic is not necessarily good in the tropics, what is good in democracy is not necessarily good under a dictatorship" (p. 288). Another geneticist who has asserted this view is James Neel (1994, p. 340), a geneticist at the University of Michigan, who has written that identifying genes as good or bad "requires massive value judgments which cannot be supported on social or scientific grounds." In a similar vein, J. D. Smith (1994), an educational psychologist at the University of South Carolina, suggests that the genes for mental retardation are just as valuable as those for normal or high intelligence because "mental retardation is a human condition worthy of being valued." Richard Soloway (1990), a history professor at the University of North Carolina, writes sneeringly that "Galton naively believed that eventually learned people would be able to agree on the social qualities of goodness of constitution, of physique and of mental capacity they want to see reproduced" (1990, p. 66).

Some of those who have adopted this position that all genes are equally valuable maintain that, although the genetic disorders have distressing effects, they may also be responsible for the desirable outcomes of high achievement and creativity. For instance, Dobzhansky (1962) suggested that epilepsy may be valuable because Dostoevsky had it, and perhaps it was an essential component of his creative genius. Perhaps, the Cambridge professor George Steiner has suggested, deafness may be valuable because Beethoven developed it in middle age, and maybe if he hadn't, he would never have composed his later works. Perhaps, he suggested further, muscular dystrophy is good because the painter Henri Toulouse-Lautrec may have had it, and if he did then perhaps his genius "sprang out of very profound physical handicaps" (Harris, 1992, p. 4).

These assertions cannot be accepted. No evidence that the vast majority of these disorders confer any kind of advantage has ever been assembled. A study of this issue by Post (1994) examined the life histories of 291 of the most eminent and creative men of the nineteenth and early twentieth centuries and concluded that they were in general considerably healthier than their contemporaries. Their average life span was 68 years, well above the average for the general population, and only 8 percent had debilitating diseases. The only possible exceptions to the conclusion that genetic diseases and disorders do not confer any advantages are the mental illnesses that are considered in the next chapter. Apart from these, genetic diseases and disorders have no value and should be regarded as defects in the physiology of the body analogous to mechanical defects in automobiles. There is no more advantage to having one of these disorders than there is to having an automobile that breaks down repeatedly.

It may be wondered why it is that otherwise rational and intelligent academics assert that there is no such thing as a bad gene. An important clue has been provided by G. W. Lasker (1991), a biologist at Wayne State University, when he writes, "The eugenics movement, based on the idea that some genes are good and others bad, has been discredited" (p. 8). Evidently these academics think along the lines that if it is conceded that there are bad genes, then it follows that the eugenicists were right to argue that it would be desirable to eliminate them. Such an admission would endorse the desirability of eugenics. So virulent is the opposition to any form of eugenics among these academics that they are unwilling to concede that any genes are undesirable. This must be the motivation of those who support the patent absurdity that the genes for genetic diseases and disorders are just as valuable as those for health and should be preserved in the population.

8. EXPERIENCES OF PARENTS OF CHILDREN WITH GENETIC DISEASES AND DISORDERS

It is sometimes said that the parents of children with genetic diseases and disorders derive great satisfaction from rearing them, and that these children are therefore valuable and should be welcomed. It is true that the parents frequently do love these children, but they also experience immense stress and dissatisfaction. A study of the lives of parents of children with genetic disorders has been published by Berit Brinchmann, a lecturer in nursing at Bodo University in Norway. She concludes that these parents have an extremely tough life and that although they love their handicapped children, at the same time they hate them. The most serious problems are the lack of rest, deprivation of sleep, and the feeding of the children. For many of these parents the home comes to seem like a prison from which it is impossible to escape.

Brinchmann (1999) describes a scene she observed of the parents of an eight-year-old girl named Katrine with a serious genetic disorder. The child is sitting at the table in the kitchen, and the family are about to have an evening snack. Katrine is tied to her chair and connected to the Nutrison tube:

The thin, tall and ungainly body jumps and shakes in the chair. She trembles and puts her fingers in her mouth like a baby. The difference between Katrine and a baby is that a baby is small and sweet. Giant babies at eight years old are unappealing and grotesque. The skin on her fingers is thick and horny from all the sucking and biting. Katrine is in a world of her own. She calls out, jumps and laughs. It is not possible to make contact. Her parents say that the worst part is that she does not develop at all. She is still a baby, at a stage of about four months, but with the body and skeleton of an eight-year-old, a big, heavy, unshapely, giant baby. (p. 139)

Brinchmann continues by describing her reactions to observing the family life of parents of children with genetic disorders. She found that she was unprepared for what she would see, even though she was a mother and a nurse. She had great problems when she was eating an evening meal with one of the families when the child was sick all over the table. It was disgusting, but these experiences are everyday occurrences and normal for families with multiply handicapped children. She describes how the exhaustion and sleepless nights experienced by the parents are like those of the first few weeks of having a newborn child; but whereas this passes quite quickly, the drudgery of rearing a severely handicapped child continues indefinitely and gets worse as the baby becomes heavier and more tiring and demanding. One mother she interviewed said that she longed to escape, but at the same time she was aware that this was impossible. She was close to breaking point: "I'm aware that if I have one more sleepless night, more sickness and washing to do, I just can't cope. It affects freedom and the like. We will never see her walk or go to school. There's a big sorrow inside us. It's there when we laugh and talk. It's there all the time, engraved independent of what's going on. The life of grief—or is it beyond grief?—that we live."

Several of the mothers reported that they would not have hesitated to have abortions if they had been told that their children would be born with multiple handicaps. One of the mothers said:

I can't see the point in putting oneself through the strain of having such a seriously handicapped child. I don't believe that anyone with their hand on their heart is really willing. I can honestly also see the pressures it puts on society. It's an enormous cost. The same resources could have been channeled elsewhere, so that others could have become well again. I don't think it's right to bring a handicapped child into the world if it's unnecessary. (p. 142)

Brinchmann also found that living with a severely handicapped child places severe strains on the whole family and that the stresses of rearing these children are so great that there is a high frequency of divorce in couples who have a handicapped child (p. 140). What Brinchmann describes is the reality of the lives of parents who have to rear children with genetic disorders. It is a very different picture from the sentimental portrayal sometimes given by those who have not had this personal experience.

Another study of the experiences of parents of children with genetic disorders involving mental retardation has been reported by Bruce, Schultz, and Smyrnios (1996). They found that the parents of mentally retarded children grieved over their children and their own misfortune and that this grief persisted with undiminished intensity through the children's childhood, adolescence, and the early adult years. To have a mentally retarded child is felt by the parents to be a personal tragedy, the impact of which does not diminish with time.

9. DYSGENIC AND EUGENIC TRENDS FOR GENETIC DISORDERS

During the twentieth century, there have been both dysgenic and eugenic trends with regard to genetic diseases and disorders. Dysgenic trends consist of medical advances that have preserved the lives of those with these disorders, enabling them to have children to whom they transmit the deleterious genes. This trend increases the prevalence of the genes in the population. These dysgenic medical advances are of three principal types: surgical treatments, pharmacological treatments, and improved treatments of critically ill newborns.

The first major dysgenic surgical treatment was developed in 1912 and consisted of an operation to correct congenital pyloric stenosis, a genetic defect in the functioning of the stomach. The next major surgical treatment was developed for retinoblastoma, a congenital eye cancer, consisting of cutting out the affected eyes. Later in the twentieth century a number of genetic disorders became surgically treatable by organ transplants, including those of the cornea, kidneys, liver, pancreas, heart, and lungs.

The development of pharmacological treatments has also contributed to dysgenics. For instance, the development of insulin in the 1920s made it possible to treat insulin-dependent diabetes, and the development of antibiotics in the mid-twentieth century made it possible to treat cystic fibrosis and other illnesses with some genetic component.

From around 1970, medicine became increasingly successful in the treatment of critically ill newborn babies. Most of these babies are either premature, born between 22 and 25 weeks gestation, with very low birth weights, or else they have congenital disabilities that previously would have been fatal. Many of these babies can now be kept alive by intensive care, but their prognosis is often poor. They are likely either to die in childhood or, if they survive, to have various impairments and a poor quality of life. In the economically developed nations, about 6 percent of newborns are so critically ill that they fall into this category.

The problem confronting physicians and the parents of these babies is whether they should be kept alive by what is sometimes called "aggressive treatment." In practice, the decision about whether to treat these babies is frequently made by the attending physician at the birth. In many Western nations physicians, sometimes alone and sometimes in consultation with parents, decide not to treat these babies and to allow them to die. In New Zealand, Australia, and the Netherlands, such decisions are "commonplace," according to Robert Blank (1995), a professor of political science at the University of Canterbury who specializes in problems of medical ethics. The same is true in Britain and in Continental Europe. In the United States, the problem of whether to provide treatment for critically ill babies came to public attention in 1971 with the case of a Down's syndrome infant with an intestinal blockage born at Johns Hopkins Hospital. The intestinal blockage could easily have been repaired by routine surgery, and this would have been done if the infant had been normal. But in this case the parents refused to give their consent to the surgery because they did not want to rear a child with Down's syndrome. The hospital did not seek a court order to carry out the operation, and the infant died. In 1973 two Yale University pediatricians admitted that they sometimes allowed critically ill newborns to die and defended this practice (Merrick, 1995). A further case occurred in 1982 with the birth of "Baby Doe" in Bloomington, Indiana, an infant with Down's syndrome and a defective esophagus. As in the Johns Hopkins case, the parents refused consent to surgery to correct the disorder, but on this occasion the physicians sought a court order for treatment. During the course of the protracted legal hearings and appeals, Baby Doe died.

United States courts have moved toward requiring physicians to keep critically ill babies alive. An important case was decided in 1994 when the U.S. Court of Appeals sitting in Richmond, Virginia, considered the problem of a 16-month-old anencephalic baby, designated Baby K, who had most of her brain missing, had never been conscious, and was only being kept alive on a respirator. The court ruled that she must be kept alive indefinitely.

While advances in medical treatments have had the dysgenic effect of increasing the numbers of those with genetic disorders, medical progress of another kind has had the eugenic effect of reducing them. This eugenic medical progress consists of the prenatal diagnosis of pregnant women for the presence of genetic disorders in the fetus and the termination of pregnancies when a disorder is identified. This procedure and its eugenic impact are discussed in Chapter 17.

10. A EUGENIC ASSESSMENT OF GENETIC DISORDERS

Genetic diseases and disorders impose heavy psychological and economic costs on the individuals who are afflicted by them, on their families, and on society as a whole. The psychological costs consist of severe and protracted suffering experienced by the individuals and their families. The economic costs consist of the considerable expenditures incurred by parents in rearing children with genetic disorders and by society as a whole.

The economic costs of the genetic disorders to society consist of the medical and welfare costs of providing care for those suffering from these disorders. It has been estimated that in the economically developed nations, about a quarter of hospital beds are occupied by patients with genetic diseases and disorders (Fletcher, 1988). Hence if these disorders could be eliminated, there would be a cost saving to society of around a quarter of all medical expenditures. In the economically developed nations, medical expenditures are about 8 percent of the gross national product (GNP), so the savings would amount to around 2 percent of the GNP. In addition, there are educational and welfare costs incurred by those with genetic disorders, which consist of the provision of special schools for children and of hostels for adults and the support of families in which neither parent is able to work because both have to look after the child. In total, these welfare costs consume approximately a further 1.5 percent of the GNP.

Although the costs incurred by genetic disorders are large, they are not so serious as the deterioration of genotypic intelligence and moral character that has taken place in the Western democracies over the course of the past century and half, a problem that will be addressed in the following chapters.

11. CONCLUSIONS

There are about 7,000 known genetic diseases and disorders. What they have in common are the distress and the costs they bring to those suffering from them and to their families and the costs of medical treatment, education, and welfare support they incur for society.

During the twentieth century there was a dysgenic trend for medical progress to preserve the lives of many of those with genetic disorders, which enabled them to have children and to transmit their adverse genes to succeeding generations. There was also a countervailing eugenic trend in the last three decades of the century for medical progress to reduce the birth incidence of genetic disorders by prenatal diagnosis and pregnancy terminations.

The argument sometimes advanced by otherwise intelligent geneticists and others that all genes are equally valuable and that the genetic diseases and disorders should be cherished cannot be accepted. The genes responsible for genetic diseases and disorders cause immense suffering, impose significant costs, and have no value. There is everything to be said for reducing the number of these genes and ultimately eliminating them.

5

⤸

Mental Illness

1. Nature and Costs of Mental Illness
2. The Classical Eugenic View of Mental Illness
3. Mental Illness and Creative Achievement in Historical Personages
4. Mental Illness and Creative Achievement in Living Writers and Artists
5. Creative Achievement in Relatives of the Mentally Ill
6. Creative Achievement and Psychoticism
7. Conclusions

⥲⥲⥲

In the preceding chapter, it was argued that genetic diseases and disorders are undesirable and that the genes responsible for them would best be eliminated. Historically, eugenicists have included the mental illnesses among the genetic disorders that they sought to eliminate. However, scientific evidence has since accumulated indicating that the genes responsible for mental illness may serve some useful purpose. Therefore, we should look critically at the former view that mental illness is wholly undesirable. This is the issue we examine in this chapter.

1. NATURE AND COSTS OF MENTAL ILLNESS

Schizophrenia, depressive psychosis, and manic-depressive psychosis are the three most common forms of serious mental illness, with rates of approximately 1 percent of the population, 3 to 5 percent, and 1 percent, respectively (Plomin, DeFries, McClearn, & Rutter, 1997; Jamison, 1993). All three are seriously debilitating conditions and cause great unhappiness to those who suffer from them and to their families. The misery of depression is so great that significant numbers of sufferers commit suicide, and it is estimated that about 70 percent of suicides have a recorded history of the disorder. Psychiatric illnesses also impose substantial costs on society in the form of medical care and welfare support. Mental illnesses are about as common as physical illnesses and consume approximately the same amount of resources.

The costs to society of schizophrenia have been estimated for the United States, Britain, Australia, the Netherlands, and Canada at about 0.3 percent of the gross national product (GNP) (Goerree, O'Brien, Goering, & Blackhouse, 1999). The costs of depression and manic-depressive psychosis are about the same. Costs arise about equally from medical and care costs and from lost economic productivity. It has been estimated that in the United States, the cost of schizophrenia exceeds even that of cancer (National Foundation for Brain Research, 1992). Schizophrenia imposes a further social cost in the form of the high rate of violent crime committed by schizophrenics, which is about four times greater than that of the general population, according to studies in Germany, Sweden, and Britain (Hafner and Boker, 1973; Lindquist & Allebeck, 1990; Coid, Lewis, & Reveley, 1993).

2. THE CLASSICAL EUGENIC VIEW OF MENTAL ILLNESS

In the first half of the twentieth century, eugenicists viewed mental illness as largely genetically determined and wholly undesirable, and they believed measures should be taken to eliminate it. To achieve this objective, programs were introduced to sterilize the mentally ill in the United States and in much of Continental Europe.

The early eugenicists were correct in their belief that mental illness has a major genetic component. Many studies have shown that identical twins are considerably more alike with respect to mental illness than are nonidentical twins. Similarly, many studies of adopted children have found that those whose biological parents had mental illness are at much greater risk of developing it (Plomin, De Fries, McClearn, & Rutter, 1997). A review of the research literature by Jones and Cannon (1998) concluded that children whose parents have mental illnesses have a 5 to 10 times greater probability of developing them than the general population have.

Further evidence for a significant genetic determination of these mental illnesses comes from studies in which they are regarded as extreme forms of traits continuously distributed in the population and measurable by questionnaire, of which the most frequently used is the Minnesota Multiphasic Personality Inventory (MMPI). A study of 119 twin pairs reared apart using this instrument has obtained estimated heritabilities of .61 for schizophrenia, .44 for depression, and .55 for hypomania (DiLalla, Carey, Gottesman, &

Bouchard, 1996). Thus there is no doubt that research in the second half of the twentieth century confirmed the eugenicists' contention that the mental illnesses have a substantial genetic basis.

3. MENTAL ILLNESS AND CREATIVE ACHIEVEMENT IN HISTORICAL PERSONAGES

Reservations about the view that mental illnesses are wholly undesirable have arisen because some evidence suggests that the genes responsible for them make a contribution to creative achievement. This claim has a long history. In the seventeenth century, John Dryden, the poet, wrote, "Great wits are sure to madness near allied, and thin partitions do their bounds divide." This contention has drawn support from four strands of evidence. First, a number of studies have reported that outstanding historical creative geniuses have had a high rate of mental illness. One of the first to make this claim was the German psychiatrist Juda (1949), who studied the lives of 294 Germans who had demonstrated great creativity. Juda concluded that they had a much higher incidence of psychoses than the general population. This conclusion has been supported by the Icelandic psychiatrist Karlsson (1978), who estimated that the rate of psychosis is 25 percent for great mathematicians, 30 percent for great novelists, 35 percent for great painters and poets, and 40 percent for great philosophers. These figures compare with a prevalence of approximately 6 percent in the general population. Karlsson concluded that typically these creative geniuses suffered from intermittent psychiatric disorders and produced their creative work during periods of remission.

A subsequent study reaching the same conclusion was made by Ludwig (1992) of 1,005 outstanding Americans of the twentieth century. He found that writers and artists had two or three times the rate of psychosis and suicide attempts as successful people in business, science, and public life and in the normal population. A similar investigation was carried out by Jamison (1993) of 36 British poets born between 1705 and 1805, in which she concluded that they were 30 times more likely to have had a manic-depressive illness than their contemporaries and 20 times more likely to have been committed to an asylum.

Post (1994) has conducted the most recent major study of mental illnesses among historical geniuses and eminent people. He examined the life histories of 291 eminent men of the nineteenth and early twentieth centuries. He concluded that they had a high rate of minor psychiatric instability, which was sufficiently severe to impair their work from time to time but not severe enough for a diagnosis of mental illness. The incidence of this minor psychiatric instability differed by occupation, being lowest among politicians (17 percent) and scientists (18 percent), and highest among composers (31 percent), artists (38 percent), and writers (46 percent). In contrast to previous researchers, Post concluded that in his sample, the lifetime incidence of depression and manic depression was only 1.7 percent, which is about one-third of that in the normal population, while he found no cases of schizophrenia.

The marked difference between Post's conclusions and those of other researchers is apparently due to the strictness with which a diagnosis of mental illness is made. Using a broad definition that includes mild forms of mental illness, the incidence of mental illness is found to be higher among highly creative people; but using the stricter definition adopted by Post, it is lower. Post, however, also concluded that mild symptoms of mental illness tend to be higher in creative indivduals. Both conclusions indicate that some degree of what can be most usefully described as "mental instability" is associated with creative achievement.

4. MENTAL ILLNESS AND CREATIVE ACHIEVEMENT IN LIVING WRITERS AND ARTISTS

A second strand of evidence for an association between mental illness and creative achievement comes from studies of living writers and artists. Several studies of this kind have found a high incidence of depression and manic-depressive psychosis. For instance, Andreason (1987) studied 30 American dramatists and novelists and concluded that 80 percent had suffered from severe or mild depression. Jamison (1993) studied 47 British writers and artists and found that 38 percent had received psychiatric treatment for depression or manic depression. Ludwig (1994) studied 59 creative writers and a matched control group of professional people and found that the writers had about five times greater lifetime prevalence of depression, mania, and anxiety states. Reviews of further research substantiating this conclusion have been published by Richards (1981), Ochse (1991), and H. J. Eysenck (1995).

5. CREATIVE ACHIEVEMENT IN RELATIVES OF THE MENTALLY ILL

A third strand of evidence for an association between mental illness and creative achievement is derived from a series of studies showing that the incidence of psychotic disorders, particularly depression and manic-depressive psychosis, tends to be higher than average among the relatives of creative individuals (Andreason, 1987; Karlsson, 1978; Isen, Daubman, & Nowicki, 1987; Ludwig, 1994). This suggests that the genes responsible for these conditions run in families, some members of which inherit many of the genes and develop a psychosis, while others inherit a few genes and have only some degree of mental instability, which, as Post (1994) concluded, makes a positive contribution to creative achievement. This is the interpretation of the results proposed by Eysenck (1993, 1995).

6. CREATIVE ACHIEVEMENT AND PSYCHOTICISM

A fourth strand of evidence for this association comes from of a body of work showing an association between creativity and the personality trait of *psychoticism* constructed by Eysenck (1993, 1995). Psychoticism is conceptualized as a continuously distributed personality trait in which psychotics, psychopaths, criminals, and aggressive, hostile, and impulsive individuals are at the high end of the scale, whereas mentally stable, conformist, empathic, conventional, and well-socialized people are at the low end.

An individual's position on the dimension of psychoticism can be measured by a questionnaire, and several studies have shown that high scores on the trait are associated with creativity. For instance, Gotz and Gotz (1979) have found that German professional artists tend to score high on psychoticism. These results are confirmed by a further set of studies showing that high scorers also tend to score high on tests of creativity, consisting of the ability to produce unusual ideas. Research finding this association has been reviewed by Rushton (1997).

The relationship between psychosis, psychoticism, and creativity has been usefully discussed by Eysenck (1993). He writes, "Our theory does not claim that psychosis as such produces creativity or that great artists and scientists are psychotic; such statements, frequently made in the past, are clearly untrue; what may be happening is that high psychoticism (P) is necessary for high creativity and that high P people may sometimes develop psychoses or at least suffer psychotic episodes during which their creative talents lie fallow" (p. 157). He cites Newton, Wagner, and Galileo as examples of highly creative people who seem to have ranked high on the trait of psychoticism.

However, high psychoticism alone is not sufficient for creative achievement. High intelligence is also required, as well as high *ego-strength*. The concept of ego-strength is borrowed from Freud and Cattell and consists of the capacity to overcome personal problems and to work persistently and single-mindedly toward long-term objectives. Eysenck's theory of creativity is that creative achievement requires "a combination of high psychoticism and high ego-strength: there is considerable evidence for the necessity of combining these two apparently antithetical properties" (1995, p. 236). Eysenck (1993) suggests that the reason the incidence of high psychoticism contributes to creative achievement is that "only aggressive, self-confident, dominant individuals can successfully show creative talents in a world full of envious mediocrities" (p. 184).

7. CONCLUSIONS

In general the genetic diseases and disorders are physiological defects that confer no advantages on those who inherit them. Eugenic measures to reduce them and the genes responsible for them would be desirable, both in the interests of the individuals concerned and their families and of society as a whole. An exception must be made, however, for mental illnesses, particularly for depression and manic-depressive psychosis. There is substantial evidence suggesting that some degree of mental instability or what can be designated "subclinical mental illness" contributes to creative achievement. There have been four kinds of study pointing to this conclusion. There has been a high incidence of these conditions in eminent historical creative individuals, in living writers and artists, and among the relatives of creative individuals. There are also high average scores on psychoticism in creative people. Taken together, these four strands of evidence constitute a reasonably strong case for an association between creativity and subclinical depression and manic-depressive psychosis, suggesting that the genes for mental illness have some positive value. Hence the complete elimination of these genes should not be an objective of eugenics.

6

Intelligence

1. Definition and Nature of Intelligence
2. Social Value of High Intelligence
3. Intelligence and Genius
4. Social Costs of Low Intelligence
5. *The Bell Curve* and Its Critics
6. The Intelligence of Populations
7. National Economic and Military Strength
8. Intelligence and Happiness
9. Potential Problems of a Highly Intelligent Society
10. Solutions to Problems of Excess Intelligence
11. Need for a Variety of Cognitive Abilities
12. Conclusions

While the reduction of genetic diseases and disorders is the first objective of eugenics, the second objective is the increase of intelligence. Two components of this objective can be distinguished. The first is to attempt to shift the whole distribution of intelligence upward such that the average intelligence level of the whole population is increased. To justify this objective, it first needs to be shown that intelligence is valuable, and this is the purpose of the present chapter. The second component of this objective is to specifically target the low end of the IQ distribution. Reduction of the numbers of the mentally retarded has been such an important objective of classical eugenics that the case for it deserves separate consideration, which we will take up in the next chapter.

1. DEFINITION AND NATURE OF INTELLIGENCE

It will be useful to begin by defining *intelligence*. A helpful starting point for the definition of intelligence was provided by a panel set up by the American Psychological Association (APA) in 1995 under the chairmanship of Ulrich Neisser and consisting of 11 American psychologists whose mandate was to produce a consensus view of what is generally known and accepted about intelligence. The definition proposed by the task force was that intelligence is the ability "to understand complex ideas, to adapt effectively to the environment, to learn from experience, to engage in various forms of reasoning, to overcome obstacles by taking thought" (Neisser, 1996, p 1.) This definition will command wide assent and is acceptable for our present purposes, except for the component of effective adaptation to the environment. A mentally retarded woman with 10 children living on welfare may be well adapted to her environment, and indeed, in the biological sense of the word, she is better "adapted" than any of the 11 distinguished members of the task force, none of whom has 10 children. A definition that avoids this difficulty was proposed by Gottfredson and endorsed by 52 leading experts and published in the *Wall Street Journal* in 1994: "Intelligence is a very general mental capacity which, among other things, involves the ability to reason, plan, solve problems, think abstractly, comprehend complex ideas, learn quickly, and learn from experience" (Gottfredson, 1997, p. 13).

It is useful to consider intelligence as a single entity that can be measured by intelligence tests and quantified by IQ (intelligence quotient). This theory was originally set out in the first decade of the twentieth century by Charles Spearman (1904), who showed that all cognitive abilities are positively intercorrelated, such that people who do well on some tasks tend to do well on all the others. To explain this phenomenon, Spearman proposed that there must be some general mental power determining performance on all cognitive tasks and responsible for their positive intercorrelation. He designated this construct *g*, for general intelligence. Spearman also proposed that in addition to *g*, there are a number of specific abilities that determine performance on particular kinds of tasks, over and above the effect of *g*. Subsequent theorists have proposed that there are also broader "group factors" or "primary abilities" which can be envisioned as aggregates of the specifics, the most important of which are verbal, comprehension, reasoning, memory, spatial, perceptual, and mathematical. This so-called "hierarchical model" of intelligence is widely accepted among contemporary experts such as the American Task Force (Neisser, 1996), Jensen (1998), Humphreys (1994), and Mackintosh (1998).

When intelligence is considered as a general ability (*g*) with a number of narrower abilities, it has generally been found that *g* is the most important determinant of task performance. For instance, performance on a test of mechanical aptitude is more sharply determined by g than by mechanical ability (Ree & Erles, 1991). Thus, although the primary abilities cannot be entirely disregarded and although societies need citizens with different mixes of these primary abilities to perform different kinds of tasks with maximum efficiency, the first objective for a eugenic society aiming to improve the intelligence of its population would be to increase the level of g.

2. SOCIAL VALUE OF HIGH INTELLIGENCE

High intelligence is socially valuable because it is a significant determinant of educational attainment, job performance, earnings, and occupational status. There is a large body of research literature supporting this conclusion (e.g., Jencks, 1972; Brody, 1992; Jensen, 1980, 1998; Eysenck, 1979; Herrnstein & Murray, 1994; Mackintosh, 1998). Table 6.1 shows correlations between these variables from Jencks (1972), who reviewed the U.S. research literature up to 1970, and from Mackintosh (1998), who reviewed more recent research drawn from several countries. (In some cases, Mackintosh gave a range of estimates that have been averaged in the table.) It will be noted that there is close agreement between the two sets of figures.

Both Jencks and Mackintosh accept that intelligence has a significant genetic basis. Jencks proposed that intelligence has a heritability of 50 percent, and Mackintosh proposed that heritability lies somewhere between 35 percent and 75 percent. Many experts regard these estimates of the heritability and the size of the correlations in Table 6.1 as too low for various technical reasons, such as the absence of corrections for unreliability of measurement and the restriction of range of test subjects. Nevertheless it is unnecessary to enter into discussions of the precise magnitude of these correlations to establish the general point that intelligence has a substantial heritability and is positively associated with these socially desirable phenomena.

It is also important to establish that intelligence is causally related to these

Table 6.1
Correlations Between Intelligence and Various Forms of Achievement,
Estimated by Jencks and Mackintosh

Achievement	Correlations with IQ	
	Jencks	Mackintosh
Educational attainment	—	.55
Years of education	.58	.60
Job performance	—	.26
Earnings	.35	.40
Occupational status	.52	.55

outcomes and is not merely a correlate. The most straightforward argument to establish this point is that intelligence can be measured in young children and that these IQs are fairly stable over time and predict subsequent achievements many years later. For instance, in a British study carried out by Yule, Gold, and Busch (1982), an intelligence test was administered to 85 five-and-a-half-year-old children. Eleven years later, the children were tested in reading and mathematics. The correlation between IQ and reading was 0.61 and between IQ and mathematics 0.72.

It can be inferred from the substantial heritability of intelligence and its causal impact on educational and occupational achievement that if the intelligence of a population could be increased, there would be concomitant increases in the educational attainment of children and adolescents, in years spent in education, in efficiency of job performance, in earnings, and in occupational status. All these things are valued highly by individuals for themselves, by parents for their children, and by governments for their populations. Enormous efforts and expenditures are incurred in trying to increase these outputs, for instance, by attempting to improve the efficiency of schools in producing higher educational standards, by encouraging young people to persevere in education, by improving job training, and by improving industrial efficiency as a means of increasing earnings. There can be no doubt that all these desirable things would be improved by an increase in the intelligence level of the population.

3. INTELLIGENCE AND GENIUS

Another argument for the desirability of raising the level of intelligence of the population is that there would be greater numbers of people with high IQs at the top end of the intelligence distribution from which geniuses come. High intelligence is an indispensable component of genius. The intellectual, scientific, and cultural advances in any civilization are made by a very small number of geniuses with exceptionally high intelligence. If the intelligence level of the population could be increased, there would be more geniuses, and this would enhance the quality of civilization.

The conclusions of decades of research on the role of intelligence and genetic factors in genius have been summarized by Sir Michael Rutter (1999) of the Institute of Psychiatry in London: "High intelligence seems a sine qua non . . . Nature or nurture? Almost certainly an interplay between the two. It seems most unlikely that people can be schooled or trained to become geniuses" (p. 23). It has to be admitted that, although it may seem obvious that people like Newton, Einstein, Beethoven, and Shakespeare must have been exceptionally intelligent, this is not straightforward to prove conclusively to sceptics. In the case of Shakespeare, he clearly had a very large vocabulary, which included words such as *incarnadine* (Lady Macbeth), *traduce* (Othello), *rotundity* (Lear), and *marjoram* (Sonnets), which are unknown to the great majority of people and may even stretch the highly intelligent readers of this book. Vocabulary size is one of the best measures of intelligence, and so it can be concluded that Shakespeare must have had an exceptionally high IQ.

Despite the obvious difficulties of assessing the IQs of geniuses of past centuries, a systematic study to solve this problem was made in the 1920s by Catherine Cox (1926). She began with a list of one-thousand geniuses compiled by an early psychologist named J. M. Cattell. From these she selected 301 for whom there were records of their intellectual development and achievements in childhood, adolescence, and early adulthood. She gave these records to psychologists experienced in the assessment of intelligence and had them estimate the IQs. She had at least two psychologists make these assessments so that she was able to check their reliability. An example illustrating the method is her assessment of the IQ of the French mathematician and scientist Blaise Pascal (1623–1662). From the historical record, it is known that at the age of 11 Pascal noticed that when he struck a plate with a knife it made a loud noise, but that if he put his hand against the plate, the noise stopped. This led him to make a number of experiments on sound, and he wrote a treatise on the subject, which was completed during his eleventh year. In his early teens he developed an interest in geometry. His father was convinced that he was attempting to understand problems for which he was not sufficiently ready and prevented him from studying Euclid's geometrical theorems. Nevertheless, the young Pascal worked out a number of these for himself, and at the age of 16 he wrote a treatise on the geometry of conic sections. At the age of 19 he invented a calculating machine, and at 25 he worked out the theory of atmospheric pressure by a barometric experiment. Cox's psychologists estimated Pascal's IQ at 185.

Another well-documented case of early intellectual precocity is the English political theorist John Stuart Mill. At the age of five, he is recorded as having had a conversation with Lady Spencer on the comparative merits of Marlborough and Wellington as generals. A year later Mill wrote a history of Rome using such phrases as "established a kingdom" and "the country had not been entered by any foreign invader." At the age of 11, he was doing mathematics at present-day college level. Using the Binet formula "mental age divided by chronological age × 100 equals IQ," the young Mill was generally performing at about the level of those twice his chronological age. His IQ was estimated at 190.

When Cox had obtained estimates of the IQs of all her geniuses, she sorted them into eight categories according to the fields in which they made their achievements. The average IQs for the categories are shown in Table 6.2. The average IQ for the entire sample is 158. In a population with an average IQ of 100, the incidence of individuals with an IQ at this level is approximately 1 in 30,000.

Table 6.2

Mean IQs of Historical Geniuses, Estimated by Catherine Cox

Category	Mean IQ	Category	Mean IQ
Artists	150	Scientists	155
Musicians	164	Soldiers	132
Philosophers	175	Statesmen	162
Religious leaders	160	Writers	164

The conclusions to be drawn from this research are that geniuses have very high IQs and are very rare. If the distribution of intelligence could be shifted upwards by eugenic interventions, there would be greater numbers at the very top end of the distribution from which geniuses are drawn. For instance, if the average IQ of the population could be raised to 115, there would be approximately 1 individual per 1,000 with an IQ over 158, a 30-fold increase over their numbers in a population with an average IQ of 100.

We should conclude by noting that high intelligence is not sufficient for genius. To produce a work of such outstanding quality that it can be described as a work of genius requires personality qualities of dedication, application, persistence, and creativity. There have certainly been people who had the requisite IQ to be geniuses, but they lacked these personality qualities. The inheritance of genes for these personality qualities, in addition to those for very high intelligence, is extremely uncommon, which is why genius is such a rare phenomenon. Nevertheless, if the intelligence of the population were raised, there would be more individuals with the intellectual capacities necessary for genius, the number of geniuses produced by such a population would be increased, and the quality of civilization would be enhanced.

4. SOCIAL COSTS OF LOW INTELLIGENCE

Just as there are social benefits of high intelligence, there are social costs of low intelligence, which consist of low educational attainment, educational dropouts, poor job performance, low earnings, and low social status. Additional social costs of low intelligence are high rates of delinquency, crime, and unemployment. The average IQs of delinquents, criminals, and the unemployed are shown in Table 6.3. The first seven entries are taken from literature reviews and show a consensus that the average IQs of delinquents and criminals is around 92. The remaining three entries are from the original studies. The study of conduct disorders consisting of persistent antisocial behavior and disobedience in young children is included because this is typically a precursor of later delinquency.

Table 6.3
Mean IQs of Delinquents, Criminals, Children with Conduct Disorders,
and the Unemployed

Criterion Group	Mean IQ	Reference
Delinquents	89	Jensen, 1980
Criminals	92	Hirschi & Hingelang, 1977
Criminals	92	Wilson & Herrnstein, 1985
Criminals	92	Quay, 1987
Criminals	92	Eysenck & Gudjonsson, 1989
Criminals	92	Raine, 1993
Criminals	92	Lykken, 1995
Conduct disorders	83	Moffit, 1993
Unemployed	81	Toppen, 1971
Unemployed	92	Lynn, Hampson, & Magee, 1984

An alternative way of expressing the relationship between crime and intelligence is in terms of the correlation coefficient. Four studies, of which the
first three are based on literature reviews, expressing the relationship as correlations are shown in Table 6.4. The figures set out in both tables indicate
that the relationship between low intelligence and delinquency is a little
stronger than the relationship between low intelligence and crimes committed by adults.

While the positive associations of low intelligence with crime and unemployment do not necessarily imply cause and effect, there are good reasons to
believe that there is some causal impact of low intelligence on these two
phenomena. In regard to crime, there are two principal explanations for its
relationship with low intelligence. These can be designated the *cognitive deficit theory* and the *alienation theory*. The cognitive deficit theory has been advanced by Wilson and Herrnstein (1985) and holds that those with low IQs
have a greater need for immediate gratification, weaker impulse control, poorer
understanding of the consequences of punishment, and a more poorly developed moral sense then do people with average or above average IQs. This
theory implies a direct causal impact of low intelligence on crime and pre-

Table 6.4
Correlations of Delinquency and Crime with IQ

Variable	Correlation/IQ	Reference
Delinquency	-.45	Jensen, 1980
Crime	-.19	Moffit, Gabrielli, Mednick, & Schulsinger, 1981
Crime	-.25	Eysenck & Gudjonsson, 1989
Crime	-.25	Gordon, 1997

dicts that an increase in the level of intelligence of the population would produce fewer individuals with the cognitive deficit of low intelligence, and hence a reduction in crime.

The alienation theory states that adolescents with low IQs tend to do poorly at school and either get only badly paid jobs or find no employment at all. This makes them disaffected and alienated from society, as a result of which they are likely to turn to crime. On the basis of this theory, an increase in the level of intelligence of the population would be somewhat less likely to produce a fall in the crime rate because no matter how great the increase of intelligence, there would continue to be a distribution from more intelligent to less intelligent, and these latter would do poorly, feel resentful and alienated, and often turn to crime. Probably both the cognitive deficit and the alienation theories are partially correct, and both theories predict that an increase in the intelligence of the population would produce some reduction in crime.

With regard to unemployment, a more intelligent population would be expected to have lower levels of unemployment because there would be fewer people incapable of working at competitive wages in the international labor market, and the reduction of unemployment would mitigate one of the major causes of alienation among the less intelligent and hence their propensity to commit crime.

5.   THE BELL CURVE AND ITS CRITICS

A new analysis of the relationship between intelligence and a number of socially important phenomena was published by Richard Herrnstein and Charles Murray (1994) in their book, *The Bell Curve*. The main body of the book consisted of an analysis of the data contained in the National Longitudinal Study of Youth, a U.S. study of a nationally representative sample of approximately 12,000 young people. In their analysis of this data set, Herrnstein and Murray confirmed the previous research showing that intelligence is positively related to educational attainment, employment, earnings, and social status and negatively related to crime. They also broke new ground in demonstrating that intelligence is related to poverty, health, single motherhood, and welfare dependency. Most of their analyses were based on whites only, making it free of possible contamination by racial differences. Their principal results are summarized in Table 6.5. To display the data, they divided the sample into five intelligence bands, which consisted of those with IQs of 126 and above, IQs from 111 to 125, from 90 to 110, from 75 to 89, and IQs of 74 and below. Then they gave the percentage of a variety of social phenomena for each intelligence band. It can be seen that there are large disparities in the incidence of these socially important phenomena. For example, 75 percent of those with IQs of 126 and above gain college degrees, whereas none of those with IQs below 74 do so; and so on down through the

Table 6.5
Incidence of Various Social Phenomena (percentages) in Five IQ Bands

Social Phenomena	126+	111–125	90–110	75–89	74–
College graduate	75	38	8	1	0
Below poverty line	1	4	7	14	26
Unemployed one month in past year (males)	4	6	8	11	14
Work impaired by poor health (males)	13	21	37	45	62
High school dropout	0	1	6	26	64
Single mother	4	8	14	22	34
Long-term welfare mother	0	2	8	17	31
Long-term welfare recipient	7	10	14	20	28
Served time in prison	0	1	3	6	13
Has child with IQ below 80	1	3	6	16	30

Source: Herrnstein and Murray, 1994.

table. The last line shows that the intelligence level of the individuals in the sample is strongly related to that of their children. Note that only 1 percent of the most intelligent group had a child with an IQ below 80, compared with 30 percent of the least intelligent group.

Herrnstein and Murray placed the heritability of intelligence somewhere between 40 percent and 80 percent. Thus, they regarded the differences in the social phenomena shown in Table 6.5 as having a significant genetic basis. They argued that social mobility has led to more intelligent individuals rising in the socioeconomic hierarchy and less intelligent individuals falling, with the result that the United States and other economically developed nations have to some degree become stratified genetically for intelligence. This view is indisputably correct, and I have set out in detail the evidence for it in my book *Dysgenics*.

The Bell Curve presented a message that most of the social science community and the media did not want to hear. Accordingly, it evoked a storm of criticism in articles and books such as *Measured Lies* (Kincheloe, Steinberg, & Gresson, 1996), the title of which betrays the strength of the hostility to Herrnstein and Murray's conclusions. Six members of the sociology faculty of the Berkeley campus of the University of California combined forces to launch a counterattack against *The Bell Curve* entitled *Inequality by Design* (Fischer et al., 1996), in which they argued that differences in intelligence, earnings, and socioeconomic status are entirely determined by the environment and that these could be mitigated by social interventions such as improvements in education and redistributive taxation. These Berkeley sociologists only succeeded in showing that they continue to inhabit a mid-twentieth-century time warp of environmental determinism. A more measured assessment of *The Bell Curve* by Devlin, Frenberg, Resnick, and Roeder (1997), *Intelligence*,

Genes and Success, was no more than a series of quibbles about the magnitudes of the effects calculated by Herrnstein and Murray. I have shown in detail that neither of these books, nor others like them, has been able to establish any coherent case against the conclusions of *The Bell Curve* and that all of Herrnstein and Murray's conclusions are essentially correct (Lynn, 1999a).

6. THE INTELLIGENCE OF POPULATIONS

We have seen that among individuals intelligence is related positively to educational attainment, efficiency of job performance, earnings, social status, and the outstanding intellectual achievements of genius and that it is also related negatively to unemployment, welfare dependency, poverty, single motherhood, ill health and crime. We can infer that the same relationships would be present among populations and that populations with a high average intelligence level would be characterized by higher rates of the desirable social characteristics and lower rates of the undesirable ones. Several studies have demonstrated that this is the case.

The first study was carried out by Maller (1933) in the early 1930s in New York City. He took as his population units the 310 administrative districts into which the city was divided. Average intelligence levels of the children in these districts were calculated from tests administered to approximately 100,000 children. Maller also collected data on several important social phenomena. Intelligence levels in the districts were correlated positively with educational attainments and negatively with delinquency, the death rate, and infant mortality.

The second major study of this issue was carried out in London later in the 1930s by Burt (1937). He took as his population units the 29 boroughs of the city and obtained a measure of the average intelligence level in each borough from tests administered to 10-year-olds. He calculated correlations between the average IQs and a variety of social and economic phenomena. His results were closely similar to those obtained by Maller in New York. The intelligence level of children in the boroughs was positively related to the level of educational attainment (indexed by the proportion of children obtaining scholarships to selective grammar schools) and negatively correlated with delinquency and infant mortality. Burt also found that intelligence levels were negatively related to the prevalence of mental retardation, poverty, and unemployment. The correlations of these social phenomena found in the New York and London studies are shown in Table 6.6.

Over the course of the next half century, several similar studies were conducted in the United States and Britain that confirmed and extended these results. Thorndike and Woodyard (1942) collected data on the IQs of 12-year-olds in 30 U.S. cities and then constructed an index of each city's "Good-

Table 6.6
Correlations of Intelligence Levels in Districts of New York
and London with Various Social Phenomena

	New York	London
Educational attainment	+.70	+.87
Delinquency	-.57	-.69
Mortality	-.43	-.87
Infant mortality	-.51	-.93
Mental retardation	—	-.91
Poverty	—	-.73
Unemployment	—	-.67

ness," based on a combination of measures of health (infant and adult mortality), average incomes, and literacy rates. The highest score was obtained by Pasadena, California, followed by Montclair and Cleveland Heights, and the lowest scores by Augusta, Meridian, High Point, and Charleston. The correlation between the cities' average IQs and their Goodness score was 0.86, showing that intelligence is a very powerful determinant of what may be called the quality of a city.

The next major investigation of this kind was carried out by Davenport and Remmers (1950), using U.S. states as population units. Average IQ levels were calculated from the test scores of over 300,000 young men. These correlated .81 with average state incomes and .67 with the proportion of adults in the state appearing in Who's Who. This was the first empirical demonstration of the eugenicists' contention that a high level of intelligence in the population would produce a large number of intellectually outstanding individuals.

In the late 1970s, I made further studies of this issue for the British Isles and for France (Lynn, 1979, 1980). The British Isles were divided into 13 regions and France into 90 departments, and mean IQs were obtained for each region and department. Data were then obtained for a variety of educational, social, economic, and health phenomena, and correlations of these with average IQs calculated. The results are summarized in Table 6.7. Educational attainment was measured for the British Isles by the numbers of first-class degrees awarded by universities as a proportion of young people; no comparable index could be obtained for France. Intellectual attainment was indexed for the British Isles by the proportion of the population who were Fellows of the Royal Society and for France by the proportion who were members of the Institut de France. It will be seen that the intelligence levels of the populations are positively associated with educational attainment,

Table 6.7
Correlations Between Intelligence Levels in the Regions of the British Isles
and France and Social Phenomena

Measure	British Isles	France
Educational attainment	.60**	—
Intellectual attainment	.94**	.26*
Earnings	.73 **	.61*
Unemployment	-.82**	-.20
Infant mortality	-.78**	-.30**

* Denotes statistical significance at $p < .05$; ** at $p < .01$.
Source: Lynn, 1979, 1980.

intellectual attainment, and earnings and negatively associated with unemployment and infant mortality.

All the aforementioned studies provide direct support for the conclusion that among populations, as among individuals, intelligence is associated with a variety of important social phenomena. They show that if the intelligence level of the population could be increased, a number of desirable social outcomes would follow, including improved educational attainment, greater intellectual achievement, and higher earnings. In addition, a number of undesirable social phenomena would be reduced, including unemployment, poverty, poor health, mortality, and crime.

7. NATIONAL ECONOMIC AND MILITARY STRENGTH

A further benefit of increasing the level of intelligence of the population would be the enhancement of economic and military strength. We saw in the preceding section that the intelligence of regional populations in the United States, the British Isles, and France is strongly related to the scientific and general intellectual achievements of the regions and to their economic efficiency, as reflected in their per capita income and rates of unemployment. The same advantages would be gained by nation states with intelligent populations.

An intelligent population has an advantage in science and technology because it has more gifted individuals able to make important scientific and technological discoveries, and it also has individuals of high average intelligence able to implement them. An intelligent population has more effective executives, managers, and operatives in its industries who are able to produce goods efficiently. An intelligent population is stronger militarily because it has a stronger scientific, technological, and economic base. It has the additional advantage of possessing more intelligent military commanders and soldiers.

The history books are filled with stories of nations with superior military technology defeating their enemies. In the year 732, the French commander Charles Martel was able to defeat the Arabs at the battle of Tours because his horses had the new technological invention of the stirrup, which enabled his soldiers to sit more securely in the saddle and provided them with the leverage to ram lances into the enemy with greater force and accuracy (Thomas, 1981). In 1415 the English defeated the French at Agincourt because they possessed the long bow, which could be fired more rapidly than the French crossbow. In the late twentieth century, the Americans and Europeans easily defeated Iraq because they had vastly superior military technology.

History also provides numerous examples of nations winning wars because they had highly intelligent military commanders. Some notable examples are Julius Caesar, who conquered France and Britain without losing a single battle, and Wellington, under whom the British won every battle against the French in the Napoleonic wars. Conversely, many wars have been lost by the blunders of military commanders lacking sufficient intelligence. For instance, Napoleon made the costly mistake of invading Russia in the autumn of 1812. If he had invaded in the spring, he would have had a much greater chance of success because he would have had several more months in which to defeat the Russians before the onset of winter. Napoleon misjudged the severity of the Russian winter, which forced him to retreat, caused the loss of about 80 percent of his army, and led ultimately to his defeat at Waterloo.

High intelligence includes the ability to make correct assessments of the probable outcomes in complex situations, such as those of warfare; and although Napoleon was undoubtedly an intelligent man, he was not sufficiently intelligent for the task of defeating Russia. Remarkably, even with this well-known historical precedent, Hitler made precisely the same mistake by invading Russia in June 1941, as a result of which he suffered a similar defeat. The recent release of British military documents from World War II revealed that in the later stages of the war, the British high command discussed various plans to assassinate Hitler, such as bombing his hilltop retreat in Bavaria; but they eventually decided, probably correctly, that Hitler had made so many blunders that it would be better for him to remain in command than to kill him and risk having him replaced with someone more intelligent.

In the twenty-first century, scientific and technological supremacy will become increasingly decisive in military conflicts. Just as in the closing stages of World War II the United States was able to defeat Japan by the use of nuclear bombs, so in the future a nation with superior nuclear weapons and delivery systems, and perhaps also biological weapons, will be able to defeat and subjugate other nations by the use, or by merely the threat, of its superior weaponry. A highly intelligent population able to produce sophisticated military hardware will become an even more important determinant of military superiority than it has been in the past.

8. INTELLIGENCE AND HAPPINESS

Although the intelligence of a population is indisputably a factor in its economic, scientific, cultural, and military strength, it is an interesting question whether intelligence is related to a population's happiness and, consequently, whether a eugenic increase in the intelligence level of a population would be expected to lead to an increase in general happiness. Being unable to find any research on this question, I have examined the results of the opinion polls carried out by the American National Opinion Research Center (NORC). Pollsters asked a representative sample of the population whether they were "very happy," "pretty happy," or "not too happy." They also gave a 10-word vocabulary test that provides a measure of intelligence. Data on both happiness and vocabulary were collected in 1974 and 1994. In both years there were small but statistically significant correlations between happiness and intelligence, suggesting that if the intelligence level of the population were raised, the sum of human happiness would also be increased.

However, this is not necessarily the case. If the three responses to the happiness question are scored 1 (very happy), 2 (pretty happy), and 3 (not too happy), the average happiness score of the respondents was 1.75 in 1974 and 1.82 in 1994. Thus people were slightly less happy in 1994 than in 1974, and this difference is statistically significant. Intelligence in the United States increased by approximately six IQ points over the 20-year period 1974–1994 (Lynn & Pagliari, 1994), so evidently an increase in the intelligence level of the population over time does not necessarily produce a concomitant increase in happiness. The positive association between intelligence and happiness in both 1974 and 1994 probably arises because the less intelligent are conscious of being relatively disadvantaged, and this reduces their happiness. As the general level of intelligence rises, the least intelligent remain in a lower social position; so despite the association between intelligence and happiness in any particular year, happiness does not necessarily increase as intelligence increases (Lynn & Lynn, in press). Thus we have to conclude that a eugenic case for measures to increase the intelligence level of the population cannot be argued on the grounds that this increase would raise the general level of happiness.

9. POTENTIAL PROBLEMS OF A HIGHLY INTELLIGENT SOCIETY

We have seen that a society that succeeded in raising the intelligence of its population would secure the benefits of higher educational standards; higher earnings; greater scientific, technological, and cultural achievements; and a stronger economy and military capability. All these are attractive outcomes. Nevertheless, it has often been argued that it would be a mistake for a society to raise the intelligence level of its population because societies need unintelligent people to do simple jobs just as much as they need intelligent people to do cognitively demanding jobs. It is asserted that if the general intelligence level of the population were to be increased, there would not be sufficient numbers of unintelligent people to do the simple, routine jobs for which nature has fitted them. This argument was noted in 1923 by Marie Stopes, the British eugenicist and birth control campaigner, in her play *Our Ostriches*.

The heroine of this play, which ran for a number of weeks in a London theater, is an earnest upper-class, eugenically minded young woman who works in a birth control clinic in the impoverished East End of London to provide contraception to the poorer and supposedly genetically less well endowed classes. When her mother learns about this, she poses a question: "But if the lower orders have fewer children, where will we get our servants from?"

This point was restated by H. J. Eysenck in an interview he gave in 1996. Asked for his views on eugenics, Eysenck replied, "Eugenicists often say that we should breed for intelligence. That is a hopeless idea, because if we all had Einstein's intelligence, who would deliver the milk? A society needs all sorts of different levels of intelligence. I think this eugenicist dream would be a great mistake." (Turner, 1996, p. 6).

This argument has also been put forward by the German geneticist Volkmar Weiss (1992). He posits a genetical system for intelligence in which there are three genetic types—the first consisting of approximately 5 percent highly intelligent individuals (IQs of 130), the second of about 27 percent average-level individuals (IQs of 112), and the third of about 68 percent dull individuals (IQs of 94). He suggests that societies need this kind of intelligence distribution to maintain social stability. Societies need a small elite to occupy the top positions, a greater number of executives to carry out the instructions of the elite, and a large number of people to do the humdrum work. As an example, he contends that modern societies need a few highly gifted individuals to invent machines, a large number of moderately gifted people to repair them, and a still larger number of quite dull people to operate them.

10. SOLUTIONS TO PROBLEMS OF EXCESS INTELLIGENCE

There are six answers to the argument that it would be a mistake to attempt to raise the population's intelligence because society needs people of low and average intelligence to carry out simple and only moderately demanding jobs.

First, even if it were true that unintelligent people are needed to carry out undemanding jobs, there are at present too many of these unintelligent people in North America and Europe for the numbers of jobs that can be done by them. This is why the unemployment rate is so high among the unintelligent. There is a mismatch between the jobs society needs done and the intelligence levels of the population such that Western societies have an excess of unintelligent people in relation to the numbers of undemanding jobs. An increase in the level of intelligence would correct this mismatch.

Second, intelligence levels of the populations of the economically developed nations rose by about 18 IQ points from the 1930s to the year 2000 (Flynn, 1984; Lynn & Pagliari, 1994), but this has not caused any problems of there being too few people to do humdrum jobs like delivering the milk. On the contrary, there are still too many people who are only able to do jobs of this kind in relation to the diminishing numbers of jobs available.

Third, many cognitively undemanding jobs have been eliminated during the nineteenth and twentieth centuries by automation. For instance, domestic servants have largely disappeared as their work of lighting fires has been replaced by central heating and washing up has been replaced by automatic dishwashers. Similarly, much of the work formerly done by farm workers, print workers, ticket vendors, and so forth, has been replaced by machines. There is every reason to expect the process of automation to continue, further reducing the numbers of cognitively undemanding jobs suitable for those with low intelligence. For example, cutting the lawn is a rather boring and cognitively simple job that many intelligent people would doubtless prefer not to have to do and are glad to pay the less intelligent to do for them. If eugenic policies reduce the numbers of the less intelligent, geneticists will no doubt produce a new species of "smart grass" that grows to a height of precisely one inch, and then stops.

Fourth, contrary to the assertions of Eysenck and Weiss, intelligent people are quite capable of doing humdrum jobs, even if they do not particularly enjoy them. Einstein worked at a fairly humdrum job as a junior scientist in the Zurich Patent Office. The philosopher Ludwig Wittgenstein, in the words of his biographers Thorne and Collocott (1984, p. 1432), advanced our understanding of the logical foundations of language by his theory that "all significant assertions can be analyzed into compound propositions containing logical constants and are truth functions of elementary propositions. An elementary proposition symbolizes a real or atomic factor or possibility or, as the Germans say, 'Sachverhat.'" Wittgenstein undoubtedly had a very high IQ; but after having established these important truths at the age of 21, he worked for the next six years as an elementary village school teacher in Austria (1920–26) and for another two years as a gardener's assistant in a monastery (1926–28). In the 1930s he took a post in the philosophy department at the University of Cambridge. Then during World War II Wittgenstein worked as a porter at Guy's Hospital in London. Returning to Cambridge after the war, he once again took up his earlier work on the logic of language, and "by means of language games," he examined the varieties of linguistic usage of certain philosophically important expressions as a means of clarifying philosophical complexity. He found that the varieties of linguistic usage of words in many cases pointed to a "family resemblance between them rather than one single essential meaning" (p. 1432).

Many similar instances could be given of highly intelligent people who have been quite capable of doing humdrum jobs. The English chemist Michael Faraday worked in his youth as a bookbinder and a laboratory assistant; the Austrian philosopher Karl Popper worked for several years as a cabinetmaker; and the American inventor Thomas Edison worked as a railroad newsboy. All these people had very high IQs and provide compelling testimony that highly intelligent people are perfectly capable of doing cognitively undemanding jobs. At a somewhat lower level, enormous numbers of intelligent students in contemporary societies work their way through college as waiters or waitresses, gas pump attendants, and the like. In a eugenic society with a high average level of intelligence and fewer unintelligent people, an intelligent citizenry could quite well do their own low-level work, such as weeding the garden, putting dishes in the dishwasher, and washing clothes.

Fifth, with a highly intelligent population, people could be paid whatever is necessary to do the humdrum jobs that need to be done. If too few people wanted to do cognitively undemanding work, they could be induced to do so if these jobs were made sufficiently remunerative. The law of supply and demand would still hold, just as it does now, such that a salary would naturally increase to the point at which someone was willing to take the job.

Sixth, in the extremely unlikely event that even high salaries failed to produce sufficient people to do certain unattractive jobs, there could be a requirement that citizens perform a certain amount of community work, analogous to the military conscription that many countries now have. This would not be unethical or an intolerable burden. The Western democracies require their citizens to serve in the military and risk their lives in times of war; so it is difficult to raise any objection to their being required to devote a few hours a week to doing such jobs as cleaning the streets or collecting the garbage, which no one wants to do.

For all the above-listed reasons, we can conclude that there should be no serious problems of the "who would deliver the milk" kind in a eugenic society with a greatly increased level of intelligence.

11. NEED FOR A VARIETY OF COGNITIVE ABILITIES

Hitherto, intelligence has been treated as a single entity that can be directed into a variety of channels. The truth of the situation is a little more complex. In addition to general intelligence, there are a number of other abilities. Different people have their own strengths and weaknesses in these, and even the exceptionally intelligent are not invariably brilliant at everything. For instance, as a young adolescent Albert Einstein was brilliant at physics and mathematics, and at the age of 12 he found an original proof of Pythagoras's theorem. However, he was weak in language-based subjects and, at the age of 16, failed the entrance examination of the Federal Institute of Technology in Zurich because of his poor performance in languages, literature, history, and art (White & Gribbin, 1993). Conversely, it is not uncommon to find that people with very strong verbal abilities are quite weak in the spatial abilities, sometimes to the extent that they are unable to learn how to drive an automobile. An example was Sir Alfred Ayer, the linguistic philosopher at the University of Oxford (Ayer, 1984). It has been shown that there are genes determining general intelligence and further genes determining primary abilities, such as verbal, spatial, and perceptual speed abilities. Five studies supporting this conclusion are reviewed by Petrill et al. (1998).

In addition to g, societies need individuals possessing a variety of these primary abilities that contribute to achievement in different occupations. There is a need for people like Einstein, with their exceptional reasoning, spatial, and mathematical abilities, to advance knowledge in physics and engineering. There is also a need for people with strong verbal abilities for professions like law, politics, economics, diplomacy, and creative writing, proficiency in which depends on strong verbal comprehension and verbal reasoning abilities. Within these broad primary abilities, there are narrower abilities that determine achievement in specific areas. For instance, in physics there are theoretical physicists and experimental physicists, each of which have their own specific patterns of reasoning, mathematical, and spatial abilities. Similarly, among those whose strength lies in verbal ability, the particular verbal abilities required for good creative writing are almost certainly rather different from the analytical verbal abilities required by the good lawyer or politician. The existence of a variety of more specific abilities, and their importance for achievements in different fields of human endeavor, means that the eugenic society should seek to foster improvements in each of these, in addition to increasing general intelligence.

12. CONCLUSIONS

Intelligence is a significant determinant of educational and occupational achievement, of earnings, and of the creative achievement of geniuses. Conversely, low intelligence is a significant determinant of delinquency, crime, long-term unemployment, welfare dependency, mortality, and single motherhood. If the intelligence of society could be raised, there would undoubtedly be social benefits in the form of a more efficient society with higher educational standards, higher earnings, more efficient work performance, a lower crime rate, and less poverty. There would be an enhancement of civilization in the form of greater scientific and cultural achievements. There would also be an increase in national economic and military strength. Objections that if the intelligence level were to be increased there would be too few unintelligent people to perform cognitively undemanding jobs were countered with a number of compelling arguments to the contrary, demonstrating that this would not be a problem.

7

❧

Mental Retardation

1. The Nature of Mental Retardation
2. Profound, Severe, Moderate, and Mild Retardation
3. Socioeconomic Status and Mental Retardation
4. Prevalence of Mental Retardation
5. Social and Work Competence of the Mentally Retarded
6. Socialization of the Mentally Retarded
7. The Mentally Retarded as Parents
8. Fertility of the Mentally Retarded
9. Conclusions

❧❧❧

In the preceeding chapter, we presented the case for the desirability of increasing the overall level of intelligence of the population. A further component of the attempt to raise intelligence is to try to reduce the incidence of mental retardation. This has long been one of the objectives of eugenics, and it is sufficiently important to deserve a chapter to itself. How undesirable is mental retardation, and how useful would it be to reduce it? These are the questions we address in the present chapter.

1. THE NATURE OF MENTAL RETARDATION

The mentally retarded are officially defined as those with IQs below 70. Those with IQs of 70 to 85 are frequently designated "borderline mentally retarded." These criteria are used by the American Association on Mental Retardation and the World Health Organization (WHO). The mentally retarded can be classified into two broad categories. The first category consists of the tail end of the normal distribution of intelligence, analogous to short stature at the tail end of the normal distribution of height. The second category of mental retardation has some qualitatively different cause, such as a disease, disorder, or injury, and is analogous to dwarfism or to short stature resulting from severe malnutrition. The most important of these causes in the second category are single chromosomal abnormalities, the effect of single Mendelian genes, birth injury to the brain, and damage to the fetus during pregnancy, caused by, for example, the mother taking excessive drugs or alcohol or contracting German measles.

The American Association on Mental Retardation classifies retardation into four groups according to the severity of the impairment: *mild*, IQ range of 50 to 70; *moderate*, IQ range of 35 to 50; *severe*, IQ range of 25 to 35; and *profound*, IQ range of 0 to 25.

A sense of the mental capacities of the mentally retarded can be gained by considering their mental ages. Among adults, an IQ of 35 is approximately equivalent to a mental age of $5^1/2$ years, that is to say that the individual has the mental capacities of the average $5^1/2$-year-old. An adult with an IQ of 50 has a mental age of 8 years and the mental capacities of the average 8-year-old. An IQ of 70, the upper limit of mild retardation, is approximately equivalent to a mental age of 11 years and represents the mental capacities of the average 11-year-old. Most people with these mental capacities are unable to function in society as independent adults responsible for the conduct of their lives, any more than this could be done by 5-, 8-, or even 11-year-old children, although a minority of the mildly retarded are able to work, generally in some form of sheltered and supervised employment.

2. PROFOUND, SEVERE, MODERATE, AND MILD RETARDATION

Profound, severe, moderate, and mild mental retardation arise from a number of causes. The most important of these have been ascertained in epidemiological studies. The results of studies in the United States (for blacks and for whites, separately), Sweden, and England are summarized in Table 7.1. In all four samples Down's syndrome is a major cause, accounting for about one-fourth of cases in Sweden and England, one-sixth of the cases among U.S. whites, and one-ninth among U.S. blacks. Those with Down's syndrome typically have IQs in the 30 to 70 range and average around 50; so their average mental age is about that of the average 8-year-old child.

The second major cause of moderate, severe, and profound mental retardation in the U.S. and Swedish surveys is cerebral palsy. This is a general term for disorders of movement and posture caused by damage to the child's brain during pregnancy or at birth. The remaining identifiable causes are largely chromosomal disorders other than Down's syndrome and multifactorial diseases.

Table 7.1
Causes of Moderate, Severe, and Profound Mental Retardation
in the United States, Sweden, and England

	United States			
Disorder	Whites (n – 92)	Blacks (n – 148)	Sweden (n – 122)	England (n – 146)
Down's syndrome	16	14	32	32
Other chromosomal and multifactorial disorders	15	7	24	14
Single-gene disorders	1	0	5	14
CNS malformations	10	2	2	8
Cerebral palsy	21	16	18	4
Other	9	2	7	16
Undiagnosed	28	59	12	12

Sources: Broman, Nichols, Shaughnessy, & Kennedy, 1987; Gustavson, Hagberg, B., Hagberg, G., & Sars, 1977; Laxova, Ridler, & Borven-Bravery, 1977.

The single-gene Mendelian disorders account for only small percentages of the severely retarded, ranging from none among U.S. blacks and about 1 percent among U.S. whites to 6 percent in Sweden and 10 percent in England. These disorders are caused by rare and mainly recessive genes for disorders such as galactosemia, amaurotic family idiocy, microcephaly, and hypertelorism. By the last decade of the twentieth century, more than a hundred rare single-gene disorders causing mental retardation had been discovered (Wahlsten, 1990). In many of these, the gene responsible for the disorder has been identified. For instance, phenylketonuria (PKU) is caused by a recessive gene on chromosome 12 (Plomin & Petrill, 1997). It will be noted that there are appreciable numbers of cases of moderate, severe, and profound mental retardation for which the cause cannot be identified, accounting for about one-third of the cases among U.S. whites, about half the cases among U.S. blacks, and about 10 percent of the cases in the Swedish and English surveys.

In the 1990s another genetic disorder responsible for mental retardation was identified and called "fragile X" syndrome. Typically, IQs of those afflicted are around 40 to 50, and they decline from adolescence onward. For instance, Dykens, Ort, and Cohen (1996) report a study of 12 fragile X males in which the mean IQ of 1- to 5-year-olds was 55 and declined steadily to 40 among adults. The genetics of the transmission of fragile X syndrome is unique. Sometimes it is passed like other X-related recessive disorders, from carrier females to affected males; but, unlike any other known disease, it is also passed from symptomless male carriers to affected females. The birth incidence is about 1 in 2,500, about the same as cystic fibrosis, as compared with about 1 in 700 for Down's syndrome (Verkerk, Peretti, & Sutcliffe, 1991).

With regard to mild retardation, there is no identifiable cause for about two-thirds of cases (McLaren & Bryson, 1987) because most of these are at the tail end of the normal distribution of intelligence. Of the remaining third of cases for which there is an identifiable cause, the commonest cause is cerebral palsy, an omnibus term for the symptoms of injury to the brain occurring to the fetus or to the baby during birth. About 70 percent of cases of cerebral palsy have IQs in the 50 to 70 range. The other identifiable cases consist of a number of Down's syndromes with higher IQs and of rare single-gene disorders, of which about 750 have been identified (Dykens & Hodapp, 1997).

3. SOCIOECONOMIC STATUS AND MENTAL RETARDATION

Children with mental retardation are born predominantly to parents of low socioeconomic status, but this relationship is substantially stronger for mild mental retardation than it is for profound, severe, or moderate retardation. One of the best studies illustrating this difference was carried out by Broman, Nichols, Shaughnessy, and Kennedy (1987) on approximately 17,000 white and 19,000 black babies born in the United States in the early 1970s and assessed for intelligence at the age of seven. Similar data are available for Scotland from a study of all seven-year-olds in the city of Aberdeen born in the years 1952 to 1954, which numbered 8,274 (Birch, Richardson, Baird, Horobin, & Illsley, 1970). These studies are analyzed for the percentages of moderate, severe, and profound mental retardation and for mild retardation coming from families in the top 25 percent of earnings, the middle 50 percent, and the bottom 25 percent. The figures are shown in Table 7.2. It will be noted that in the U.S. sample, the prevalence of profound, severe, and

Table 7.2
Prevalence (percentages) of Profound-Severe-Moderate and of Mild Retardation in Relation to Earnings in the United States (Whites and Blacks) and in Scotland

	Profound-Severe-Moderate Retardation			Mild Retardation		
	United States		Scotland	United States		Scotland
	Whites	Blacks		Whites	Blacks	
Top 25 percent	.40	.44	.33	.30	1.19	.37
Middle 50 percent	.61	.56	.42	1.31	3.59	2.64
Bottom 25 percent	.83	.94	.29	3.34	7.75	5.93

Sources: Broman, Nichols, Shaughuessy, & Kennedy, 1987; Birch, Richardson, Baird, Horobin, & Illsley, 1970.

moderate retardation among the lowest socioeconomic groups is about double that of the highest; whereas in Scotland, there are no significant earnings differences. For mild retardation, the socioeconomic differentials are much greater. Among whites in the United States and in Scotland, the lowest socioeconomic group produces more than 10 times as many mild retardates as the highest group. Among U.S. blacks the differential is smaller, but it is still appreciable.

The results of the three studies can be considered from the point of view of the proportion of mild retardates coming from parents in the bottom 25 percent of earnings. The figures are 53 percent among U.S. whites, 48 percent among U.S. blacks, and 51 percent among the Scots. Parents in the bottom 25 percent of earnings produce about half the mildly mentally retarded because these parents tend to be of low intelligence. Profound, severe, and moderate retardation is less closely associated with parental earnings than mild retardation is because most of the more severe retardation results from genetic disorders or congenital accidents, such as the spontaneous appearance of a chromosomal disorder as in Down's syndrome; from an injury occurring to the fetus during pregnancy or at birth; or from the chance coming together of two parents who happen to be carriers of the same recessive gene, a double copy of which produces severe retardation such as galactosaemia. Accidents of this kind occur with about equal frequency in all social classes. The reason that parents in the bottom 25 percent of earnings produce about double the numbers of babies with profound, severe, and moderate mental retardation, as do parents in the top 25 percent of earnings is that some of those babies with IQs in the 0 to 50 range are the very low tail end of the normal distribution of intelligence, so they come disproportionately from parents with low intelligence who have low earnings.

These studies show that parents with low intelligence, indexed by low earnings, have disproportionately large numbers of mentally retarded children. Direct evidence confirming this conclusion is available in the largest study of mental retardation, consisting of some 80,000 individuals in Minnesota who were either retarded or the descendants of the retarded (Reed & Reed, 1965). This data set has been examined by Anderson (1974), who calculated that among the normal population, approximately 2.2 percent of children are mentally retarded. If one parent is mentally retarded but not the other, 17 percent of the children are mentally retarded; if both parents are mentally retarded, 48 percent of the children are mentally retarded. These estimates indicate that if none of the mentally retarded were to reproduce, the frequency of mental retardation in the next generation would decline by about 25 percent (Reed & Anderson, 1973).

4. PREVALENCE OF MENTAL RETARDATION

In order to assess the seriousness of the problem of mental retardation, we need to know its prevalence. From the early years of the twentieth century, studies have been carried out in various countries to ascertain the prevalence of the different categories of mental retardation. The general consensus emerging from these investigations is that among Caucasian populations of Europe and North America, around 2.7 percent of the population are mentally retarded. About 70 percent of these, representing 2.2 percent of the total population, are the tail end of the normal distribution of intelligence. In the normal distribution of intelligence, the mean IQ of the population is set at 100 and the standard deviation at 15. The effect of this is that 2.2 percent of the population have an IQ of 130 and above (two standard deviations above the mean), while another 2.2 percent have an IQ below 70 (two standard deviations below the mean). Most of these fall into the mildly retarded category, with IQs in the 50–70 range. In addition, about 0.5 percent of the population have mental retardation caused by special adverse factors. These are predominantly the moderately retarded, wih the IQs in the 35–50 range, who comprise about 25 percent of the total retarded population. The remaining 5 percent are severely or profoundly retarded, with IQs in the 25–55 and 0–25 ranges (Burack, Hodapp, & Zigler, 1997).

One of the most extensive studies of the prevalence of mental retardation is the Epidemiologic Catchment Area Study, carried out in the United States in the late 1980s (Robins & Regier, 1991). This study was based on a sample of almost 20,000 individuals and reported the prevalence of profound, severe, and moderate retardation and of mild retardation by the Mini-Mental State Examination, an intelligence test. The results were that among 18 to 34-year-olds, 0.32 percent had profound, severe, or moderate retardation and 2.31 percent had mild retardation. Among older age groups the percentage was higher, but much of this should be ascribed to the onset of senile dementia and other disorders of aging.

5. SOCIAL AND WORK COMPETENCE OF THE MENTALLY RETARDED

We have noted that the upper threshold of mental retardation consisting of an IQ of 70 represents the mental abilities of the average 11-year-old child. These retardates can read and write, but they are not competent to lead independent lives as adults. An IQ of 50, the lower limit for mild retardation and the upper threshold for the more severe forms of mental retardation, represents the mental abilities of the average 8-year-old and these are still less able to look after themselves. With these mental capacities, their social competence in looking after themselves, in employment, and as parents is inevitably quite limited. The more severely retarded are almost always looked after by their parents or in institutions. As adults, hardly any of the profound, severe, and moderate mentally retarded work in normal employment, although some of them can be trained to do simple jobs. The mildly retarded are generally brought up by their parents. In adulthood many of them live with other retardates in hostels or community residences (group houses) run by house parents or wardens (residence managers). A number of others marry nonretardates who support them and live in the community. Children with mild mental retardation frequently attend regular schools and learn to read simple material, but few of them are able to master basic mathematical skills (Baroody, 1988).

One of the first large-scale studies of the work capacity of the mentally retarded was carried out in the United States by Weaver (1946) on approximately 8,000 retarded conscripts in the U.S. army. Most of these had IQs lower than 75; 44 percent of them, with an average IQ of 68, were found to be unsatisfactory for work in the military. The remaining 56 percent, with an average IQ of 72, were assessed as satisfactory for low-level jobs, such as orderlies. The average IQ of the satisfactory group was above the threshold of 70 for mental retardation, so most of these would be regarded as borderline retarded. The more impaired of the mildly retarded would not be represented in this sample because the army would have screened them out during the draft induction procedures.

Four recent studies have examined how many of the mentally retarded are employed. Wehman, Kregel, and Seyforth (1985) found that 60 percent were unemployed and a further 10 percent worked part-time or in sheltered workshops. Only 30 percent worked in normal, full-time employment. Almost exactly the same figures were found by Wagner and Blackorby (1996) in an analysis of a nationally representative sample of approximately 8,000 young Americans with disabilities in the period 1987–90. They found that among those aged 22–25 who were mentally retarded, 71 percent were unemployed. In a third study, Polloway, J. Smith, Patton, and T. Smith (1996) found that 18 percent of the mildly mentally retarded worked in normal employment. In Britain a study of 404 mentally retarded individuals in Wales carried out in the late 1980s found that only 5 percent of them worked for wages, largely in sheltered schemes (Evans, Todd, Beyer, Felce, & Perry, 1994).

In most Western countries, the mentally retarded who are unemployed obtain welfare incomes. When the mentally retarded do find employment, they tend to make unsatisfactory workers because of their high rate of absenteeism and poor work performance (Brickley, Browning, & Campbell, 1982; Rudred, Ferrara, & Ziarnik, 1980). A study of the poor work capacities of the mentally retarded has been carried out in Northern Ireland by Donnelly, McGilloway, Mays, Perry, & Lavery (1997). They examined 114 mentally retarded people discharged from a mental hospital into the community during the years 1987 to 1990. Ninety-seven of them lived in homes requiring paid staff to look after them. Only one worked full time and four worked part time.

The social competence of the mentally retarded has been summarized by Patton and Polloway (1992) as "characterized by unemployment or underemployment, low pay, part-time work, frequent job changes, non-engagement with the community, limitations in independent functioning, and limited social lives" (p. 413).

6. SOCIALIZATION OF THE MENTALLY RETARDED

The mentally retarded not only have low intelligence. Typically, they also have poorly socialized behavior that is expressed in a variety of ways, including a high incidence of conduct disorders, aggression, inappropriate sexual behavior, and crime (Harris, P., 1993; Sigafoos, Elkins, Karr, & Attwood, 1994). A number of investigations of institutional populations of mentally retarded people have documented high rates of aggressive behavior, and violence is reported to be one of the major management problems within such institutions (Crocker & Hodgins, 1997). It has been estimated by Bruininks, Hill, and Morreau (1988) that 36 percent to 45 percent of the mentally retarded exhibit socially unacceptable aggression and destruction of property. Other researchers have concluded that about half of the severely mentally retarded have "poor impulse control" and frequently make violent attacks on others (Cherry, Matson, & Paclawskwj, 1997).

Numerous studies have shown that the mentally retarded have a greater tendency to commit crime than those in the normal range of intelligence. It has been estimated that about 10 percent of the prison population in the United States is mentally retarded (Brown & Courtless, 1967), about four times as many as would be expected on the basis of their prevalence in the population. The earlier research literature showing a high rate of crime committed by the mentally retarded has been reviewed by Wilson and Herrnstein (1985). A more recent study substantiating this effect has been published by Crocker and Hodgins (1997). They took a birth cohort of 15,117 people born in Stockholm in 1953 and assessed them at the age of 30 for mental retardation and their criminal records. They found that 2.5 percent of the men in the cohort had been assessed as mentally retarded and that 56.7 percent of these had been convicted of a crime, as compared with 31.7 percent of those not mentally retarded. Among the women, 1.7 percent had been diagnosed as mentally retarded, and 31.7 percent of these had had a criminal conviction, as compared with 5.8 percent of those not mentally retarded. Taking men and women together, the mentally retarded had about double the crime rate of those whose intelligence is in the normal range.

It is generally considered that there are three reasons for the overrepresentation of the mentally retarded among the criminal population: first, that they have a poor understanding of the law and the consequences of breaking it; second, that they have a poorly developed moral sense; and third, that they have less to lose by criminal acts and imprisonment because most of them do not have satisfying jobs or even any jobs to forego if they are imprisoned.

The poor socialization of the mentally retarded is also expressed in their inappropriate sexual behavior. One of the problems frequently encountered by nurses who care for mentally retarded men is being subjected to various forms of sexual harassment and intimidation. A study of this issue by Thompson, Clare, and Brown (1997) found that two-thirds of mentally retarded men in an institution posed problems of this kind, of which the principal expressions were "continually touching other people's genitals," "verbal threats to others of aggressive sexual acts," "sexual activity with willing women but in inappropriate places," "continued masturbation to the point of penis bleeding," and "formation of strong inappropriate attachments to a female" (p. 589).

7. THE MENTALLY RETARDED AS PARENTS

A number of the mentally retarded have children, but generally they make inadequate parents. Because they have difficulty in looking after themselves, they have difficulty in rearing children. One of the first studies of this question was made by Mickelson (1947) in an investigation of 90 families in Minnesota in which the wife (74 percent), the husband (9 percent), or both (17 percent) were mentally retarded. Only 42 percent of the couples were regarded as providing satisfactory care. This was assessed in terms of whether the children were kept clean; were adequately fed, clothed, and supervised; and attended school regularly. It has sometimes been argued that mentally retarded parents can be given training to make them into competent child rearers; but in a later paper Mickelson (1949) reported that it was difficult to do this.

A more recent study of the parenting skills of the mentally retarded has been reported by Accardo and Whitman (1990). They identified a sample of 79 families in St. Louis, Missouri, in which the women were mentally retarded with an average IQ of 52 and a range from 35 to 69. These women had among them produced a total of 226 children, representing an average of 2.8 and a range of 1 to 9. Of these children, 103 (46 percent) had been removed from their families and taken into care by the social services because the parental care was inadequate. Of the 123 children remaining with their parents, 71 had suffered child abuse, sexual abuse, and/or neglect. Thus, of the total of 226 children, 150 (66 percent) were being unsatisfactorily brought up. The authors provide an illuminating vignette of their experience of the quality of child rearing provided by these parents. They found that typically their mothers could not remember their children's birth dates. Often they could not remember their children's names or distinguish one child from another, feed their children adequately, or administer medicines.

A review of the literature by Feldman (1994) concludes that the mentally retarded, defined in this review as those with IQs below 80, are generally inadequate in ensuring that their children's physical, nutritional, cognitive, emotional, health, and safety needs are properly met. Feldman notes that their children have a high prevalence of developmental delay, learning problems, and emotional and social maladjustment; and that "these families often have multiple problems related to poverty, parental psychopathology, child abuse, and the lack of social supports, which may adversely affect the parent's capacity to adequately raise children," and as a consequence of which, "these families are considered to be the most difficult and time-consuming cases for social service workers" (pp. 300–301). Many references are cited to support these conclusions to the effect that the mentally retarded do not make good parents and their children manifest a high prevalence of a range of social pathologies. In a later paper, Feldman (1998) provides further evidence for this conclusion and wrote that the children of the mentally retarded are "at risk for neglect, developmental delay, mental retardation, cerebral palsy, and behavioral and psychiatric disorders" (p. 2).

A recent study in Norway reports that 40 percent of the children of mentally retarded parents were inadequately cared for and that 43 percent of those children were themselves retarded (Morch, Skar, & Andersgard, 1997). The research literature indicates that between one-third and two-thirds of mentally retarded women do not make fit parents and that 40 to 50 percent of their children are also retarded.

8.   FERTILITY OF THE MENTALLY RETARDED

Few of the profound, severe, and moderate mentally retarded have children, but some of them do. A representative study of the fertility of the more severely retarded was carried out by Scally (1973) in Northern Ireland. He identified all the adult profound, severe, and moderate mentally retarded in the province in the mid-1960s. These numbered 4,631, representing a rate of 0.32 percent, the typical prevalence rate of those with an IQ of less than 50. These had had 791 children, representing a fertility rate of 0.34 percent as compared with a general population fertility rate of 3.0 and, therefore, about 10 percent of the normal average fertility.

The mildly retarded are much more likely to have children. In the United States, it is estimated that about 120,000 children are born to mentally retarded parents each year, constituting about 3 percent of all births (Keltner, 1992). Mentally retarded women produce slightly greater numbers of children than do women of normal intelligence. This difference is found in data collected by Vining (1982) for a representative sample of white and black women aged 24 to 34 in 1978 and shown in Table 7.3. The high fertility of the mentally retarded is an expression of the generally dysgenic nature of fertility in the economically developed nations in the twentieth century.

Females with mild mental retardation are generally fertile and have a relatively high incidence of unintended pregnancies. They have difficulty using contraception effectively, and they are deficient in the social skills required

Table 7.3
Numbers of Children of Mentally Retarded and of Normal White and Black American Women in 1978

	Retarded Women	All Women
White	1.59	1.46
Black	2.60	1.94

Source: Vining, 1982.

to ward off unwanted sexual advances. A study of young women with mild mental retardation living in community houses in Australia carried out by McCabe and Cummins (1996) found that they had a generally low level of understanding about conception. For instance, only 4 percent knew that semen is required for pregnancy; 61 percent of these young women had been pregnant, although only 48 percent said they had had sexual intercourse.

9. CONCLUSIONS

Approximately 2.7 percent of the populations of Western countries are mentally retarded, 2.2 percent being mildly retarded and about 0.5 percent being either profoundly, severely, or moderately retarded. The mentally retarded are an economic and social burden on society because the great majority are unable to work and because they impose large medical and welfare costs. Most of the mildly mentally retarded are fertile but they typically make poor parents. When both parents are mentally retarded, about half their children are also mentally retarded. Despite the assertions of certain geneticists that "all genes are equally valuable" and that "there is no such thing as a bad gene," the great majority of sensible people should have no problem in accepting that mental retardation is undesirable for society and a tragedy for couples who have a mentally retarded child and that it would be best prevented. In 1976, the U.S. Presidential Committee set a national goal for the reduction of severe mental retardation by 50 percent by the year 2000 (Comptroller General, 1977), thereby showing a degree of national consensus on the desirability of this eugenic objective. The case for a reduction of mild mental retardation is hardly less strong.

8

⤣

Personality

1. Neuroticism
2. Introversion-Extroversion
3. Openness to Experience
4. Agreeableness
5. Conscientiousness
6. Psychopathic Personality
7. Conclusions

↜↜↜

We turn now to the issue of whether personality and personality disorders should be the object of eugenics. The early eugenicists favored the improvement of personality qualities quite as strongly as the improvement of physical and intellectual qualities. Galton (1909) wrote of the value of "character," "manliness and courteous disposition," and "worth" (pp. 37, 105) and of the desirability of encouraging those who possessed these qualities to have more children and those who were deficient in them to have fewer children.

The understanding of personality and its disorders was quite weak in the late nineteenth and early twentieth centuries when eugenicists were considering the application of eugenics to personality. In the second half of the twentieth century considerable progress was made in the analysis of the nature of personality and its disorders. There developed a widespread consensus that personality should be conceptualized as consisting of a number of continuously distributed and independent traits, with personality disorders occupying the most extreme positions on these distributions. For instance, neurotics have come to be seen as occupying an extreme position on the trait of "neuroticism."

By about 1990 a widespread consensus emerged among personality theorists that there are five major personality traits. These are neuroticism, introversion-extroversion, openness to experience, agreeableness, and conscientiousness. This has become known as the "Big Five" personality model, useful descriptions of which have been given by Costa and McCrae (1992a, 1992b), Barrick and Mount (1991), Zuckerman (1992), and MacDonald (1995), and which will be used in this chapter as the framework in which to discuss the desirability of an application of eugenics to personality.

1. NEUROTICISM

The first of the Big Five personality traits to be identified was neuroticism, which has also been known as anxiety or emotionality. In the second decade of the twentieth century, this trait was identified as "general emotionality" by Burt (1915). The term *anxiety* has been employed by a number of personality theorists, such as Cattell (1965). The term *neuroticism* was introduced by H. J. Eysenck (1947), who has been largely responsible for its widespread usage. Neuroticism has been usefully analyzed by Costa and McCrae (1992a) into five components that are described by pairs of adjectives denoting opposite poles of the characteristic. These are: calm-worrying; even-tempered–temperamental; self-satisfied–self-pitying; comfortable–self-conscious; and unemotional-emotional. At the high end of the trait are found neurotic personality and anxiety disorders (Eysenck, H. J., 1947; Widiger & Trull, 1992). This has led Cattell (1987, p. 213) to propose that a low level of neuroticism is desirable and that eugenic measures could usefully be taken to lower the average level of the trait. This proposal may seem attractive on the grounds that a eugenic intervention of this kind would reduce the numbers of emotional, temperamental, and self-pitying neurotics. Nevertheless, caution is required. Early in the twentieth century, Yerkes and Dodson formulated a general principle known as the "Yerkes-Dodson law," which states that intermediate levels of anxiety (neuroticism) are more effective for efficient work performance on tasks of moderate difficulty than is either high or low anxiety. The evidence for this theory has been summarized by M. W. Eysenck (1982), who showed that the theory holds for a variety of animals as well as for human beings.

In addition to the general principle that a moderate level of neuroticism is optimum for efficient work performance, there are several studies showing that in certain populations and for some tasks, the relationship is slightly positive, such that high neuroticism appears to facilitate efficient work performance. For instance, it was shown by Kelvin, Lucas, and Ojha (1965) that there is a positive association between neuroticism and the academic performance of students in British universities, suggesting that a higher than average level of neuroticism is an asset for academic achievement among students. A study carried out in Britain by Mughal, Walsh, and Wilding (1996) on 75 insurance sales consultants found that neuroticism was positively correlated with effectiveness (.28) and sales (.29). A meta-analysis by Barrick and Mount

(1991) of the U.S. research literature consisting of 117 studies and approximately 20,000 individuals found that neuroticism is positively associated with effective work performance among professionals (.13), but negatively associated among skilled and semiskilled workers (-.12). The inference to be drawn from these studies is that neuroticism may have a slight positive relationship with work performance in white collar occupations.

An important clarification of the relationship between neuroticism and efficiency of performance has been made by McKenzie (1989). He found that if the trait of ego-strength (control over the emotions) is factored into the relationship, students with high neuroticism and high ego-strength perform well in university examinations, while those with high neuroticism and low ego-strength perform poorly. Probably the mechanism responsible for these differences is that students with high ego-strength are able to direct their emotional energies into productive work, whereas those with low ego-strength are rendered disorganized by their high emotionality. Thus, the evidence points to the conclusion that although a lowering of the level of neuroticism in the population would produce a reduction in the number of neurotics, this would also entail a cost of reducing work efficiency among some in the upper-middle range of the trait, especially among those with high ego-strength. For this reason Cattell's proposal for eugenic measures to reduce neuroticism cannot be endorsed, and a neutral position should be adopted with regard to this trait.

2. INTROVERSION-EXTROVERSION

The trait of introversion-extroversion was first conceptualized in the 1920s by Jung to designate the inward looking (introverted) and outward looking (extroverted) personality types. In contemporary personality theory, the trait is described by Costa and McCrae (1992a) in terms of the following pairs of objectives: reserved-affectionate; sober-fun loving; quiet-talkative; loner-joiner; passive-active; and unfeeling-passionate. Probably there would be general agreement that neither extreme of this dimension is attractive and that we prefer people who are neither excessively reserved, sober, or quiet nor excessively affectionate, talkative, or fun loving.

Neither extroverts nor introverts appear to make a greater contribution to society as assessed by their work efficiency. This is the conclusion reached by Barrick and Mount (1991) from their meta-analysis of studies of the relationship between the Big Five personality factors and work performance. They found that the relationship between introversion-extroversion and efficiency of work performance among professional people, managers, sales personnel, or skilled and semiskilled workers was close to zero, and that the overall correlation across all occupations was a negligible .08. Possibly, introversion may be an asset for some occupations, such as that of research scientist, as proposed by Cattell (1965); but this may be counterbalanced by extroversion being an asset in other occupations, such as, possibly, sales representatives. Among occupations as a whole, the research evidence suggests that neither introverts nor extroverts make a greater contribution to the well-being of society and that there is no reason to attempt to change this trait by eugenic intervention.

3. OPENNESS TO EXPERIENCE

The openness to experience trait is described by Costa and McCrae (1992a) as comprising the characteristics of imaginative-down to earth; creative-uncreative; original-conventional; preferring variety-preferring routine; curious-incurious; and liberal-conservative. The trait has its origin in Cattell's (1965) factor of radicalism-conservatism, in which the more radical have stronger intrinsically motivated curiosity, interest in intellectual matters, imagination, and creativity. Cattell showed that the trait is moderately strong among creative scientists, artists, and writers.

At first sight, openness to experience may seem like a desirable characteristic, which it would be useful to strengthen; but once again caution is needed. Barrick and Mount (1991) in their meta-analysis of studies on the relationship of the Big Five personality traits to efficient work performance found no association between efficiency and openness to experience. The reason for this is that it is only in a few occupations, such as those of scientists, artists, and some entrepreneurs, that a high level of openness to experience is an asset. For many more routine occupations, it is of no relevance or is even a hindrance. Society needs creative people to promote innovation, but it also needs noncreative people to carry out routine administrative work and skilled trades. Society does not need creative tax officials, police officers, electricians, and plumbers. It is preferable that those working in these and many similar occupations work according to the rule book. Hence no eugenic measures should be attempted to alter this trait.

4. AGREEABLENESS

Agreeableness is described by Costa and McCrae (1992a) in terms of its components and their opposites as follows: softhearted-ruthless; trusting-suspicious; generous-stingy; acquiescent-antagonistic; lenient-critical; and good-natured–irritable. Other personality theorists have described the positive components of the trait as including friendliness (Guilford & Zimmerman, 1949), social conformity (Fiske, 1949), and compliance (Digman & Takemoto-Chock, 1981). Barrick and Mount (1991) write that "traits associated with this dimension include being courteous, flexible, trusting, good natured, cooperative, forgiving, soft hearted, and tolerant" (p. 4). This trait is not well described by the term *agreeableness* and might be better labeled *altruism*, which better captures the sense of responsibility and the capacity for working selflessly for others that this trait represents. However, *agreeableness* has become so widely used that we are stuck with it. Equally inadequate is its opposite, *disagreeableness*. M. Zuckerman (1991) calls this "aggression-hostility," which conveys more powerfully the other extreme of the trait.

Although agreeableness appears to be a socially desirable characteristic, there is no case for attempting to improve the trait on the grounds that it makes any contribution to work efficiency. Barrick and Mount's (1991) meta-analysis concluded that the overall correlation between agreeableness and work efficiency across all occupations is a negligible .04. Nevertheless, there is clearly a persuasive case that agreeable, friendly, generous, and socially concerned people are preferable to those who are disagreeable, hostile, mean, and self-ish. This was probably the trait of which Galton was thinking when he wrote of the value of courtesy and worth. A similar view was taken by Hermann Muller (1939) who listed for eugenic action "those temperamental qualities which favor fellow-feeling and social behavior" (p. 64).

In addition to a general preference for agreeable and altruistic people over the disagreeable and selfish, at the disagreeable end of the trait are found those in whom the sense of social obligation is weak or absent. These are the de-linquent, the criminal, and the psychopathic personalities. For instance, it has been shown by Heaven (1996) in a study of Australian high school students that there is a negative correlation of -.28 between agreeableness and self-reported violence and vandalism. We should note also that those who are high on agreeableness are happier than those who are low. This has been shown by Brebner (1998) in a study of 143 Australian university students, among whom happiness was correlated .29 with agreeableness. Possibly one of the explanations for this is that those who are strong on agreeableness obtain a degree of happiness in the fulfillment of social obligations.

There is a strong case that agreeableness is preferable to disagreeableness and that the quality of social life would be improved if the mean and the distribution of the trait could be shifted upward. This would result in fewer psychopaths at the lower end of the distribution and in a general improve-ment in the civility of social life. We conclude, therefore, that an increase in the personality trait of agreeableness should be included among the objec-tives of eugenics.

5. CONSCIENTIOUSNESS

The last of the Big Five personality traits is conscientiousness. It is de-scribed by Costa and McCrae (1992a) in terms of the following pairs of ad-jectives: conscientious-negligent; hardworking-lazy; well organized-disorga-nized; punctual-late; ambitious-aimless; persevering-quitting. MacDonald (1995) describes the trait as consisting of "the ability to defer gratification, persevere in unpleasant tasks, pay close attention to detail, and behave in a responsible, dependable manner" (p. 534).

A number of studies have shown that individuals who are high on conscientiousness are efficient and productive workers. In the United States it has been found that conscientiousness correlates around .5 with educational attainment (Smith, G. M., 1967; Digman & Takemoto-Chock, 1981). A Netherlands study on a representative sample of adults carried out by Duijsens and Diekstra (1996) obtained a similar result.

There is also extensive evidence that conscientiousness is positively related to efficiency in the workplace. Barrick and Mount's (1991) meta-analysis found that conscientiousness is consistently associated with three criteria of efficiency, which are described as "job proficiency" (performance ratings and productivity), "training proficiency," and "personnel data" (salary levels and turnover rates). These positive associations between conscientiousness and efficient work performance are present for professionals, managers, salespeople, and skilled and semi-skilled workers. The overall correlation between conscientiousness and efficient work performance was .22. A meta-analysis of 36 European studies based on 3,300 individuals has been carried out by Salgado (1997), who calculated a closely similar correlation of .25. In further recent studies, it has been found that conscientiousness is associated with low absenteeism ($r = .24$), apparently because much absenteeism is a form of malingering and conscientious individuals feel a moral responsibility to work (Judge, Martocchio, & Thoresen, 1997). High conscientiousness also contributes to the efficiency of team performance in group-decision-making tasks ($r = .18$) (Le Pine, Hollenbeck, Ilgen, & Hedlund, 1997). Finally we note that conscientiousness is positively associated with happiness, as found in the study by Brebner (1998) of Australian students, among whom the correlation between conscientiousness and happiness was .38. We conclude, therefore, that it would be desirable to increase the overall level of conscientiousness by eugenic measures.

6. PSYCHOPATHIC PERSONALITY

Psychopathic personality consists of a weakness or an absence of moral sense and is expressed in persistent antisocial behavior and crime. Psychopathic personality is a function of very low agreeableness and very low conscientiousness. The association between low agreeableness and psychopathic personality has been reported in several investigations. Wiggins and Pincus (1989) obtained a correlation between the two characteristics of .61, and Widiger and Trull (1992) obtained a correlation of .35. This association has been confirmed in the Netherlands by Duijsens and Diekstra (1996) and Hendriks, Hofstee and De Raad (1999).

The association between low conscientiousness and psychopathic personality has been found by Widiger and Trull (1992), who reported a correlation between the two of - .41, and by Costa and McCrae (1990), who reported a correlation of - .42. Further studies showing low conscientiousness in psychopathic personalities have been published by Wiggins and Pincus (1989); Costa and McCrae (1990); Yeung, Lyons, Waternaux, Faraune, and Tsuang (1993); Clark, L. A., and Livesley (1994); and Duijsens and Diekstra (1996).

It appears therefore that psychopathic personality is associated with a low level of the two traits of agreeableness and conscientiousness. There are two possible explanations. The first is that psychopathic personality is a joint product of low values of the two personality traits of agreeableness and conscientiousness. The second is that there may be two types of psychopathic personality: those who are callous and aggressive in their interpersonal relationships and are excessively low on agreeableness, and those who lack the capacity for sustained work effort and are excessively low on conscientiousness. The existence of these two types, sometimes designated "aggressive psychopaths" and "inadequate psychopaths," has long been known and has been demonstrated by factor analytic studies of the characteristics of psychopaths, which have shown the presence of these two factors (Harpur, Hare, & Hakstian, 1989; Frick, O'Brien, Wooton, & McBurnett, 1994). These are ideal types in the sense that most psychopaths are both aggressive and inadequate or, in terms of the five-factor personality theory, a product of low agreeableness and low conscientiousness.

Psychopathic personalities are a serious social problem, and reducing their numbers should be an objective of eugenics. This would entail raising the levels of the personality traits of agreeableness and conscientiousness, low levels of which are expressed in psychopathic personality. In subsequent chapters I shall write of the desirability of reducing the prevalence of psychopathic personality as a convenient shorthand for the desirability of raising the underlying personality traits of agreeableness and conscientiousness. I shall also continue to use Galton's term *moral character* as a useful antonym for *psychopathic personality*.

7. CONCLUSIONS

In this chapter, we have considered the Big Five personality traits from the point of view of the desirability of eugenic improvement. We have concluded that there are no persuasive arguments for attempting to alter the traits of neuroticism, introversion-extroversion, or openness to experience. There are, however, sound arguments for increasing the levels of agreeableness and conscientiousness. Both of these are socially desirable traits, and there is a strong case that society would be improved if its citizens became more agreeable and conscientious.

With regard to agreeableness, an upward shift in the mean and the distribution of the trait in the population would produce a society with a higher level of courtesy and civility. The population would also be happier. With regard to conscientiousness, an upward shift in the mean and the distribution of the trait would produce similar benefits and the additional advantage that work would be performed more efficiently.

Furthermore, low agreeableness and low conscientiousness are the underlying traits responsible for criminals and psychopathic personalities, and an upward shift in the traits would result in fewer of these. Criminals and psychopathic personalities make no positive contribution to society. On the contrary, they are a social menace, and there are strong arguments for having fewer of them. Psychopathic personalities are a sufficiently serious social problem that they deserve further consideration, which is the subject of the next chapter.

9

⌣

Psychopathic Personality

1. Nature of Psychopathic Personality
2. Prevalence of Psychopathic Personality
3. Crime
4. Unemployment
5. Drug Abuse
6. HIV and AIDS
7. Teenage Mothers
8. Teenage Fathers
9. Costs of Psychopathic Personality
10. Psychopathic Personality and Creativity
11. Psychopathic Personality and Intelligence
12. The Underclass
13. The Cycle of Deprivation
14. Conclusions

⌣⌣⌣

In the Chapter 8, we saw that extremely low levels of the personality traits of agreeableness and conscientiousness are associated with psychopathic personality. This is a socially undesirable condition that it would be desirable to reduce. In this chapter we examine the problems of psychopathic personality in more detail. We look at that personality's relationship with crime, unemployment, drug abuse, and sexually transmitted diseases, and we discuss its social costs. Finally, we consider whether psychopathic personality may have some socially desirable features associated with creative achievement.

1. NATURE OF PSYCHOPATHIC PERSONALITY

We noted in the preceding chapter that psychopathic personality is a product of the low extremes of the personality traits of agreeableness and conscientiousness. The essential features of psychopathic personality are persistent antisocial behavior, arising from low agreeableness, and a lack of moral sense, arising from low conscientiousness. The first identification of psychopathic personality in the medical literature was given in the early nineteenth century by the American physician Benjamin Rush, who described a personality type characterized by "innate preternatural moral depravity". The use of the term *innate* shows that Rush recognized the genetic nature of the disorder. A little later in the nineteenth century the British physician John Pritchard (1837) proposed the concept of *moral imbecility* for those deficient in moral sense, analogous to the cognitively impaired who at the time were described as "imbeciles". The term "psychopathic personality" was first used by the German psychiatrist Emile Kraepelin (1915). In the course of the twentieth century the condition has frequently been described, notably by Hervey Cleckley (1941) in his book *The Mask of Sanity*, so-called because psychopaths are not easily recognized and appear superficially to be sane. A recent account of the disorder has been given by Robert Hare (1994) in his book *Without Conscience*, a title that aptly describes a central feature of psychopathic personality. Hare writes that the psychopath is "a self-centered, callous, remorseless person profoundly lacking in empathy and the ability to form warm emotional relationships with others, a person who functions without the restraints of conscience; what is missing in this picture are the very qualities that allow human beings to live in social harmony. It is not a pretty picture" (pp. 2–3).

Psychopathic personality is also known as personality disorder, antisocial personality disorder, and sociopathic personality. Frequently these terms are used as alternatives to psychopathic personality and sometimes to make subtle distinctions between different variants of the condition. The American Psychiatric Association (APA) dropped the term *psychopathic personality* in 1984 and substituted antisocial *personality disorder* as an alternative. Lykken (1995) treats these terms as synonymous, so far as the symptoms are concerned, but distinguishes between psychopathic personality as genetically caused and sociopathic personality as environmentally caused.

Another concept closely related to psychopathic personality is *deviance*. Rowe (1986) has proposed the concept of *d* (deviance) as a general factor analogous to *g*, the factor of general intelligence. Rowe's *d* is a useful concept; but it has not yet come to be widely accepted, and I shall stick with the term psychopathic personality, which was used for most of the twentieth century.

The amoral, antisocial, and aggressive nature of the psychopathic personality has been elaborated by the APA in its 1994 edition of *Diagnostic and Statistical Manual of Mental Disorders (DSM)*. It lists eleven features of the condition, now renamed *antisocial personality disorder*. These are: (1) inability to sustain consistent work behavior; (2) failure to conform to social norms with respect to lawful behavior; (3) irritability and aggressivity, as indicated by repeated physical fights or assaults; (4) repeated failure to honor financial obligations; (5) failure to plan ahead, or impulsivity; (6) no regard for truth, as indicated by repeated lying, use of aliases, or "conning" others; (7) recklessness regarding one's own or others' personal safety, as indicated by driving while intoxicated or recurrent speeding; (8) inability to function as a responsible parent; (9) failure to sustain a monogamous relationship for more than one year; (10) lacking remorse; (11) the presence of conduct disorder in childhood. It may be useful to note that among these characteristics, numbers 3, 8, 9, and 11 are moral failures in regard to social relationships, whereas the remainder are moral failures in regard to the self and to self-discipline.

The centrality of a deficiency of moral understanding in psychopaths has been usefully documented in a meta-analysis of 15 studies of moral reasoning capacity of delinquents carried out by Nelson, Smith, and Dodd (1990). All the studies showed that delinquents have immature moral reasoning capacity. More recently O'Kane, Fawcett, and Blackburn (1996) have confirmed a significant correlation between psychopathic personality and defective moral reasoning.

The symptoms of psychopathic personality virtually always appear quite early in childhood, when they are expressed as conduct disorders consisting principally of aggressiveness, hyperactivity, and persistent disobedience. A summary of the typical life history of the psychopath has been given by Moffitt (1993):

> At the age of 3–4 years they begin to display serious conduct disorders such as biting and hitting; by the age of 10 they are shoplifting and truanting; by the age of 16 they are selling drugs and stealing cars; by the age of 22 they are into robbery and rape; by the age of 30 they are committing fraud and child abuse. . . . The underlying disposition remains the same, but its expression changes form as new social opportunities arise at different points in development. (p. 679)

No one has ever disputed that psychopaths are a serious social problem. This has been recognized throughout the course of history, during most of which they have been routinely executed in substantial numbers.

2. PREVALENCE OF PSYCHOPATHIC PERSONALITY

Psychopathic personality is not qualitatively distinct from normal personality; rather it is a condition at the low tail end of the normal distribution of the personality traits of agreeableness and conscientiousness. Nevertheless, for an appreciation of the magnitude of the problem of psychopathic personality, it is useful to regard psychopaths as a type whose prevalence in the population can be estimated. The results of six epidemiological studies of the prevalence of psychopathic personality in Caucasian populations are summarized in Table 9.1. It will be noted that the prevalence for males and females combined is around 3.5 percent. This is a little higher, although of the same general order of magnitude, than the prevalence of mental retardation, which is approximately 2.7 percent. However, psychopathic personality is much more common in men, among whom its prevalence is around 6 percent, than in women, among whom its prevalence is only around 1 percent.

3. CRIME

There is a strong association between psychopathic personality and crime. Many psychopaths are criminals, and many criminals are psychopaths. The reason for this is that psychopaths lack the conscience and the capacity to control their behavior and their impulses, which together restrain most people from committing crimes. In addition, many psychopaths are highly aggressive, and this predisposes them to commit violent crimes. A recent review by Moran (1999) estimates that around 60 percent of male prisoners are psychopaths. Some studies have produced lower figures and some higher. The proportion of imprisoned criminals who are psychopathic varies from around 30 percent, as estimated by Hare, Hart, & Harpur (1991), to around 75 percent, as estimated in a report issued by the Correctional Service of Canada (1990). One of the explanations for the discrepancy between these estimates is that it is difficult to diagnose the presence of psychopathic personality in criminals. They normally try to conceal their psychopathic personality because they believe, generally correctly, that they are more likely to secure privileges, parole, and early release from prisons or mental hospitals if they can establish

Table 9.1
Lifetime Prevalence Rates (percentages) of Psychopathic Personality

Country	Both Sexes	Males	Females	Reference
Canada	3.7	6.5	0.8	Bland, Orne, & Newman, 1988
Finland	6.0	8.1	3.8	Lehtinen, Lindholm, Veijola, & Vaisanen, 1990
New Zealand	3.1	—	—	Oakley-Browne, Joyce, Wells, Busnhell, & Hornblow, 1989
New Zealand	5.0	—	—	Moffit, 1993
United States	2.5	—	—	Compton, Helzer, & Hiou, 1991
United States	3.5	5.8	1.2	Kessler, McGongale, Zhao, & Nelson, 1994

that they are not psychopaths. It is fairly easy for psychopaths to conceal their psychopathic personality because the condition can only be diagnosed by questioning them about their past behavior and their feelings, such as whether they got into fights at school or were ever suspended, whether they feel remorse for their crimes, and so forth. Psychopathic criminals have a vested interest in lying about their past psychopathic behaviors and frequently do so, as Hare (1994) has noted; and this makes it easy to underestimate the extent of psychopathic personality among criminals.

In considering the issue of the proportion of criminals who are psychopaths, a distinction needs to be made between recidivists (habitual criminals who commit crime after crime and are undeterred by punishment) and nonrecidivists (who may simply have made one serious mistake). For instance, a person who makes an error of judgment in driving an automobile can kill a pedestrian and can be imprisoned for manslaughter or dangerous driving. Such persons are not normally psychopaths. Similarly, drug barons frequently employ young women to import drugs into North America and Europe; and from time to time these young women are apprehended, convicted, and imprisoned for long periods. Many of these young women are not psychopaths but simply gullible. They provide further evidence for the association between low intelligence and crime reviewed in Chapter 6.

It is arguable that all recidivists should be regarded as psychopaths from their track record, even though the presence of psychopathic personality is difficult to establish by questioning them about their past psychopathic behavior. This position is supported by a study by Herrnstein (1983), which showed that recidivist criminals in prison were approximately three standard deviations above the normal population, as assessed by the psychopathic deviate scale of the Minnesota Multiphasic Personality Inventory (MMPI). This is a very high score, obtained only by approximately 1 per 1,000 of the population; and it suggests that virtually all recidivists are psychopaths. The mean psychopathic deviate score of all criminals in prison in the Herrnstein study was approximately two standard deviations above the mean, comprising the top 2.2 percent of the population and indicating a very high prevalence of psychopathic personality among criminals. The close association between psychopathic personality and serious crime is further supported by epidemiological studies showing that these have about the same prevalence in the population. As noted in Table 9.1, the prevalence of psychopathic personality among men is approximately 6 percent. It has been found that approximately 6 percent of the male population are recidivists. Tracy, Wolfgang, and Figlio (1990) studied of two cohorts of boys in Philadelphia, the first born in 1945 and the second born in 1958. They found that in the first cohort, 6 percent became recidivists, and in the second cohort, 8 percent became recidivists. Approximately half of these were violent, and the other half were nonviolent burglars. Other criminologists who have also estimated the percentage of recidivists among males at approximately 6 percent include Stattin and Magnusson (1991) and Raine (1993). The reason that these percentages are about the same as the percentage of the male population who are psychopaths is that recidivists and psychopaths are largely the same people.

The majority of crimes are committed by this quite small number of recidivist psychopaths. For instance, in the Philadelphia study, the 6 percent of the first cohort who were recidivists were responsible for 51 percent of the crimes committed in the city; and in the second cohort the 8 percent of recidivists were responsible for 68 percent of the crimes (Tracy, Wolfgang, & Figlio, 1990). A study of 30,000 males in Denmark found that more than half of all crimes were committed by about 1 percent of psychopathic men who commit hundreds of crimes (Mednick & Christiansen, 1977). In Britain, Farrington (1994) reviewed a number of studies and concluded that a small number of serious psychopathic recidivists commit about half of all known crimes.

4. UNEMPLOYMENT

As noted in the first section of this chapter, the first of the APA's criteria for antisocial personality disorder is "inability to sustain consistent work behavior." As there is no compulsion to work in the Western democracies and those who dislike working can subsist on welfare and can obtain supplementary incomes from crime, many psychopaths opt not to work. There is research evidence that criminals, many of whom are psychopathic, have poor employment records. Raine (1993) cited five studies of this kind and concluded that "many studies have shown that adult offenders are much more likely to be unemployed and have difficulty in obtaining employment than non-offenders" (p. 283).

However, to say that psychopathic criminals have "difficulty" in obtaining employment is a euphemistic expression. The truth has been put more bluntly by David Lykken (1995), one of the leading experts in psychopathic personality, who wrote in his book *The Antisocial Personalities,* "Many unsocialized young people have never even known anyone who held a full-time job, and jobs are not on their personal agendas. Having fun and hanging out and crime are on their agendas. Working for a living is for socialized people" (p. 221).

5. DRUG ABUSE

Because psychopaths lack discipline and the capacity to control their desire for immediate gratification in the light of their knowledge of long-term adverse consequences, many of them become drug abusers. A number of studies have investigated the proportion of psychopaths among drug abusers undergoing treatment. The results of eight such studies of opiate abusers in several countries are summarized in Table 9.2. It will be noted that the proportion of

Table 9.2
Percentage of Opiate Abusers Diagnosed as Psychopathic Personalities

Location	Percentage Psychopaths	Reference
Australia	60.8	Darke, Hall, & Swift, 1994
Italy	61.0	Clerici, Carta, & Cazzullo, 1989
Netherlands	91.0	De Jong, van den Brink, Harteveld, & van der Wielen, 1993
United States	53.9	Rounsaville, Weismann, Kleber & Wilber, 1982
United States	68.0	Kosten, Rounsaville, & Kleber, 1982
United States	65.0	Khantzian & Treece, 1985
United States	100.0	Craig, 1988
Yugoslavia	62.5	Vukov, Baba-Milkic, Lecic, Mijalkovic, & Marinkovic, 1995

psychopaths among drug abusers ranges between 60.8 percent and 100 percent. Furthermore, these are the drug abusers who are undergoing treatment
and are less psychopathic than drug abusers who do not undergo treatment.

6. HIV AND AIDS

A further cost imposed on society by psychopaths is that they have a high
prevalence of HIV and AIDS and make a substantial contribution to spreading these infections throughout the population. This is principally accomplished by narcotic-addicted psychopaths sharing needles; by psychopathic
females trading sex for money to finance their drug abuse; by a high rate of
psychopaths' engaging in casual and unprotected sex; and by the widespread
occurrence of homosexual rape perpetrated by psychopaths in prisons
(Robertson, 1999). Studies of those with HIV infection have typically found
that about a third of them are psychopaths. The results of three such studies
are summarized in Table 9.3. Further evidence showing that needle-sharing
drug addicts are psychopaths has been reported by Kleinman et al. (1994).

Table 9.3
Percentage of Psychopathic Personality among Those with HIV Infection

Location	Percentage Psychopaths	Reference
United States	36	James, Rubin, & Willis, 1991
United States	33	Perkins, Davidson, Leserman, Liao, & Evans, 1993
United States	37	Jacobsberg, Frances, & Perry, 1995

Much of the spread of the HIV and AIDS epidemic is attributable to the antisocial behavior and irresponsibility of psychopaths.

7. TEENAGE MOTHERS

Male psychopaths inflict damage on society principally by their high rates of crime. Female psychopaths inflict damage on society principally by becoming unmarried teenage mothers. Hardly any teenagers become pregnant and bear children intentionally. In the United States it has been found that approximately 98 percent of teenage pregnancies are unintended (Harlap, Kost, & Darroch, 1991), and similar results have been found in Britain (Crosier, 1996). Teenage pregnancies occur through casual, unplanned, and unprotected sex, all of which are characteristic of psychopaths. Teenage mothers also have other typically psychopathic characteristics. A review of 49 studies of U.S. teenage mothers published during the years 1984–1994 by Goodson, Evans, and Edmundson (1997) concluded that teenage mothers came predominantly from low socioeconomic status families with poor education, from parents who had physically and sexually abused them during childhood, and from mothers who were themselves teenage mothers. They are typically drug and alcohol abusers, cigarette smokers, and delinquents; have poor educational records; have negative attitudes to school; are school dropouts; and have below-average intelligence.

Among teenage single mothers, there is a hard core who have repeated pregnancies and give birth to more than one child. A study of these has been made by Stevens-Simon, Kelly, and Singer (1996). In Colorado, they interviewed 200 low-income, pregnant, unmarried teenagers during their third trimester of pregnancy and found that 12 percent of these had become pregnant on at least one previous occasion. Compared with the others, these were more likely to be school dropouts, to have used illicit drugs, to say they had no intention of using a contraceptive implant in the future, and to express no regrets about having a child. Most of these irresponsible young women are psychopaths.

8. TEENAGE FATHERS

Teenage fathers have the same psychopathic characteristics as teenage mothers. Nationally representative surveys in the United States have shown that in two recent decades, approximately 7 percent of 12- to 18-year-old males had fathered a child (Marsiglio, 1987). These adolescent fathers have a high prevalence of early sexual experience, of multiple sexual partners, of non-use of contraception, of drug abuse, and of delinquency (Elster, Lamb, & Tavare, 1987). A study of 125 incarcerated adolescent males with an average age of 15 years carried out in 1984 by Nesmith, Klerman, Kim, and Feinstein (1997) found that 32 (26 percent) had impregnated a girl and 13 had been responsible for more than one pregnancy. The total of 47 pregnancies resulted in 25 live births, 19 miscarriages, and 3 abortions. Of this 125 males, 67 had been previously incarcerated, and 80 percent had failed at least one grade. The average age of their first sexual intercourse was 11.9 years; the average number of sexual partners was 15; and 96 (77%) had a family history of teenage pregnancy occurring among mothers, fathers, sisters, or brothers.

Despite their social and educational failures, these teenage fathers were by no means lacking in self-esteem, another characteristic of psychopaths—63 percent said they would be pleased if they got a girl pregnant, 78 percent said they would be a good role model for a child, and 89 percent said that a child would be proud of them.

Other studies of teenage fathers have found that, like teenage mothers, they are characterized by poor school attainment, high rates of school dropout, and low aspirations for academic achievement. This has been shown in the United States (Robbins, Kaplan, & Martin, 1985; Elster, Lamb, & Tavare, 1987) and in Britain (Dearden, Hale, & Alvarez, 1992). Teenage fathers have been found to have high rates of delinquency, drug abuse and disruptive behavior in school (Ketterlinus, Lamb, Nita, & Elster, 1992; Resnick, Chambliss, & Blum, 1993). A recent study confirming a number of these associations has been published by Thornberry, Smith, and Howard (1997). Their sample was the Rochester Youth Development Study, a socially representative panel study of one thousand 15- to 19-year-olds in the city of Rochester, New York. They found that 28 percent of the sample had become fathers by the age of 19 and that fathering a child was significantly correlated with low parental education and receipt of welfare, poor reading and math scores, early sexual experience, drug use, and criminal records. The research evidence shows that teenage fathers, like teenage mothers, are predominantly psychopaths or have strong psychopathic tendencies. They inflict damage on society by their own psychopathic behavior and by the transmission of their psychopathic tendencies to future generations.

9. COSTS OF PSYCHOPATHIC PERSONALITY

The costs that psychopaths impose on society arise from six principal sources: their high rate of crime, unemployment, drug abuse, HIV and AIDS, teenage parenthood, and welfare dependency. Estimates of the financial costs of psychopaths in the United States have been made by Westman (1994). He calculates that the annual cost of keeping one psychopathic recidivist in prison is approximately $34,000 a year and that the total cost of recidivism is approximately $18 billion. The financial cost to society of crimes committed by psychopaths can be broadly estimated for the United States for the mid-1990s from data given by Levitt (1996), who found that the annual cost of government expenditure on prisons was approximately $40 billion. Thus, if we assume that approximately 60 percent of prisoners are psychopaths, the annual cost of keeping psychopaths in prison is about $24 billion. This is reasonably close to Westman's (1994) estimate of $18 billion as the annual cost of re-cidivism. In the mid-1990s there were about one million men in U.S. pris-ons, representing about 0.8 percent of the male population. The prevalence rate of psychopaths is about 6 percent, so only about 13 percent of psycho-paths were in prison. The remaining 87 percent were at large, and many of them were actively committing crime. These impose the further costs of maintaining a large police force; and a criminal justice system; insurance against crime; private security; and the psychological costs of being robbed, burgled, raped, assaulted, and murdered. Westman estimated the welfare de-pendency costs of psychopaths at approximately $5,000 a year per individual and $12 billion in total. He estimated the annual cost of social service sup-port for the abused and neglected children of psychopathic parents at approxi-mately $12,000 per individual and $8 billion in total. These large costs do not include the medical costs of psychopathic drug addiction and sexually transmitted diseases. Taking all these costs together, psychopaths impose a huge burden on society.

10. PSYCHOPATHIC PERSONALITY AND CREATIVITY

Although the large costs imposed by psychopaths add up to a persuasive case that society would be improved if the numbers of psychopaths could be reduced, some caution needs to be exercised because of a possible association between psychopathic personality and creativity. An association of this kind has quite frequently been claimed. If these claims are correct, a reduction in the numbers of psychopaths might entail a reduction in the numbers of cre-ative individuals. This is a possible cost that needs to be considered.

The case for an association between psychopathic personality and creative achievement rests on two lines of evidence. The first of these regards psycho-pathic personality as a continuously distributed trait in the general popula-tion and examines its relationship with creativity. A study of this kind has been published by MacKinnon (1978) using the psychopathic deviate scale of the MMPI. He found that creative writers and artists scored above the popu-lation average on the scale.

Most of the work of this kind has been done using Eysenck's personality trait of "psychoticism" as a measure of psychopathic personality. The case that this trait is largely a measure of psychopathic personality has been persua-sively made by M. Zuckerman (1991). Research showing that psychoticism is positively associated with creativity has been carried out by Rushton (1997). He found that among college professors, creativity as assessed by citation counts (the extent to which their work is cited by others) was correlated .26 with psychoticism, suggesting that psychoticism makes significant contributions to creativity. He also found that among students, performance on a creativity test correlated .17 with psychoticism, again indicating that the trait contributes to creative achievement. Further evidence obtained by others for a positive association between psychoticism and creativity has been reviewed by Rushton (1997). This conclusion has also been reached by Eysenck (1993, 1995), who proposed that the evidence points to a positive association between creativity and a high average level of psychoticism rather than the very high level of psychoticism expressed in severe psychopathic personalities.

The second source of evidence for an association between some moderate degree of psychopathic personality and creativity consists of studies of the personality of eminent creative individuals. The most thorough study of this kind has been made by Post (1994). He examined the life histories of 291 eminent men of the nineteenth and early twentieth centuries and assessed them for psychopathic personality. He conceptualized psychopathic personality as a personality continuum on which three types should be distinguished. These are fully expressed psychopathic personality, subclinical psychopathic personality, and mild traces of psychopathic personality. He classified the eminent men into six groups, which consist of scientists, thinkers, politicians, artists, composers, and writers. Post then assessed the extent to which these men possessed these three psychopathic characteristics.

Post's first conclusion was that none of the eminent men had sufficiently pronounced psychopathic characteristics to be diagnosed as full psychopaths, although he considered that Hitler and Kierkegaard came close to being included in this category. We have previously noted that in the general population, about 6 percent of males are psychopaths; so the conclusion that there were no full psychopaths in the group of eminent men indicates that psychopathic personality must be a serious handicap for outstanding political and creative achievement.

Post's second category consisted of subclinical psychopathic personality, comprising psychopathic tendencies sufficiently strong to have an adverse impact on personal relationships and careers. Post estimated the prevalence of this condition at about 10 percent of the male population and at 14 percent among his eminent men, suggesting that some element of subclinical psychopathic personality may be an advantage for achievement. However, there were considerable variations in the incidence of subclinical psychopathy among the different categories of achievement: 2 percent among scientists, 11 percent among politicians and composers, 14 percent among thinkers, 20 percent among writers, and 25 percent among artists. Thus, only among the thinkers, writers, and artists was the prevalence of subclinical psychopathic personality appreciably greater than in the general population, and among scientists it was much lower.

The third category comprised a still weaker form of psychopathic personality, consisting of some antisocial characteristics but not of a seriously disruptive nature. Post estimated these as present among 54 percent of his eminent men, ranging from a low of 40 percent among scientists to a high of 70 percent among writers. He estimates that this condition is present in about 16 percent of the general male population, considerably less than among all the six categories of eminent men.

The conclusions reached by Post from his biographical studies of creative persons are essentially the same as those reached by Rushton (1997) from his studies of the relationship between psychopathic personality (psychoticism) and creativity in the general population, namely that high average psychopathic personality (psychoticism) or some element of subclinical psychopathic personality (but not fully expressed psychopathy) is positively associated with creativity.

The explanation for this association has been discussed by M. Zuckerman (1994). His thesis is that those who are high on psychopathic personality have a strong need for sensation and this leads them into a variety of risk-taking activities, including dangerous sports, crime, sexual adventures, and sometimes into creative achievement: "Sensation seeking is directly related to various tests of cognitive innovation, variety and originality" (p. 369). The creative scientist, artist, writer, or musician has to be something of a risk taker. Such individuals are likely to be attacked by the conventional establishment, who dislike having their ideas or methods challenged. Socrates, Galileo, Darwin, Rembrandt, and Monet are examples of creative risk takers who were ferociously attacked by their conventional contemporaries. Some people are willing to take the risks of creative endeavor and may even derive some enjoyment from the sensation provided by the controversies they stir up. We can regard these people as a certain type of subclinical creative psychopath, or, in Eysenck's terminology, those who are moderate to high on psychoticism.

11.   PSYCHOPATHIC PERSONALITY AND INTELLIGENCE

In assessing the contribution of psychopathic personality to creative achievement, we need to consider the mediating role played by intelligence. It is only those with some psychopathic tendencies who are also intelligent who are capable of significant creative achievement because intelligence is a necessary component for all forms of achievement. In addition, many of those who have the combination of average intelligence with some psychopathic tendencies are able to find a useful niche in society as small-scale entrepreneurs, as street traders, as artists, as pop musicians, or as workers at similar occupations that are comparatively free from external discipline and routine and that provide stimulation and variety while not being highly intellectually demanding. This group is not a serious social problem and adds color to social life.

The real problem group is that of the full-blown psychopaths and those with psychopathic tendencies combined with low intelligence who are only able to work in cognitively undemanding jobs, such as laborers, waiters, bartenders, and handymen. Their psychopathic personality makes them intolerant of unstimulating work of this kind and discontented with the poor remuneration and low status. Intolerance of work in such low-status employment draws them into crime, which provides the stimulation and the excitement they crave, satisfies their aggressiveness and hostility to others, provides instant gratification and financial rewards, and is not intellectually demanding. This is why habitual criminals are typically characterized by low intelligence and high psychopathic personality.

12. THE UNDERCLASS

The two dysgenic problems of low intelligence and psychopathic personality coalesce in a sector of society that has become known as the underclass. This subculture is typically located in impoverished inner-city districts and is characterized by poor educational attainment; high levels of long-term unemployment; and high rates of crime, drug addiction, welfare dependency, and single motherhood. Low intelligence and psychopathic personality are the underlying psychological determinants of this syndrome of social pathology.

The underclass is a new term for an old phenomenon. In the eighteenth and nineteenth centuries, this class was known in the United States and Britain as the "undeserving poor." These were people who were physically healthy and able to work, but who were work-shy, feckless, criminal, and violent and whose poverty was a consequence of their personal inadequacies. A distinction was drawn between these and the deserving poor who worked and were poor because they earned low wages and had large families to support. The general view was that only the deserving poor should be given charitable help; the undeserving poor needed to be discouraged and contained by social disapproval and stigma. Society provided those regarded as the undeserving poor with the bare essentials of accommodation and food in workhouses, which were deliberately designed to be spartan to discourage their use. The Victorians understood with a clarity that became lost in the second half of the twentieth century that rigorous social control was necessary to contain the growth of a subclass of undeserving poor. The contemporary concept of the underclass is a sanitized term for the undeserving poor and was first used by Gunnar Myrdal (1962). It came into wide circulation in the early 1980s, following Ken Auletta's (1982) use of the term in three articles published in The New Yorker in 1981 and in book form a year later.

The underclass is perpetuated through the transmission of low intelligence and psychopathic personality, from generation to generation, from parents to children. This transmission takes place through genetic and environmental processes. The genetic processes consist of the inheritance of low intelligence and psychopathic personality. The environmental processes consist of the poor childrearing techniques of parents with low intelligence and psychopathic personality and of the social influence of the psychopathic subculture. The intergenerational continuity of the underclass was first shown in the 1870s by Richard Dugdale (1877) in a study of criminals in the prisons of New York State. Dugdale noticed that many of these criminals were related and were members of extended criminal families. One of the largest of these was the Jukes family. Dugdale traced its pedigree through five generations, back to five sisters living in the second half of the eighteenth century. He found that the descendants of these sisters formed an extended family network of criminals, school dropouts, illiterates, alcoholics, unemployables, and prostitutes.

A follow-up study of this family was carried out some 40 years later by A.H. Eastabrooke (1916). He was able to trace 748 descendants of the original five sisters and obtained accurate information on the lives of 658 of them. He categorized these into three groups: The first consisted of 323 individuals, approximately half of the sample. Eastabrooke described these as "the scum of society . . . inefficient and indolent, unwilling or unable to take advantage of any opportunity which offers itself or is offered to them"; this group consisted essentially of psychopaths of various kinds.

The second group comprised 255 individuals who were marginally adequate semi-skilled and unskilled workers. The third group comprised 76 individuals (11 percent) who were described as "socially adequate" and "good citizens." Thus, after six to seven generations, 89 percent of the descendants of the original five sisters were either seriously psychopathic or only marginally adequate citizens.

About the same time as Eastabrooke was working on the Jukes family, another investigation of the intergenerational transmission of social pathology was made by H. H. Goddard (1912). This study consisted of seven generations of the Kallikak family, which was comprised of two branches descended from Martin Kallikak in the mid-eighteenth century. One branch consisted of responsible citizens descended from Kallikak's wife. The other branch consisted of the descendants of an extramarital liaison between Kalikak and a mentally retarded woman. For this branch, Goddard traced 450 descendants over seven generations. Of these 450, 143 were mentally retarded, and there were also numerous alcoholics and prostitutes. Thus, by the time of World War I, Dugdale, Eastabrooke, and Goddard had identified what was later to become known as the underclass and had shown that this group perpetuates itself down the generations in certain sociopathological families.

13. THE CYCLE OF DEPRIVATION

The studies of Dugdale, Eastabrooke, and Goddard first identified what was to become known as "the cycle of disadvantage" or "the cycle of deprivation." In these so-called cycles, the social pathology of the underclass is transmitted from generation to generation. During the twentieth century the existence of this phenomenon was confirmed by numerous studies, many of which have been summarized by Rutter and Madge (1976). For instance, Robins (1966) carried out a study in St. Louis of five hundred adolescents referred to a psychiatric clinic for a range of psychological problems over the years 1922–32. The parents were investigated for criminal behavior. Examination of the relationship between psychopathic personality in the adolescents and criminal behavior in the parents showed that 35 percent of psychopaths had criminal fathers and 32 percent had criminal mothers. In 34 percent of the psychopaths the fathers were long-term unemployed, and in 31 percent they were alcoholics. Only 16 percent of the fathers and 18 percent of the mothers were entirely free of psychopathic symptoms.

A study extending the parent-child resemblances for crime to three generations has been published by Robins, West, and Herjanic (1975). They obtained data in St. Louis on rates of serious crime among adolescents and then investigated the crime rates among their parents and grandparents. They found that adolescents who had criminal records were disproportionately likely to have parents and grandparents who also had criminal records. For instance, among white adolescents with criminal records, 33 percent had a grandparent with a criminal record; whereas of those without a criminal record, only 3 percent had a grandparent with a criminal record.

Another feature of the underclass that is transmitted from parents to children is single teenage motherhood. Fustenberg, Levine, and Brooks-Gunn (1990) reported a study of young black women in Baltimore, Maryland, who had sexual intercourse before the age of 16 and teenage pregnancies and births up to the age of 19. They also examined the National Longitudinal Study of Youth (NLSY) data and examined these experiences in the daughters of black teenage mothers and of black women who had not been teenage mothers. The results are shown in Table 9.4. Notice that the daughters of teenage mothers were substantially more likely to have had sexual intercourse before the age of 16 and to have become pregnant and given birth before the age of 19.

Table 9.4
Sex, Pregnancies, and Births (percentages) of the Children of Teenage Mothers and Nonteenage Mothers

| Daughters | Teenage Mothers | | Control |
	Baltimore	NLSY	NLSY
Intercourse before age 16	52	26	18
Pregnancy before age 19	51	46	31
Birth before age 19	36	33	21

The authors of the study also examined the characteristics that distinguished the daughters who became teenage mothers from those who avoided becoming teenage mothers. The results were that those who became teenage mothers were more likely to have failed a grade at school (57 percent versus 28 percent), less likely to have graduated from high school (43 percent versus 82 percent), less likely to be employed (29 percent versus 67 percent), more likely to receive public assistance (60 percent versus 3 percent), and less likely to use birth control (24 percent versus 42 percent). In addition, 31 percent of the teenage mothers were happy to become pregnant and had two or more children by the age of 19. This study illustrates the tendency of the underclass to perpetuate itself from one generation to the next, producing a second generation of teenage mothers characterized by poor educational attainment and a high incidence of unemployment and welfare dependency and a significant proportion of whom are glad to become teenage mothers themselves.

The intergenerational perpetuation of the underclass has also been found in Britain. One of the leading studies was carried out by Essen and Wedge (1982). They examined a sample of all babies born in one week in 1958 and followed them up over subsequent years, from which a variety of data has been collected. They divided the sample at the age of 16 years into those children whose family was "deprived" (underclass), as defined by unemployment of the father or male head of household, low income, poor quality housing, and five or more children, and "nondeprived." The results for the two groups of 16-year-olds, summarized in Table 9.5, show that the children from the deprived group had 16 times more mental retardation than those in the nondeprived group and were 12 times more likely to have been taken into foster care because their parents did not look after them properly. They performed poorly on tests of reading and mathematics. They were 20 times more likely to be illiterate, 12 times more likely to be innumerate, and twice as likely to have behavior problems in school.

Table 9.5
The Cycle of Disadvantage in Britain

Children	Family Background	
	Nondeprived	Deprived
Number	5,962	148
Mental retardation—percent	0.5	8.1
Children taken into foster care—percent	1.6	17.0
Reading quotient	104	85
Mathematics quotient	104	89
Illiterate—percent	0.6	11.4
Innumerate—percent	1.1	12.2
Behavior problems—percent	1.4	2.8

The intergenerational transmission of crime in Britain has been studied by West and Farrington (1997) of the Institute of Criminology at Cambridge University. In the late 1970s they wrote that "the fact that crime tends to be concentrated in certain families and that criminal parents tend to have criminal children has been known for years" (p. 12). Their own work added further evidence on this issue. They investigated a sample of 389 youths and their families born in a working-class area of London from 1951 to 1954 and followed these youths over subsequent years when they were aged 18 and 19. The criminal records of the youths were investigated together with those of their fathers, mothers, and sisters. This made it possible to examine to what extent young men with criminal records had fathers and mothers who also had criminal records and also to what extent female criminals had fathers and mothers with criminal records. The results, given in Table 9.6, show that among the young men, crime is approximately twice as common among those with criminal fathers as among those with noncriminal fathers (36.8 percent as compared with 18.7 percent), and about two and a half times as common among those with criminal mothers. In females the intergenerational continuity is stronger. Among female criminals, 17.2 percent had criminal fathers, as compared with 6.1 percent who had noncriminal fathers.

14. CONCLUSIONS

Psychopathic personalities inflict great costs on society in the form of their high rates of crime, antisocial behavior, unemployment, drug abuse, welfare dependency, and the spreading of HIV and AIDS and other sexually transmitted diseases. The social costs imposed by psychopaths are greater than those imposed by those with genetic disorders of health, mental retardation, and low intelligence, which consist largely of the financial burden to society and the unhappiness experienced by the individuals concerned and their families. Psychopaths inflict greater social damage. It is this group more than any other of whom H. G. Wells (1905) wrote in A Modern Utopia, "They spoil the world for others" (p. 85), and for whom he proposed a eugenic solution.

Table 9.6
Percentage of Male and Female Criminals in England with Criminal and Noncriminal Fathers and Mothers

Parents	Male Criminals	Female Criminals
Father criminal	36.8	17.2
Father noncriminal	18.7	6.1
Mother criminal	54.0	27.8
Mother noncriminal	23.1	5.1

Psychopathic personality and low intelligence are the psychological characteristics underlying the underclass, a subculture characterized by low educational achievement, crime, drug addiction, unemployment, welfare dependency, and single motherhood. The underclass is perpetuated from generation to generation by a combination of genetic inheritance, poor socialization of children by parents, and social learning of the norms of the underclass culture. The underclass has been allowed to grow in Western societies through a relaxation of the social controls of punishment and stigma used in previous centuries to contain it, and its elimination must be one of the objectives of eugenics.

The problem posed by psychopathic personality for a eugenic society is more complex than the problems of those with genetic diseases, mental retardation, and low intelligence because some element of psychopathic personality serves a useful social purpose by contributing to creative achievement. If a eugenic society were able to reduce the incidence of psychopathic personality, there might be some loss of creative achievement. There would nevertheless be substantial gains in social order, civility, work efficiency, and the reduction of crime. This would be a desirable trade-off, but the impact would need to be monitored. In the longer term it should be possible for a eugenic society to produce a population free of psychopathic personalities but with some individuals with the right mix of subclinical psychopathic personality and high intelligence and strong ego-strength that appears to be conducive to creative work.

III

⌒

The Implementation of
Classical Eugenics

We have now completed our examination of the objectives of eugenics and concluded that they consist of the reduction or elimination of genetic diseases and disorders, mental retardation and psychopathic personality, and the increase of intelligence and the personality traits of agreeableness and conscientiousness. It is now time to consider how these objectives could be achieved. There are two broad strategies for the promotion of eugenics. These consist of the classical eugenics of selective reproduction and the new eugenics of human biotechnology. We discuss classical eugenics first and turn to the new eugenics in Part IV.

Classical eugenics attempts to apply to humans the selective breeding techniques used for centuries on animals and plants to produce improved strains by breeding from the best individuals. The way this should be applied to humans was set out by Galton (1908). He proposed that the population should be divided into three categories which he designated the "desirables," the "passables," and the "undesirables" (p. 322). The "desirables" were those endowed with the qualities of health, intelligence, and moral character and, therefore, whose fertility it would be the objective of eugenics to increase. Galton called this "positive eugenics." The "undesirables" were those particularly poorly endowed with the qualities of health, intelligence and moral character and, therefore, whose fertility it would be the objective of eugenics to reduce. Galton called this "negative eugenics". In contemporary terms, Galton's "undesirables" are those with genetic disorders, the mentally retarded, the unintelligent, and those with low agreeableness and conscientiousness consisting of criminals and psychopaths. Galton's "desirables" are the healthy, the intelligent and those with the personality traits of high agreeableness and conscientiousness.

The implementation of a program of classical eugenics raises three general problems. The first consists of the genetical processes involved. It has so often been asserted that a eugenics program would not be feasible genetically that this contention needs to be considered. The second consists of formulating the specific policies and the political feasibility of introducing them. The third consists of the ethical acceptability of eugenic policies designed to alter the fertility of different sections of the population. We are now ready to consider these problems.

10

⤳

The Genetic Foundations
of Eugenics

1.   Basic Genetic Processes
2.   Single Additive Genes
3.   Intergenerational Transmission of Additive Genes
4.   Traits Determined by Additive Genes
5.   Dominant Genes
6.   Recessive Genes
7.   X-Linked Recessive Genes
8.   Polygenetic Inheritance
9.   Conclusions

⤳⤳⤳

Galton and the classical eugenicists believed that it would be both possible and desirable to improve the genetic quality of human populations by using the methods of selective breeding that had been used successfully for centuries by plant and animal breeders to produce improved strains. When eugenics came under attack in the later decades of the twentieth century, it was frequently asserted that these methods would not work. It was asserted also that the eugenicists did not understand genetics. Steve Jones (1993), for instance, a geneticist at the University of London, has written that the eugenicists' "views were taken seriously, although in retrospect it is obvious that they knew almost nothing about human inheritance" (p. 12). Milo Keynes (1993), a Cambridge University biologist, has written of Galton's eugenic proposals for increasing the numbers of children of the more desirable citizens that "through his ignorance of Mendelism, Galton was unbiological when he thought that eugenic policies could be achieved by encouraging the fertility of families to which eminent men belonged" (p. 23). Sir Walter Bodmer, a British geneticist, sneered at "the mindless practice of eugenics" (Bodmer & McKie, 1994, p. 236), despite welcoming the prospect of a number of eugenic advances, such as the elimination of baldness and myopia. J. Testart (1995), a biologist, has asserted that "we know that the methods of classical eugenics are without effect" because "positive eugenics could allow the birth of defective babies" (for instance, by the operation of recessive genes and mutations), while "negative eugenics precludes the birth of normal babies" (for instance, because some of the mentally retarded have children of normal intelligence) (pp. 305–6).

One of the criticisms frequently made of classical eugenics was that it assumed that characteristics like intelligence and personality are under the control of single genes, that this is incorrect because it is now known that these characteristics are determined by a number of genes (polygenetically), and that this means that the classical eugenics of selective reproduction would not work. Thus, Richard Soloway (1990), a historian at the University of North Carolina, writes that "polygenetic inheritance, gene-gene interaction, and gene-environment interaction undermined the scientific foundations of eugenics" (p. 353).

In the 1990s, eugenics was frequently dismissed as a "pseudoscience" by critics such as Paul (1995, p. 18), a political scientist at the University of Massachusetts; Billig (1998, p. 8), a psychologist at the University of Loughborough in Britain; and Koenig (1997, p. 892), a German historian. A pseudoscience is a false science whose basic principles are incorrect and that therefore will not work, like medieval alchemy, which attempted to turn base metals into gold, or contemporary astrology, which claims to predict the future from the signs of the zodiac. Eugenics was placed in the same category.

These criticisms that the classical eugenics of selective reproduction would not work have been made so frequently that they have to be tackled. Accordingly, we devote this chapter and the next to showing that eugenics is not a pseudoscience and that, on the contrary, Galton and his successors were right in their assertion that eugenic policies would work. In this chapter we examine the various types of genes and gene action responsible for the characteristics that it is the objective of eugenics to improve. The understanding of these makes it possible to assess the extent to which it would be possible to effect improvements by the classical eugenic methods of selective reproduction. In the next chapter, we consider in more detail the methods of selective breeding and the impact of the application of these to human populations.

1. BASIC GENETIC PROCESSES

The basic genetic processes were discovered in the 1860s by the Czechoslovak monk Gregor Mendel, but his discoveries were so far ahead of his time that no one understood them until 1900. Mendel worked out the principles of inheritance on plants, but the principles apply quite generally to plants, animals, and humans. The first essential point of Mendel's discoveries was that genes normally come in pairs. Each member of these pairs is called an *allele*; so normally everyone has two alleles for each genetically determined characteristic. People frequently write of "genes" when, strictly speaking, they mean "alleles." There are four kinds of alleles: additive, dominant, recessive, and X-linked. *Additive alleles* both contribute equally and therefore additively to the expression of the characteristics they affect. *Dominant alleles* are so-called because they dominate, or overpower, the recessive allele with which they are paired, so that the recessive has no effect or, sometimes, just a small effect. Conversely, *recessive alleles* are dominated, or overpowered, by the dominant genes with which they are paired. *X-linked alleles* are an exception to the general rule that genes come in pairs. They come singly on the X chromosome. Males have only one X chromosome, and thus only one of any X-linked alleles.

The two alleles that determine a genetic characteristic may be the same or they may be different. In the case of additive alleles, there are often a number of different alleles for any particular gene, so the two that people have are likely to be different. In the case of dominant and recessive alleles, people can have two dominants, two recessives, or one dominant and one recessive.

When people mate, they transmit one of their two alleles to their children. The children therefore receive one allele from their father and one from their mother, giving them two alleles. This process and its consequences can be illustrated as follows. Suppose that for a particular characteristic there are two alleles, labeled A and B, in the population. In one kind of mating, both father and mother have two A alleles. These are designated AA and are both said to be "homozygous" for this gene. Their children will inherit one A from their father and one A from their mother, so they will also be AA. Thus, all the children of this kind of mating are the same as the parents, and all the siblings are the same as each other. A parallel result would happen if both parents were BB, in which case all their children would be BB. This genetic process explains why children often resemble their parents and why siblings often resemble each other. If all genetic inheritance were like this, eugenics would be easy. If the A allele is desirable and the B allele is undesirable, perhaps because it is responsible for a genetic disorder, all that a eugenic program would have to do would be to sterilize the BBs. The BB genes would be eliminated from the population, and the undesirable characteristic caused by the B allele would also be eliminated.

However, eugenics programs are not so straightforward as this. In another kind of mating, both parents will have one A and one B allele—they are both AB. These couples will produce four allele combinations in their children. (1) One child will inherit an A from the father and an A from its mother and be an AA; (2) the second will inherit a B from its father and a B from its mother and be a BB; (3) the third will inherit an A from its father and a B from its mother and be an AB; (4) the fourth will inherit a B from its father and an A from its mother and be a BA. But a BA is exactly the same as an AB, so there are only three different genotypes in the population; the AAs, the BBs, and the ABs. From AB-AB matings, the children fall out, on average, in the proportions of 25 percent AA, 25 percent BB, 25 percent AB, and 25 percent BA. Because AB and BA are the same, they can be combined, giving 50 percent AB.

The effect of these genetic processes is that children often do not resemble their parents. Two AB parents can produce AA and BB children, who are unlike the parents, as well as AB children who are like the parents. This process explains why children are often different from their parents and why children of the same parents often differ from one another. This mode of inheritance makes eugenic programs more difficult, although it does not make them impossible.

2. SINGLE ADDITIVE GENES

The simplest genetic process determining characteristics for which there is variability in the population consists of a single gene that has two alleles (alternative forms). The German psychologist Volkmar Weiss (1992) proposed that intelligence is determined in this way, subject to the qualification that in addition to the major gene for intelligence with its two alleles, there are also some minor genes having small effects.

Weiss's model proposes that one of the intelligence alleles, which he designates M1, confers a genotypic IQ of 130 and that the other allele, which he designates M2, confers a genotypic IQ of 94. Thus in the population there are three genotypes: (1) the M1M1s, with two high-intelligence alleles and a genotypic IQ of 130; (2) the M2M2s, with two low-intelligence alleles and a genotypic IQ of 94; and (3) the M1M2s, who have one high- and one low-intelligence allele and therefore have a genotypic IQ of 112 (the average of 130 and 94). Weiss proposes further that the M1 alleles are less common than the M2 and that they are present in Caucasian and Mongoloid populations in approximately 20:80 ratios. About 4 percent of the population consists of the genetic elite of M1M1s, about 32 percent of genetically intermediate M1M2s, and about 64 percent of the genetically inferior M2M2s. Assortative mating (the tendency for people to mate with those who are similar to themselves) increases slightly the proportion of M1M1s to around 5 percent, and of M2M2s to around 68 percent.

Because there are three genotypes in this model, it would be expected that these three types would show up in the distributions of IQs measured by intelligence tests as three distinct phenotypes, that is, types of individual. However, intelligence tests do not show the expected trimodal distribution but rather a continuous bell curve. It may be considered that this disproves the model. However, Weiss has four answers to this objection. First, the normal distribution of intelligence tests is an artifact imposed on the scores by psychometricians. Second, the genotypic trimodal distribution is blurred by the action of other genes with small effects. Third, the genotypic trimodal distribution is also blurred by environmental effects; for example, a child with a genotypic IQ of 130 might suffer oxygen deprivation at birth, causing some brain damage sufficient to reduce its measured intelligence to around 120, whereas a child with a genotypic IQ of 112 might have an exceptionally favorable environment sufficient to raise its measured intelligence to around 118. Fourth, the genotypic trimodal distribution would be further blurred by measurement errors in the assessment of IQs. Taken together, these four factors would produce the standard normal continuous distribution of intelligence shown in psychology textbooks.

The consensus view among behavior geneticists is that Weiss's model is an oversimplification and that intelligence, and also personality traits, are determined by many additive genes (Plomin, DeFries, McClearn, & Rutter, 1997). Nevertheless, Weiss's model is a useful simplification, and the principle of determining the transmission of genotypic intelligence from parents to children and the impact of eugenic interventions are the same, whether one gene or many genes are involved.

3. INTERGENERATIONAL TRANSMISSION OF ADDITIVE GENES

The application of Weiss's model to the transmission of genotypic intelligence from parents to children is shown in Table 10.1. The two left-hand columns give the six possible combinations of matings between the M1M1 and the M2M2 genotypes; the three right-hand columns give the distribution of the children of those matings. Row 1 of the table shows that if an

Table 10.1
Mating Combinations of Parents with Two Additive Alleles M1 and M2 and Frequencies of Genotypes of Their Children

Genotypes of Parents		Genotypes of Children		
Father	Mother	M1M1	M1M2	M2M2
M1M1	M1M1	100	—	—
M1M1	M1M2	50	50	—
M1M2	M1M2	25	50	25
M1M1	M2M2	—	100	—
M1M2	M2M2	—	50	50
M2M2	M2M2	—	—	100

M1M1 mates with another M1M1, all their children are M1M1. Both parents transmit one of their two alleles to their children. As both the parents have two M1 alleles, these are all they can transmit; so all their children have to be M1M1. This would represent the genetic elite that transmits its high genotypic intelligence down the generations and whose existence was first shown by Galton (1869, 1874) for elite British families and by Woods (1913) for similar families in the United States.

Row 2 shows the distribution of the children of an M1M1 father and an M1M2 mother. Half the children are M1M1, resembling their father (M1M1), while the other half are M1M2, resembling their mother (M1M2); so the average of the children is the same as the average of the parents. The same distribution of children results from an M1M2 father and an M1M1 mother (not shown in the table). Row 3 shows the distribution of the children where both parents are M1M2. A quarter of the children inherit the M1 allele from both parents and therefore belong to the M1M1 genetic elite. Another quarter inherit the M2M2 allele from both parents and belong to the M2M2 low intelligence group. The remaining half inherit one M1 and one M2 and are M1M2s, like their parents. The distribution of the children of these parents explains why siblings frequently differ quite appreciably from one another. The average difference in intelligence between siblings is about 14 IQ points, and the average correlation between their IQs about .5 (Bouchard, 1993). The distribution also explains how a pair of average parents can produce a range of children, some of whom are more intelligent their parents, some of whom are less intelligent, and some of whom are average, like their parents. This explains why outstandingly gifted individuals are quite commonly born to quite ordinary and average parents.

For instance, the German geneticist Otmar von Verschuer (1957) examined the family pedigree of the composer Robert Schumann. He found that the Schumann family had been quite ordinary artisans for many generations and that neither Robert's father nor mother, nor any of his 136 ancestors or relatives, had displayed any musical talent. This illustrates the principle that highly gifted individuals not infrequently appear as a result of the inheritance of an unusually favorable combination of genes carried by average parents.

Row 4 shows the children of a mating of a high IQ M1M1 father with a low IQ M2M2 mother (the results would be the same for an M2M2 father and an M1M1 mother, not shown in the table). In this case, all the children are average M1M2 because they all inherit an M1 from one parent and an M2 from the other.

Row 5 shows the mating of an average M1M2 father with a low IQ M2M2 mother. Half the children are M1M2 and the other half are M2M2. This mating is the mirror image of that in Row 2.

Finally Row 6 shows the results of matings between two low IQ M2M2s. All the children are M2M2. They have to be because they can only inherit the M2 alleles. These are the mirror image of the M1M1 genetic elite, who only produce M1M1 children. The children of the M2M2s provide a model of the transmission of low intelligence from generation to generation.

This model is a simplification of what actually occurs. In the real world neither the genetic elite nor the genetic underclass breeds entirely true. The genetic elite produces some children whose intelligence is lower than that of the parents, while the genetic underclass produces some children whose intelligence is higher than that of the parents. This phenomenon is known as regression to the mean and will be considered in the next chapter. Nevertheless, Weiss's model provides an approximation of the fact that societies do have elite families that produce highly intelligent individuals in successive generations, such as the Darwin family in England, four members of which, descendants of Charles Darwin, are on the faculty of the University of Cambridge. Societies also have an underclass of low intelligence, which also perpetuates itself from generation to generation in the process known as the "cycle of deprivation" or "the cycle of disadvantage" (Rutter & Madge, 1976), which was discussed in Chapter 9, Section 13.

4. TRAITS DETERMINED BY ADDITIVE GENES

Traits determined by additive genes can be improved by selective reproduction. This can be readily understood by considering the mating types and the characteristics of their offspring shown in Table 10.1. A number of eugenic scenarios can be constructed from the table. For instance, a program of positive eugenics would encourage the genetic elite to have more children. If a policy of this kind were targeted on the M1M1 genetic elite and succeeded in persuading them to double their number of children, then the number of M1M1s in the child generation would also double because the genetic elite only produce M1M1 children. A broader eugenic program that encouraged M1M1-M1M2 couples to have more children would also be effective because the average of the children would be above the average of the population, like that of their parents.

A program of negative eugenics would reduce the numbers of children produced by the M2M2s. This was the thinking behind the program for the sterilization of the mentally retarded that was introduced in North America and Europe in the first half of the twentieth century. In terms of the mating model shown in Table 10.1, the program could be implemented by the sterilization of the M2M2s. This would prevent the reproduction of the mating combinations in the bottom three rows of the table and would prevent the birth of a lot of M2M2 children. The effect of this would be that a much higher proportion of children would be M1M1 or M1M2 and that the intelligence level of the child generation would rise. Some M2M2 children would continue to be born from M1M2-M1M2 matings (from parents of average intelligence) because a quarter of the children of these are M2M2s, as shown in Row 3 of Table 10.1. Nevertheless, sterilization of all M2M2s would have a large eugenic impact in reducing the numbers of low-intelligence children. Although some of these would continue to be born from M1M2-M1M2 matings, a continuation of such a policy over several generations would progressively weed out the M2 allele, and the intelligence level of the population would increase in each generation.

It may be useful to note that there are no regression-to-the-mean effects in the children of any of the parental mating combinations shown in Table 10.1. Each mating combination breeds true; that is, the average of the children is precisely the same as the average of the parents. This is because genetic regression does not occur for traits determined by additive genes, but only for traits determined by dominant and recessive genes. Regression to the mean can also occur through environmental effects. (We consider more fully the issue of regression effects in Chapter 11.)

Although this is a simplified model of the genetics of intelligence and its transmission from parents to children, it illustrates the essentials of the process. By the end of the twentieth century, there was a universal consensus among behavior geneticists that intelligence is largely determined by additive genes and conforms broadly to the genetic model of which a simplified version is shown in Table 10.1. Personality traits also appear to be largely determined genetically by additive genes (Plomin et al., 1997).

The potential effectiveness of a eugenic program remains the same whether intelligence and personality traits are determined by one gene with two alleles, as in the model we have been considering; by two genes each of which has two alleles; by 20 genes each of which has three alleles; by 100 genes each of which has four alleles; or whatever. Nor does it make any significant difference to the effectiveness of eugenic intervention if a few of the genes determining intelligence are dominant and recessive, as appears to be the case. Whatever numbers of alleles and genes are involved in the determination of intelligence and personality, there can be no doubt about the effectiveness of a eugenic program of the traditional kind used by stock breeders of increasing the reproduction of the most desirable individuals (M1M1) and of reducing the reproduction of the least desirable (M2M2). This has been known and demonstrated for centuries. The model shown in Table 10.1 illustrates how the process works genetically.

5. DOMINANT GENES

One of Mendel's most important discoveries was that sometimes one of a pair of alleles, called the dominant, overpowers the other, called the recessive. In this type of gene action only the dominant allele determines the characteristics expressed (the phenotype), and the recessive has no effect. An individual with one dominant and one recessive allele is known as a *carrier* of the recessive. Sometimes dominance is incomplete, and the recessive does have a small effect.

As with additive alleles, dominant and recessive alleles combine to produce three genetic types. These can be designated the DDs, who have two dominant alleles; the RRs, who have two recessives; and the DRs, who have one dominant and one recessive. The different combinations of matings between these genotypes, and the distribution of the genotypes of the children of these matings, are shown in Table 10.2. The mode of inheritance of dominant and recessive alleles is the same as that of additive alleles shown in Table 10.1. However, because the dominant allele overpowers the recessive, a single recessive has no effect on the phenotype. There are therefore only two phenotypes, those who have the dominant allele and those who do not. This is in contrast to the case with additive alleles for which there are three phenotypes, with one that is intermediate between the two extremes.

Examining Table 10.2, we see in Row 1 that when both parents are DD, all the children are DD because they can only inherit the D allele. Row 2 shows that when the parents are DD and DR, half the children are DD and the other half DR; all the children have the condition determined by the D allele. Row 3 shows that when the parents are DR and DR, a quarter of the children are DD, half are DR, and a quarter are RR. The quarter that are RR have the condition determined by the R allele, which could be a genetic disorder. Row 4 shows that when the parents are DD and RR, all the children are DR. Row 5 shows that when the parents are DR and RR, half the children are DR and the other half are RR. Finally Row 6 shows that when both the parents are RR, all the children are RR.

A common mating combination is the DR-RR. If the dominant gene is adverse, half the children are DR and have the adverse condition, while the other half are RR and are free of it. The effect of this is that in pedigrees of families in which the gene is present, the condition is expressed in successive generations in half the children. When this pattern of inheritance is found, it can be inferred that a dominant gene is responsible for the condition. One of the first dominant genes to be identified in this way was the gene for

Table 10.2
Mating Combinations of Parents for Dominant and Recessive
Alleles (D and R) and Genotypes of Their Children

Genotypes of Parents		Genotypes of Children		
Father	Mother	DD	DR	RR
DD	DD	100	—	—
DD	DR	50	50	—
DR	DR	25	50	25
DD	RR	—	100	—
DR	RR	—	50	50
RR	RR	—	—	100

Huntington's chorea, which was identified in the second decade of the twentieth century by C. B. Davenport (1916). In 1983 the gene was located by James Gusella on the short arm of chromosome 4. The gene is involved in the production of two neurotransmitters, acetylcholine and gamma amino butyric acid, and the defective gene reduces their production, causing the physical and mental deterioration that typically begins to develop in middle age among people afflicted with this disease. A number of rare genetic diseases are transmitted by dominant genes, some examples of which are given in Chapter 4, Section 2.

Dominant genes for diseases and disorders are, in principle, easily eliminated from the population by eugenic intervention. All that is required is to prevent those with the disorders from having children, either by genetic counseling or sterilization. This would prevent the genes from being transmitted to future generations. If all those with the disorder caused by the gene are prevented from having children, the gene can be eliminated from the population in one generation, except for a few cases appearing as a result of new mutations. In the rare cases in which the disorder does not express itself until middle age, as is the case with Huntington's chorea, there is a problem because affected individuals are likely to have children before the disease appears. This problem could be overcome if all the children of those with Huntington's disease, each of whom has a 50 percent risk of carrying the gene, remain childless or have prenatal genetic testing for the presence of the harmful gene and pregnancy terminations in cases where the gene is identified.

Most dominant genes are advantageous, or at least not harmful. There are some dominant genes that enhance intelligence (Jensen, 1998). Beneficial dominant genes, such as those for enhanced intelligence, are not so easy to increase by eugenic intervention because they cannot as yet be identified. Nevertheless, eugenic measures to increase the fertility of the more intelligent would increase the number of dominant genes that contribute to high intelligence. These measures would be effective because those carrying dominant genes transmit them to half their children.

6. RECESSIVE GENES

The genetic process of the transmission of recessive genes is also shown in Table 10.2. Because a single recessive allele has little or no effect, it cannot be detected, except by biochemical tests. There are a large number of potentially harmful single recessives in the population, and it is generally considered that virtually everyone has at least one of these and many people have several. This makes it more difficult to eliminate harmful recessives by eugenic intervention. Those who have a harmful double recessive manifest the disorder or disease. These could be sterilized, and this would have some impact on reducing the transmission of the gene to future generations. However, this impact would be relatively small because there would still be so many individuals with a single recessive remaining in the population. It has been estimated by Stern (1973) that if individuals with a double recessive comprise 1 percent of the population, the sterilization of these would reduce the birth incidence of the recessives by 9 percent in the next generation, and if this policy were pursued for 10 generations, the birth incidence would be reduced to a little less than half of the initial frequency. Therefore, as the recessive became rarer, it would become increasingly difficult to eliminate, and it would take thousands of generations to eliminate it completely.

Opponents of eugenics have often made the point that it could take thousands of years for eugenic measures to eliminate all harmful recessives. This is true, but nevertheless Stern's calculations show that the birth incidence of the disorder or disease caused by the recessive could be reduced by 9 percent in one generation and more than halved in about 250 years (10 generations). This would be a significant gain.

There are some recessive genes that depress intelligence (Jensen, 1998). This is inferred from the phenomenon of inbreeding depression, which consists of impairments of various kinds, including intelligence, arising in the children of parents who are closely related. Because individuals who are closely related have a high probability of being carriers of the same recessive genes, their children have an increased probability of inheriting the double recessive and manifesting the impairment. However, not all recessive genes are harmful. In the case of Huntington's chorea, and a number of other disorders, it is the dominant allele that is harmful and the recessive that is for health. Similarly, in regard to intelligence there are some recessives that enhance spatial intelligence (Jensen, 1998). This is shown by the fact that spatial ability is high among the children of close relatives, who have an increased probability of having children with the double recessive responsible for the expression of the condition.

7. X-LINKED RECESSIVE GENES

X-linked recessive genes are carried by females but usually only express their effects in males. The genetic mechanism responsible for this is that males have one X chromosome and one Y chromosome, whereas females have two X chromosomes. A recessive gene on the X chromosome of females is normally suppressed by the dominant gene on the other X chromosome. Males have no second X chromosome with a dominant gene, so the recessive expresses itself. Females only express the condition in the unusual case where they inherit two of the recessive genes.

Malfunctioning X-linked recessive genes are responsible for many genetic disorders and diseases that normally occur only in or disproportionately in males. Some of the commonest of these have been described in Chapter 4 and include hemophilia, color blindness, fragile X syndrome, and Duchenne's muscular dystrophy. There may also be X-linked recessives having advantageous effects. A theory has been proposed that spatial ability or a component of spatial ability is increased by an X-linked recessive and that this contributes to the higher spatial ability normally present in males (Stafford, 1961; Thomas & Krail, 1991), although the theory has not won universal acceptance.

The prevalence of harmful X-linked recessive genes in the population can be reduced by selection. A number of those with an X-linked disorder are born to parents where the father has the disorder and the mother is normal. Thus, if males with the disorder do not have children, the gene is not transmitted, and its frequency in the population is correspondingly reduced. However, most children with X-linked disorders are born to parents where the father is normal and the mother is a carrier. These children are hard to select out because the gene in the mother cannot be detected. This makes it difficult to eliminate the X-linked disorders by conventional selection methods, although they can be reduced.

8. POLYGENETIC INHERITANCE

Hitherto we have considered the modes of action of the four types of single genes. However, many characteristics, including a number of the multifactorial disorders and diseases described in Chapter 4 and also intelligence and personality, are carried by the action of a number of genes, each of which makes a relatively small contribution to the condition. There is frequently also an environmental component to the expression of the condition. For instance, there is a genetic predisposition to contracting lung cancer that is exacerbated by the environmental impact of air pollution and cigarette smoking. Similarly, intelligence and personality are determined partly by genes but also by environmental factors.

Where a number of genes determine a characteristic, the mode of inheritance is called *polygenetic*. It was shown by the British geneticist Sir Ronald Fisher (1918) in a classical paper that in this mode of inheritance each of the genes involved operates in accordance with the Mendelian principles of single-gene action. Fisher showed that polygenetic inheritance normally causes continuously distributed characteristics, such as height, intelligence, and personality traits. He also showed that the degree of similarity of relatives for the characteristic depends on how closely they are related genetically. For instance, fathers and mothers have half their polygenes in common with their children and so also do siblings. For intelligence, the parent-child correlation derived from numerous studies is .42 and that for siblings is .47 (Bouchard, 1993). The predicted correlations if intelligence were entirely inherited polygenetically are .50, so the obtained correlations are very close to those predicted. First cousins have one-eighth of their genes in common, and the correlation for intelligence is .125, again very close to that predicted from their genetic relationship. This is one of the several kinds of data indicating that intelligence is largely determined polygenetically.

Polygenetic traits can be changed by selective breeding using the same methods as those for single genes. These consist of measures to increase the numbers of children of those carrying the genes for the characteristic regarded as desirable and to reduce the numbers of children of those carrying the genes for the characteristic regarded as undesirable.

9. CONCLUSIONS

When Galton first proposed the concept of eugenics in the 1860s, and for some decades thereafter, the genetic processes of inheritance were not understood, except by Mendel, whose work no one knew about until it was discovered in 1900. Despite the fact that the mechanisms of gene action were not known to the early eugenicists, this does not mean that eugenic selective breeding programs would not work, as asserted by many of the critics of eugenics. On the contrary, selective breeding for improved strains of plants and animals has been used for centuries, and there is no doubt that it has been effective and that it would be equally effective applied to human beings.

Nevertheless, the understanding of the various gene processes described in this chapter makes it possible to think more precisely about the effectiveness of classical eugenics programs of selective reproduction. Characteristics determined by single dominant genes and by multiple additive genes are relatively easy to change by selective reproduction, while those determined by recessives and X-linked recessives are more difficult, but not impossible, to change. In the next chapter, we consider the effectiveness of eugenic intervention for characteristics determined polygenetically, that is to say, by a number of genes interacting with the environment.

11

⤳

The Genetic Principles
of Selection

1. Selective Breeding of Plants and Animals
2. The Breeding of Thoroughbreds
3. Polygenetic Disorders
4. Heritability of Intelligence
5. Heritability of Personality
6. Heritability of Psychopathic Personality
7. Quantification of Effectiveness of Selection Programs
8. The Problems of Regression to the Mean
9. Understanding Regression to the Mean
10. Conclusions

⤳⤳⤳

In this chapter we consider the principles governing the effectiveness of eugenic programs for improving characteristics determined by several genes (polygenetically) interacting with the environment. We begin by outlining how this technique has been used successfully for centuries by plant and animal breeders, who have bred from the best specimens and produced improved stocks. The effectiveness of a selective breeding program depends on the heritability of the characteristic for which the breeding is done and on the stringency of the selection of the couples chosen for breeding. If a characteristic has no heritability, it is impossible to change it by selective reproduction. Hence it is necessary for the eugenicist to demonstrate that the characteristics to be bred for improvement have some heritability. This is shown for a number of polygenetically determined disorders and diseases and for intelligence and personality. We consider next the quantification of the effectiveness of selective breeding programs for intelligence. We conclude by considering the problem of regression to the mean.

1. SELECTIVE BREEDING OF PLANTS AND ANIMALS

The eugenicists believed that it would be possible to improve the genetic quality of human populations by adopting the methods used by horticulturalists and animal breeders for improving the quality of plants and animals. The technique consists simply of selecting individuals with desirable qualities and breeding from them. People began to do this in the early civilizations shortly after they made the transition from a nomadic way of life to a settled life based on agriculture about ten thousand years ago. In the fourth century B.C., Plato referred to the selective breeding of horses and cattle in *The Republic* as if this were common knowledge, and he suggested through his spokesman Socrates that the same principle should be applied to humans in the Utopian state.

During the past two thousand years numerous species of animals and plants have been bred for a variety of characteristics. In Roman times beekeepers bred bee colonies for calmness by destroying the more aggressive ones (MacKenzie, 2000). In medieval Europe, horses were used in warfare and these needed to be larger and heavier to carry knights wearing armor and carrying lances and to bear down more powerfully on the enemy. Accordingly, by the eighth century A.D., in the words of a leading historian, "Horses were being bred to be as big as possible" (Thomas, 1981, p. 89).

For many centuries people have bred sheep and cattle for better quality meat. In the middle years of the eighteenth century, an English farmer, Robert Bakewell, bred the modern strain of sheep known as New Leicesters. People also bred different strains of dogs with a variety of useful characteristics, such as spaniels that would creep up on birds and then spring to frighten them into the hunter's net and sheepdogs that would herd sheep. In France, 17 different breeds of sheepdogs and stock dogs were bred to perform different kinds of farm work, and in England 26 breeds of hunting dogs were bred for a variety of hunting purposes. Other dogs, such as cocker spaniels, were bred primarily as house pets.

Selective breeding has also been used for many centuries to obtain improved strains of plants. The numerous varieties of roses in present-day gardens have been produced by selective breeding from the wild dog rose, which grows naturally in woods and hedges. Virtually all our fruits, vegetables, and plants have been bred selectively for improved strains. For example, the modern strawberry was bred from the small wild strawberry by an English plant breeder in the early nineteenth century (Farndale, 1994). Corn has been bred to increase its oil content. In the 1890s the oil content was 5 percent; it has been increased by selective breeding over generations to 20 percent (Crow, 1986).

In the course of the twentieth century, a number of psychologists have shown that it is possible to breed animals for intelligence and temperament. The classical study of breeding for increased intelligence in rats was carried out by Tryon (1940). He measured rats' intelligence by their ability to learn how to run through a maze without making errors of turning into cul-de-sacs. He began by selecting a group of "brights," who did well at this task, and a group of "dulls," who did poorly. He bred from these two groups over 21 generations and obtained two genetically different strains. A similar study was carried out by W. R. Thompson (1954). He found that over six generations the strains of bright and dull rats progressively diverged. The bright rats became progressively brighter and reduced their average errors in the maze from 190 to 142, while the dull rats became progressively duller.

Rats have also been bred for the trait of emotionality, a rat analogue of the trait of anxiety in humans. The experiment was carried out by Broadhurst (1975). Two groups designated "reactive" (emotional) and "nonreactive" (unemotional) were selected and interbred for 15 generations. This produced two genetic strains. In all these experiments the two selected strains bred true, that is to say that there was no regression to the original population mean, which some critics have asserted would nullify a selective breeding program.

2. THE BREEDING OF THOROUGHBREDS

An instructive example of selective breeding for improved stocks is the breeding of thoroughbred racehorses. All thoroughbreds are descended from horses listed in the *Stud Book* of 1791, and their pedigrees have to be approved for registration as authentic thoroughbreds. The crucial characteristic of a thoroughbred is how fast it can run, and it has been shown that this depends on several factors, including its body size, leg length, lung capacity, temperament, competitiveness, and endurance. All these characteristics are to some degree genetically determined. The heritability of running speed has been estimated at between 15 percent and 35 percent (Langlois, 1980; Budiansky, 1997).

The owners of thoroughbreds breed them in the hope of obtaining an exceptionally fast horse that will win races and earn lots of money. They use the traditional method of selective breeding and typically breed from the fastest 10 percent of stallions and the fastest 50 percent of mares. These selective breeding programs have produced an improved population of thoroughbreds whose average running speeds increased by about 1 percent a year from 1952 to 1977 (Budiansky, 1997). What has happened is that the genes (alleles) contributing to fast running speeds have been increased in the thoroughbred population and that at the same time the alternative forms of these genes (alleles) for poorer running speeds have been reduced. This has led to an improvement in the average running speed of the entire population.

However, although the average running speed of thoroughbreds has increased, there has been no improvement in the fastest running speeds, which have remained the same for about a century. Records are not broken virtually every year, as they are in Olympic events. The fastest horse ever was Sovereign, who lived between the two World Wars. The reason the fastest running times have not improved is that all the genes (alleles) for the fastest running speeds must have been present in the 1791 thoroughbred population. These have been increased by selective breeding, while at the same time the alleles that reduce speed have been reduced. It is very unlikely that new mutant genes for faster speeds have appeared. Because running speeds are determined by a number of characteristics, each determined by a number of genes, the chances of a horse inheriting all the best genes for running speed are very low; and it is a matter of chance when these, together with optimum environmental factors, appear in a particular horse.

The experience of the breeding of thoroughbreds over the past two centuries serves as a useful model for what could be anticipated if eugenic measures were introduced for humans. In the case of intelligence, there would not be any increase in the highest intelligence hitherto achieved. The highest IQs ever recorded are about 200, the intelligence level estimated for Blaise Pascal (Cox, 1926) and Francis Galton (Terman, 1917a). An IQ of 200 means that a child of a particular age is at the intellectual level of the average child of twice this age (e.g., a four-year-old is at the level of the average eight-year-old). We should not expect that a eugenic program would increase the highest achievable IQ to 300 or 400. This is because all the right genes and the most favorable environmental conditions have already appeared from time to time and produced people like Pascal and Galton. What a eugenic program would accomplish would be the reduction or elimination of the genes for low intelligence. The average intelligence level of the population would be improved, just as the average running speed of thoroughbreds has been improved; but there would be no increase of the highest IQs, just as there has been no improvement in the running speeds of the fastest thoroughbreds.

3. POLYGENETIC DISORDERS

The great majority of genetic diseases and disorders are not caused by single genes but by the action of a number of genes and adverse environmental effects. The prevalence of these disorders and diseases can only be changed by eugenic intervention if they have some heritability, and this needs to be demonstrated. The genetic contribution to the development of these disorders has been shown by studies of the similarity for them among monozygotic (identical) and dizygotic (nonidentical, or fraternal) twins. The degree of similarity between twin pairs is frequently calculated as concordance rates. Twins are described as concordant if they both have the disease. If the disease has a genetic component, the concordance rate is greater for monozygotic than for dizygotic twins because monozygotic twins have all their genes in common, whereas dizygotic twins have on average only half their genes in common. The effect of greater genetic similarity is to make monozygotic twins more similar to each other than dizygotic twins are. The concordance rates of monozygotic and dizygotic twins for 25 common multifactorial diseases assembled from a number of studies have been estimated by Connor and Ferguson-Smith (1988) and are shown in Table 11.1. Where the concordance for monozygotic twins is greater than that for dizygotic twins, indicating a genetic susceptibility to the diseases, the importance of the genetic contribution is indicated by the degree of monozygotic twin concordance. If a disease is wholly determined genetically, the concordance rate for monozygotic twins

Table 11.1
Concordance Rates for Common Multifactorial Diseases and Disorders

	Concordance (percent)	
	Monozygotic	Dizygotic
Cleft lip +/- cleft palate	35	5
Cleft palate alone	26	6
Spina bifida	6	3
Pyloric stenosis	15	2
Congenital dislocation of the hip	41	3
Talipes equinovarus	32	3
Hypertension	30	10
Diabetes mellitus (insulin-dependent)	50	5
Diabetes mellitus (insulin-independent)	100	10
Ischaemic heart disease	19	8
Cancer	17	11
Epilepsy	37	10
Schizophrenia	60	10
Depression	40	11
Manic depression	70	15
Mental retardation	60	3
Leprosy	60	20
Tuberculosis	51	22
Atopic disease	50	4
Hyperthyroidism	47	3
Psoriasis	61	13
Gallstones	27	6
Sarcoidosis	50	8
Senile dementia	42	5
Multiple sclerosis	20	6

will be 100 percent, as is the case with insulin-independent diabetes mellitus. If the concordance rate is low, as in the 6 percent concordance for spina bifida, the genetic contribution is small.

The feasibility of selection to eliminate polygenetic disorders depends on their heritability. If the heritability is high, as is the case for insulin-independent diabetes mellitus, the birth incidence of the disorder can be substantially reduced if those with the disorder do not have children. If the heritability of the disorder is low, a reduction in the fertility of those with the disorder has little impact.

4.   HERITABILITY OF INTELLIGENCE

Selective breeding programs will only work for characteristics that have some heritability (some genetic variability in the population). With regard to intelligence, Galton and later eugenicists believed that the heritability is substantial. During the middle decades of the twentieth century, this view came under attack, notably from Kamin (1974), but by the last two decades of the twentieth century, no serious students of this issue disputed that genetic factors are a significant determinant of intelligence.

There are two principal kinds of evidence pointing to the conclusion that intelligence is substantially genetically determined and from which its heritability can be calculated. The first of these consists of studies of monozygotic twins reared apart. It has been found that these have highly similar IQs, represented by a correlation between adult twin pairs of .72 (Bouchard, 1993). This figure needs to be corrected for the unreliability of measurement. Assuming the test has a reliability of .9, the corrected correlation between the twin pairs is .80. This correlation is a direct measure of heritability.

The second method consists of comparing the degree of similarity between identical twins and same-sex, nonidentical twins brought up in the same families. Because identical twins are genetically identical, whereas nonidentical twins have only half their genes in common, if genetic factors are operating, identical twins should be more alike than nonidenticals. The simplest method for quantifying the genetic effect was proposed by Falconer (1960) and consists of doubling the difference between the correlations of identical and same-sex nonidenticals. Studies of the intelligence of adult twin pairs have been summarized by Bouchard (1993, p. 58). He found correlations of .88 for identical twins and .51 for same-sex, nonidenticals. The difference between the two correlations is .37, and doubling this difference gives a heritability of .74. This figure needs to be corrected for the imperfect reliability of the tests. Using a reliability coefficient of .9, the corrected correlations become .98 for identicals and .56 for same-sex nonidenticals. The difference between the two correlations is .42, and doubling this difference gives a heritability of .84. This is very close to the heritability of .80 derived from the first method. This is why most experts on this issue estimate the heritability of intelligence as approximately .80, or 80 percent (Eysenck, 1979, p. 102; Jensen, 1998, p. 78). The heritability of intelligence among children is somewhat lower, probably because there are environmental effects acting on children that wear off during adolescence.

The two twin methods for estimating the heritability of intelligence have been augmented by genetic modeling techniques based on data collected for the resemblance of other family relationships, such as siblings, half siblings, unrelated adopted children reared in the same family, and the like. These technologies produce broadly similar estimates of heritability. Some students of this issue prefer to give a range of heritabilities. For instance, Herrnstein and Murray (1994) propose a range of 40 to 80 percent and Mackintosh (1998) of 30 to 5 percent. The heritability varies somewhat from one population to another, especially when estimated from children and adolescents as well as from adults. It should be noted that the precise heritability does not matter for the eugenic objective of raising the intelligence level of the population. So long as a characteristic has some heritability, it can be improved genetically.

It is important to distinguish between what are called broad heritability and narrow heritability. *Broad heritability* is a measure of the heritability attributable to all kinds of genes. *Narrow heritability* is a measure of the action of additive genes only. The figure of .8 is an estimate of broad heritability. Narrow heritability has been estimated by Jinks and Fulker (1970) and Jensen (1972) at .71, or 71 percent, leaving a .09, or 9 percent, heritability attributable to the effects of dominant and recessive genes. Thus, intelligence is largely determined by additive genes and only to a small extent by dominant and recessive genes. Narrow heritability is used to calculate the effects of selection programs of varying degrees of severity, as we shall see in Section 7 of this chapter. The first gene determining intelligence in normal populations was discovered by Chorney et al. (1998). It lies on chromosome 6, and possession of the gene or, more strictly, allele, confers about 4 IQ points to an individual's intelligence.

5. HERITABILITY OF PERSONALITY

By the closing decades of the twentieth century, it had become clear that personality traits have a significant heritability. All five of the big personality traits have shown evidence of heritability (Costa & McCrae, 1992b, p. 658), and most traits yield heritabilities in the range of 40 to 50 percent (Zuckerman, M., 1992, p. 675).

Heritabilities of the Big Five personality traits have been estimated from the similarity between adopted children and their biological and adoptive parents and between identical and non-identical twins reared in the same and in different families and from the similarities between full siblings, half siblings, and biologically unrelated children reared in the same family. The results of nine of these studies have been integrated by Loehlin (1993), and his calculations have shown that the heritabilities fall between .39 and .49. A subsequent study by Jang, Livesley and Vernon (1996) used 123 pairs of identical and 127 pairs of nonidentical twins to calculate heritabilities for the Big Five personality traits. Their results are closely similar to those of Loehlin. Both sets of estimates are shown in Table 11.2.

6. HERITABILITY OF PSYCHOPATHIC PERSONALITY

We noted in Chapter 9 that psychopathic personalities are located at the low extremes of the personality traits of agreeableness and conscientiousness. Hence, because both agreeableness and conscientiousness have significant heritabilities, we should expect that this would also be the case for psychopathic personality. Eight studies of the degree of similarity of identical and same-sex nonidentical twins for psychopathic personality have been summarized by Mason and Frick (1994). In all the studies, identical twins were more concordant than nonidenticals, and the studies as a whole yielded a heritability of .41. Three subsequent studies have confirmed this conclusion. Nigg and Goldsmith (1994) found correlations for psychopathic personality of .50 for identical twins and .22 for nonidenticals, yielding a heritability of .56. Lyons, Tone, Ersen, and Goldberg (1995) reported an unusually large sample of more than 3,000 adult male twins in which correlations for psychopathic personality were .47 for identical twins and .27 for nonidenticals, giving a heritability of .40. In the third recent study, Silberg et al. (1996) examined the incidence of conduct disorder in adolescents, a precursor of psychopathic personality among adults. Data were obtained from 389 male twins aged 11 to 16, assessed by parents and teachers for conduct disorder and attention deficit hyperactivity disorder (ADHD). The assessments were combined to provide an index of a syndrome of "hyperactivity-conduct disorder," the adolescent precursor of psychopathic personality and present in 14 percent of the sample. The genetic analysis concluded that 54 percent of the variance

Table 11.2
Heritabilities of the Big Five Personality Traits Estimated by Loehlin (1993) and Jang, Livesley, and Vernon (1996)

Trait	Loehlin	Jang et al.
Neuroticism	.41	.41
Extroversion	.49	.53
Openness to experience	.45	.61
Agreeableness	.39	.41
Conscientiousness	.40	.44

was attributable to the action of additive genes and 34 percent to the action of dominants, producing a combined broad heritability of 88 percent. The remaining 12 percent was attributed to unique environmental effects—those effects unique to each twin, such as one twin suffering a birth injury damaging the behavioral control mechanisms or one twin being drawn into bad company. There was no effect of shared environment, such as the way in which parents socialize and discipline their children. This is consistent with other studies suggesting that style of child rearing has little or no impact in producing psychopathic children.

Psychopathic personality is frequently expressed in crime, and crime also has a significant heritability. Raine (1993) summarized 13 studies of crime in adult twins, for which the average concordances were 52 percent for identical twins and 21 percent for nonidenticals. This conclusion is corroborated by studies of adopted children, who resemble their biological parents for criminal activities more closely than they resemble their adoptive parents (Mednick, Gabrelli, & Hutchings, 1984; Brennan, P., Mednick, & Jacobsen, 1996). Surveys of the literature by Eysenck and Gudjonsson (1989), Gottesman and Goldsmith (1995), Lykken (1995), and Raine (1993) all conclude that the heritability of crime is about 50 percent or a little higher.

In the 1990s the genes responsible for personality traits with significant heritabilities began to be discovered. In the case of psychopathic personality, in which high aggressiveness is a major component, the first gene was located in 1993 by H. G. Brunner and his colleagues at the Nijmegan University Hospital in the Netherlands. They investigated a large Dutch family pedigree spanning four generations, in which there were 14 males with borderline mental retardation and impulsive-aggressive-psychopathic behavior, including the criminal offenses of arson, attempted rape, exhibitionism, and physical assault. The pedigree analysis showed that the syndrome appeared in different branches of the family in some males but not in their brothers. This tends to rule out shared family and other environmental effects on the abnormal behavior, which should affect both of a pair of brothers reared in the same family and environment. The appearance of the syndrome only in males suggests that the gene is transmitted through normal female carriers and is located on the X chromosome. Biochemical analysis showed that affected males had a defect in their neurotransmitter metabolism for the breakdown of 5 HT (5-hydroxytryptamine), causing abnormally high 5HT. Further analysis showed the presence of a defective MAOA (monoamine oxidase A) gene, causing elevated 5 HT levels (Brunner, Nelen, Breakfield, Ropers, & van Oost, 1993; Brunner, Nelen, & van Zandvoort, 1993).

This research indicating that this gene defect produces psychopathic behavior does not mean that all, or even most, psychopathic behavior is caused by a defective allele of this particular gene. It is probable that a number of defective genes are responsible for psychopathic personality or for a predisposition toward the development of psychopathic personality. In fact, Goldman,

Lappalainen, and Ozaki (1996), in a review of this question, list 23 "candidate genes" for which there is some evidence that they may be involved in the development of psychopathic personality.

7. QUANTIFICATION OF EFFECTIVENESS OF SELECTION PROGRAMS

The effectiveness of programs of selective breeding can be quantified. The effectiveness depends on the stringency of the selection of the individuals chosen to be the parents and on the heritability of the characteristic to be bred for. With regard to the stringency of selection, if a small, exceptionally well endowed percentage is selected as the parents of the next generation, as is generally the case in the breeding of domestic animals and racehorses, considerable improvements are obtained in the offspring. If the parents of the next generation are less stringently selected, the breeding program is less effective.

The second factor determining the effectiveness of a selective breeding program is the narrow heritability of the trait being bred for, which, as has been noted, is the heritable component ascribable to the action of additive genes. We have noted in the preceding sections of this chapter that most of the heritability of intelligence and personality is due to the action of additive genes.

The effectiveness of selective breeding, taking into the account the stringency of the selection of the breeding pairs and the narrow heritability of the trait being bred for, can be calculated using the formula for computing the effects of selection given by Cavalli-Sforza and Bodmer (1971):

$$x1 = x - mh^2 + m$$

where $x1$ is the mean of the first generation of children of selected parents, x is the mean of the selected parents, h^2 is the narrow heritability of the trait or characteristic, and m is the mean of the population. This formula can be used to calculate the effectiveness of selective breeding programs for enhanced intelligence with different degrees of stringency of selection. Suppose, for instance, that all the mentally retarded were sterilized. Everyone else was permitted to have children. The mentally retarded with IQs below 70 comprise about 2 percent of the population. The remaining 98 percent of the population would constitute the breeding group for the next generation, and these would have a mean IQ of approximately 101. If we assume a narrow heritability for intelligence of .71, as estimated by Jinks and Fulker (1970), the formula allows us to calculate that the mean IQ of the children of the selected parents will be 100.71. Such a sterilization program would therefore produce a .71 IQ point gain in the intelligence level of the child generation.

It will be noted that the children do not have such a high mean IQ as the selected parents, whose mean IQ is 101.0. The children show some regression to the mean, which we shall consider in the next section. Nevertheless, there is some gain. The improvement of .71 IQ point provided by such a program may not seem strikingly impressive. Nevertheless, even an increase of this apparently small size would have some significant impact at the extremes of the distribution. As the intelligence distribution shifted upward, the proportion of the population with IQs above 130 would increase from 2.28 percent to 2.56 percent. At the same time, the proportion of the population with IQs below 70 would decrease from 2.28 percent to 2.02 percent. Both of these would be modest but worthwhile gains. Furthermore, if this program were to be implemented over several generations, it would increase the average intelligence level in each successive generation of children.

A program of this kind would not be at all stringent in its selection of the breeding population. It would not adopt the stringent methods of stockbreeders who breed for improved strains of livestock by selecting small numbers of the best males for breeding purposes and typically mate them with a larger number of selected females. We can quantify the effect of using this procedure for humans by applying the Cavalli-Sforza–Bodmer formula. With regard to intelligence, suppose that the top 25 percent of both sexes were selected as the breeding population. The lower threshold for inclusion in the group would be an IQ of approximately 110, and the average IQ of the group would be approximately 118. We can calculate from the formula that the children of this group would have an IQ of 112.8. This would be a substantial increase, and further increases could be obtained by continuing the program over several generations. Broadly similar gains could be achieved for conscientiousness and agreeableness, although, as the narrow heritabilities of these are lower than that of intelligence, a selective breeding program would not be quite so effective. Nevertheless, carried out over several generations, conscientiousness and agreeableness could certainly be improved substantially by such a program of positive eugenics.

8. THE PROBLEMS OF REGRESSION TO THE MEAN

It has often been asserted that eugenic policies would not work because of regression to the mean. The phenomenon of *regression to the mean* is the tendency of parents at the extremes of continuously distributed traits, such as intelligence, to have children who are less extreme than themselves. In other words, the children regress toward the mean. Thus highly intelligent parents have children who are, on average, less intelligent than themselves; and mentally retarded parents have children who are on average less retarded than themselves.

As an empirical phenomenon, the existence of regression to the mean was first shown by Francis Galton (1869) in his *Hereditary Genius*. Galton examined the descendants of one hundred highly eminent men and concluded that a comparable degree of distinction was achieved by only 36 percent of their sons. There is, however, a problem with this argument because it takes no account of the quality of the wives of the highly eminent men. Children inherit half their genes from their fathers and half from their mothers. Eminent men normally marry women with less outstanding qualities than themselves because women with the same qualities are so rare. The result of this will be that their sons inherit fewer of the genes responsible for eminence than were present in their fathers. In other words, the wives of highly eminent men tend to dilute the genes for distinction in the children.

Thus to provide a convincing demonstration of regression to the mean and to assess its magnitude, we have to have data for both parents and their children. Three studies providing these data for intelligence are Terman's (1925) study of highly intelligent parents and their children, Scarr and Weinberg's (1978) study of moderately intelligent parents and their children, and Reed and Reed's (1965) study of retarded parents and their children. The results of these are summarized in Table 11.3.

The Terman study began with 1,528 highly intelligent California children with IQs of 140 and above. They were followed up at intervals, and when they were in their fifties, Oden (1968) collected data for the IQs of their spouses and the average IQs of their children. The average IQ of the spouses of this group was 125, and the average of the couples was 138.5. These couples produced 1,571 children, the average IQ of whom was 133.2 (Oden, 1968). Thus, the average IQ of the children was 5.3 points below that of their parents, showing a small regression to the mean.

The second study illustrating regression to the mean is that of Scarr and Weinberg (1978). Their sample consisted of 71 fathers and mothers with IQs of 120.2 and 117.7, averaging 118.9. These parents had a total of 143 children whose average IQ was 116.7. The regression effect is present but very small, amounting to only 2.2 IQ points.

The leading study showing regression to the mean in the children of mentally retarded parents is that of Reed and Reed (1965). They reported on

Table 11.3
Three Studies of Regression to the Mean from Parental to Children's IQs

Number Parents	IQ Parent 1	IQ Parent 2	IQ Parent Average	Number Children	IQ Children Average	IQ Regression	Reference
1,528	152.0	125.0	138.5	1,571	133.2	5.2	Terman, 1925; Oden, 1968
71	120.2	117.7	118.9	143	116.7	2.2	Scarr & Weinberg, 1978
53	—	—	74.0	177	82.0	8.0	Reed & Reed, 1965

53 couples with an average IQ of 74, and on 177 of their children, who had an IQ of 82, showing an upward regression of 8 IQ points.

A number of critics of eugenics have contended that regression to the mean would make eugenic programs ineffective. The argument is that regression to the mean over a number of generations has the effect that the descendants of the very intelligent have progressively lower intelligence until they end up at the mean of the population. Conversely, the descendants of the mentally retarded become progressively more intelligent in successive generations, and they also end up at the mean of the population. Thus, there is no point in attempting to increase the numbers of children of the very intelligent or to reduce the numbers of children of the mentally retarded, because after a few generations regression to the mean ensures that the descendants of these two groups are indistinguishable.

This thesis has been proposed by Scarr (1984). She put forward a model of intelligence in which there are five classes, numbered from 0 (lowest IQ) to 4 (highest IQ), and asserted that "by the eighth generation, the descendants of classes 0 and 4 are distributed about equally in all five classes" (p. 26). The same argument was advanced by Eysenck in a criticism of Herrnstein's thesis of the development of a genetic elite in the United States. In his book *IQ in the Meritocracy*, Herrnstein (1971) argued that the social classes in the United States were becoming increasingly genetically stratified into castes, consisting of a genetic elite of the highly intelligent, a genetic underclass of the unintelligent, and other intermediate groups. Eysenck (1973) argued that this could not happen because of regression to the mean: "This is precisely what cannot happen upon genetic considerations; regression makes it quite impossible that castes should be created which will breed true—that is where the children will have the same IQs as their parents. Within a few generations, the difference in IQ between the children of very bright and very dull parents will have been completely wiped out" (p. 219). The most recent variant of this thesis, that regression to the mean makes it impossible to breed selectively for enhanced intelligence, comes from Preston and Campbell (1993). They present a model in which there is a negative association between intelligence and fertility. The effect of this is that the children of this population have reduced intelligence. However, they argue that in subsequent generations, the intelligence of the population gradually recovers through regression to the mean until it reverts to the mean of the original population.

9. UNDERSTANDING REGRESSION TO THE MEAN

These assertions that regression to the mean would render a eugenic program ineffective or, in the case of the variant proposed by Preston and Campbell (1993), that it would correct the impact of dysgenic fertility, are obviously wrong because if they were true the selective breeding of plants and animals, of which a number of examples have been given in the first two sections of this chapter, would not work. If these assertions were correct, improvements brought about by selective breeding would disappear after several generations through regression to the mean. Furthermore, natural selection would not work either because evolution by the survival of the fittest could not take place through the regression of the descendants of the fittest back to the population mean.

The essential error in the arguments of Scarr, Eysenck, Preston and Campbell, and others who have argued that regression to the mean would negate eugenic programs lies in the assumption that regression to the mean of the descendants of extreme groups continues for a number of generations until they reach the mean of the population. Contrary to this assumption, regression only occurs in the first generation of children. In a selective breeding program, the children of the selected parents establish a new mean, which is maintained in subsequent generations. The genetics of this is explained by D. S. Falconer (1960) in his *Introduction to Quantitative Genetics*, where he states that in selective breeding programs, "provided there is no other reason for the gene frequency to change, the population mean will be the same in the generations following as in the F2," that is, the first generation (p. 259). This has been explained more recently by Loehlin (1998), who points out that this is the error in the thesis of Preston and Campbell (1993), that dysgenic fertility has no long-term effect because although the child generation has a lower IQ than the parental generation, subsequent generations gradually regress back to the original mean.

Selective breeding, therefore, operates like a ratchet with minor slippage. With each generation of selection, the trait is improved; but there is some slippage backward. Continued selection over successive generations results in improvement in the trait. This is what is actually obtained in the selective breeding experiments for intelligent and emotional rats described in the first section of this chapter, and it is what would happen with a eugenic program designed for humans.

10. CONCLUSIONS

In the last decades of the twentieth century, a number of the critics of eugenics dismissed eugenics as a pseudoscience, that is to say, a false science that would not work. In the previous and the present chapters, we have seen that this assertion is incorrect. On the contrary, it has been known and shown empirically for thousands of years that the selective breeding of animals and plants to produce improved strains does work. Selection for improved strains of plants and livestock is not a pseudoscience but a genuine science, the genetic basis of which is understood and the effectiveness of which can be estimated for various levels of stringency of selection for reproduction and for different heritabilities of the characteristics being selected for. There is no doubt whatsoever that these methods would be effective for human populations.

For a program of selective breeding to work, the characteristics bred for have to have some heritability. There is overwhelming evidence that this is the case for numerous multifactorial genetic disorders, for intelligence, for the personality traits of agreeableness and conscientiousness, and for psychopathic personality. These are all amenable to eugenic improvement by selecting against those with undesirable characteristics (i.e., reducing their fertility) and selecting for those with desirable characteristics. We begin our consideration of how this might be done in practice in the next chapter.

12

∽

Negative Eugenics: Provision of Information and Services

1. Reduction of Unplanned Pregnancies and Births
2. Adverse Effects of Teenage Motherhood
3. Magnitude of the Problem of Unplanned Births
4. "Just Say No" Campaigns
5. Sex Education in Schools
6. School-Based Clinics
7. Promoting the Use of Contraception
8. Emergency Contraception
9. Unreliability of Contraception
10. Need to Overcome Impediments to Research for Better Contraceptives
11. Sterilization
12. Eugenic Impact of Abortion
13. The Facilitation of Abortion
14. Conclusions

∽∽∽

Classical negative eugenics consists of measures designed to reduce the fertility of people with genetic disorders, low intelligence, and psychopathic personality. Programs to achieve this objective are of two general kinds. The first consists of the provision of information and services on contraception, abortion, and the like to these people to enable them to control their fertility more effectively. These are discussed in the present chapter. The second kind of classical negative eugenics consists of the provision of incentives and the application of coercion or compulsion and is taken up in Chapter 13.

1. REDUCTION OF UNPLANNED PREGNANCIES AND BIRTHS

The first objective of a program of negative eugenics should be the reduction of unplanned pregnancies and births. There is substantial evidence that these occur disproportionately among those with low intelligence and psychopathic personality. Although there is dysgenic fertility for intelligence in the United States, it has been shown by Vining (1995) in an analysis of a large sample of 34- to 45-year-old women that the relationship between intelligence and the ideal number of desired children among both blacks and whites is essentially zero (correlations of - .09 and - .03, respectively). Thus if all women had their ideal number of children, dysgenic fertility would cease.

Therefore, one of the objectives of eugenics should be to enable all women to have their ideal number of children, and a component of this objective is to help the less intelligent and more psychopathic to limit their number of children to the ideal by helping them to avoid unplanned pregnancies and births. Many of these unplanned births occur in single women without stable partners. This is particularly true for single teenage mothers, few of whom plan to have children. Single mothers without stable partners come predominantly from the unemployed and the least educated, who are also the least intelligent and most psychopathic (Leibowitz, Eisen, & Chow, 1986; Ermisch, 1991). In a summary of the research literature, Nock (1998) wrote that "unmarried motherhood is associated with poverty, low income, low educational attainment and increased welfare receipt" (p. 250). A further study by Moore, Manlove, Glei, and Morrison (1998) analysed 7,459 18-year-old females drawn from the American National Educational Longitudinal Study. Of these, 471 were single mothers, and these had poor school grades and scores on educational tests.

Numerous studies have found that single mothers tend to have psychopathic personality or tendencies. In a recent review of the literature, Moore et al. (1998) noted that single motherhood is associated with "varied problem behaviors, ranging from school behavior problems to early substance abuse, delinquency, and violence" (p. 434). A study by Woodward and Fergusson (1999) has shown that girls with conduct disorders and antisocial behavior in childhood, the precursors of psychopathic personality in adulthood, had a significant tendency to become teenage mothers. The reason why single mothers without stable partners tend to be more common among the less intelligent and the more psychopathic is that they do not use contraception consistently and efficiently. This is illustrated by a British study of a nationally representative cohort on whether contraception was used during the past year in relation to educational level, which can be regarded as a proxy for intelligence and psychopathic personality (Johnson, Wadsworth, Wellings, & Field, 1994). The results showed that college and high school graduates used contraception most, while those with no educational qualifications used contraception least. The results obtained in this study are shown in Table 12.1.

It is easy to understand why single mothers tend to have low intelligence and psychopathic tendencies. Those with low intelligence are less likely to use contraception because they do not fully understand how to use it or how to obtain it; and if they incur an unplanned pregnancy, they are less likely to have it terminated. All these pose cognitive problems that those with low intelligence are less able to solve. Those with psychopathic tendencies are more likely to take the risk of unprotected sexual intercourse and, if they become pregnant, are less likely to have the pregnancy terminated.

2. ADVERSE EFFECTS OF TEENAGE MOTHERHOOD

In addition to its dysgenic effect, teenage motherhood has adverse consequences for both the mothers and their children. Most teenagers mothers in the last decade of the twentieth century in the United States and in Europe kept their babies rather than, as in previous decades, giving them up for adoption (Morgan, 1999). In a review of research on the effects of teenage motherhood, Card, Petersen, and Greeno (1992) concluded that "this research has documented a long list of negative consequences: truncated education, lower paying jobs, greater unemployment, greater likelihood of poverty, larger families and closer spacing between children, greater likelihood of mental disruption or out of wedlock childbearing, children who are slow to develop and who do more poorly in school when they begin their education" (p. 3). Ruch-Ross, Jones, and Musick (1992) and Morris, Warren, and Aral (1993) documented the research evidence showing that teenage mothers have a high probability of providing inadequate parental care and of having low-birth-weight infants with physical and mental impairments; that they are typically high school dropouts and welfare recipients; and that children born to teenagers have an above-average probability of having low IQs, and being school failures, substance abusers, and delinquents.

For all these reasons the eugenic objective of reducing the numbers of unplanned pregnancies and births among single teenagers and single women

Table 12.1
Nonuse of Contraception (percentages) in Relation to Educational Level in Britain

	Educational Level				
	College	A Level	O Level	Other	None
Males	12.9	12.3	14.8	21.9	34.4
Females	13.2	16.2	15.9	22.3	34.0

commands widespread assent. So strong is the general consensus that single teenage motherhood is undesirable that the governments of several Western nations, including those of the United States and Britain, have made it official government policy to attempt to reduce their numbers.

3. MAGNITUDE OF THE PROBLEM OF UNPLANNED BIRTHS

There are large numbers of unplanned births to teenagers and to single mothers in many Western nations. Statistics on these for 11 major nations were collected by Clearie, Hollingsworth, Jamison, and Vincent (1985) and are shown in Table 12.2. The table gives figures for the percentages of single teenagers (those below the age of 20) giving birth and also for the percentage of abortions calculated as a percentage of births. The figures for blacks and whites in the United States are taken from Westoff, Calot, and Foster (1983). It will be noted that there is considerable variation among countries and that the United States, England, and Canada have larger numbers of teenage births than do Continental Europe and Japan. The highest birth rate to teenagers is in the United States, in which 28 percent of teenagers have babies. This high figure is to some degree inflated by the large numbers of blacks (51 percent) becoming teenage mothers; but even among whites the percentage of teenage mothers at 22 percent is substantially higher than in the European countries and Japan.

The figures for the abortion rate as a percentage of the birth rate enable us to make some inferences about why the rates of single motherhood differ

Table 12.2
Births to Single Teenage Women as Percentages of Population and Abortions as Percentages of Births to Single Teenagers (1981)

Country	Births	Abortions
Canada	14	60
Denmark	7	148
England	15	58
Finland	9	106
France	9	58
Japan	3	143
Netherlands	4	49
Norway	11	97
Sweden	7	39
United States	28	145
Whites	22	—
Blacks	51	—
West Germany	10	32

among countries. For instance, Japan has a very low birth rate and a very high abortion rate, from which we can infer that the birth rate is kept down by the abortion rate. The Netherlands has a very low birth rate and a very low abortion rate, from which we can infer either that Dutch adolescents are not sexually active or, more probably, that they are efficient users of contraception. The United States has high birth rates and high abortion rates from which it can be inferred that U.S. adolescents are particularly inefficient users of contraception.

It is not only among teenagers that substantial numbers of pregnancies and births are unplanned and unintended. This is also true of all women. For instance, a survey carried out in the United States in 1982 by Westoff (1988) found that slightly over half of all conceptions were unintended and that an estimated 47 percent of these were terminated by abortion. From this it can be inferred that approximately 25 percent of births are unintended. There is a broadly similar pattern in Britain. Surveys carried out over the years 1972–1989 by the Royal College of Obstetricians and Gynaecologists (1991) found that approximately 30 percent of births were unplanned and unwanted.

The principal reasons for these large numbers of unintended births is that contraception is not used at all or is used inefficiently and that when unplanned pregnancies occur they are not terminated by abortion. A survey carried out in the United States in 1982 found that 21 percent of U.S. women at risk of unintended pregnancy were not using contraception. The percentage among teenagers was 56 percent, falling in successive age groups to 12 percent (Westoff, 1988). Similarly, in Britain an analysis of the General Household Survey of 1991 found that 20 percent of sexually active women not wishing to become pregnant were not using contraception (Goddard, 1993).

The general conclusion to be drawn from all these surveys is that in the economically developed nations, and particularly in the United States, large numbers of births are unintended, especially among the less intelligent and the psychopathic, because women do not use contraception efficiently or do not have their unplanned pregnancies terminated. From this it can be inferred that if contraception and abortion were used more efficiently, there would be fewer unplanned pregnancies and births and a reduction of dysgenic fertility. The least controversial way to increase the efficient use of contraception is by the provision of information and family planning services.

4. "JUST SAY NO" CAMPAIGNS

There are three broad strategies for attempting to reduce the number of unplanned births. The first is to try to persuade teenage girls not to have sexual intercourse; the second is to try to induce women who do not wish to become pregnant to use contraception more efficiently; and the third is to try to induce women who become unintentionally pregnant to have their pregnancies terminated. We consider these three strategies in this and in the next two sections.

So far as the promotion of sexual abstinence among teenage girls is concerned, attempts to promote this were launched in the United States in the 1980s and 1990s in the form of "Just Say No" campaigns. The president of the Family Research Council, Gary Bauer (1994), has been a leading advocate of these campaigns. He contends that there should be school instruction "in basic things like discipline, self-control, delay of gratification, self-respect, and how to handle relationships" (p. 59).

A program to promote the Just Say No philosophy was introduced in the later 1980s in Atlanta, Georgia, schools by Marion Howard of Emory University in Atlanta. She gave instruction in "decision-making skills" to 14-year-old girls in several schools. She found that over the next year 5 percent of girls who had taken her course were sexually active as compared with 15 percent of those who did not take the course; so it seems that the instruction had some effect (Dryfoos, 1990). A further study finding a positive impact of Just Say No education has been reported by Zabin (1992) on 106 teenage girls at a school in Baltimore, Maryland. In this study it was found that the program delayed the onset of sexual activity by around six months. By the age of 15.5 years, 35 percent of girls exposed to the Just Say No philosophy were sexually active, as compared with 50 percent not exposed. However, 50 percent of the exposed group had become sexually active by the age of 16, six months after the control group. It may be considered that delaying the onset of sexual activity by six months is only a small gain, but on the other hand any gain is worthwhile.

Evidently impressed by these studies, President Bill Clinton in 1997 authorized a $50 million federal government "Abstinence until Marriage" program that targeted particularly teenage girls and young women. Not everyone, however, has been persuaded that campaigns of this kind are likely to be effective. Jane Fonda, the movie actress, has opposed President Clinton's initiative on the grounds that "abstinence until marriage is based on an unreal world that isn't out there" (Rees-Mogg, 1997, p. 20). Despite Ms. Fonda's reservations, it does not seem improbable in the light of the research evidence that Just Say No teaching may have some effect in reducing teenage sex and pregnancy. There is everything to be said for governments sponsoring programs of this kind. Nevertheless, as Ms. Fonda has observed, instruction in Just Say No skills cannot realistically be expected to provide a complete solution to the problem of teenage pregnancies and births to single women. Programs of this kind need to be supplemented with the provision of information and services on the use of contraception and abortion.

5. SEX EDUCATION IN SCHOOLS

It is widely believed that unplanned pregnancies and births can be reduced by the provision of sex education in schools, which includes instruction in the use of contraception. There is no doubt that many sexually active female adolescents and adult women who do not wish to become pregnant nevertheless use contraception inconsistently and are therefore at risk of becoming pregnant. For instance, in the United States, studies carried out in the 1980s found that only about one-third of sexually active single adolescents reported regular and consistent use of contraception (Trussell, 1988).

An account of the provision and the effectiveness of sex education in schools in the United States and Europe has been provided by E. F. Jones et al. (1986). They conclude that countries that have provided the most effective education on contraception are the Netherlands, Denmark, and Sweden, which have achieved a teenage birth rate about one-third of that of white girls in the United States, as shown earlier in Table 12.2. In the Netherlands, sex education in schools includes information on contraception and abortion and is provided universally under the provisions of the national curriculum laid down by the government. In the United States, decisions about what topics to cover in sex education classes are normally left to local school districts, and many schools do not provide instruction on contraception. In the 1980s it was estimated that only about 55 percent of U.S. teenagers had received lessons on contraception in school.

There seems a reasonable case that sex education in schools, which includes instruction in contraception and the adverse effects of teenage motherhood on getting a good education and rewarding employment, is likely to reduce teenage pregnancies and births. Nevertheless, it has to be noted that the research evidence for this is conflicting, and several of those who have examined the studies on this issue have concluded that sex education in schools is ineffective (Kirby, 1984; Dawson, 1986; Stout and Rivara, 1989; Dryfoos, 1990). Dawson examined the data of the American National Survey of Family Growth and concluded that adolescents who had taken a sex education course in school were more likely to use contraception during their first sexual intercourse; but most of them did not continue to use contraception consistently thereafter. Sipe, Grossman, and Milliner (1988) reported a study of the effects of summer schools for adolescents that provided educational and vocational training, personal counseling, and instruction on contraception in five urban communities. The program had no effect on increasing the use of contraception. Dryfoos (1990, p. 192) concluded an extensive review of the literature on this issue with the statement "Sex education courses per se have never been proven to have had any direct effect on pregnancy rates"; and the same conclusion is reached by Stout and Rivara (1989). It has even been found by Marsiglio and Mott (1986) that teenage mothers were more likely to have had sex education in school than comparable teenagers who were not mothers.

All these studies and conclusions call into question whether instruction in contraception as part of sex education in schools is likely to have much impact on reducing pregnancies and births among teenagers and single women. Furthermore, a problem with instruction in the use of contraception in schools is that it involves an implicit assumption that teenagers are going to have sexual intercourse and appears to accept this as natural and inevitable and to condone it. Many parents in the Moral Majority group in the United States, the Responsible Society in Britain, and similar organizations in Continental Europe take the view that instruction in contraception is likely to encourage teenage girls to engage in sexual intercourse; and for this reason they oppose it. Nevertheless, the low pregnancy rates in the Netherlands, Denmark, and Sweden, which are apparently attributable to a considerable extent to sex education in schools, suggest that it would be desirable to adopt this instruction more widely in the United States, Britain, and other high-teenage-pregnancy-rate countries as a component of sex education in which strong emphasis is placed on the undesirability of teenage pregnancies.

6. SCHOOL-BASED CLINICS

From about 1970 it became evident that giving teenagers information and advice on contraception was not sufficient to ensure that all teenagers took this advice, obtained contraception, and used it efficiently on all occasions. Part of the problem was that getting contraceptives involves foresight and effort. An additional problem was that young girls wishing to take the contraceptive pill or to have an intrauterine device fitted generally had to consult their physicians. This caused embarrassment and the apprehension that physicians would inform the girls' parents, and many young girls did not want their parents to know they had begun a sexual relationship.

As these problems became apparent, a number of concerned adults in the United States decided that a solution might be to set up school-based clinics that would provide advice on contraception. Typically these operated in trailers parked on the school campus. They were staffed by nurses and physicians who frequently were available for any medical consultation, for instance, for diagnosis and treatment of sexually transmitted diseases, as well as for advice on contraception and for issuing prescriptions for contraceptives and sometimes, although much less frequently, actually giving out contraceptives. These school-based clinics began to be established in several U.S. cities in the 1970s. By 1991 around 200 of them were operating in high schools across 32 states and in most major cities.

Some of the studies of the effectiveness of school-based clinics have been encouraging. Dryfoos (1990) reviewed the research literature and concluded that they have some impact on increasing the use of contraception, although whether they have any significant effect on reducing the incidence of pregnancy has not been firmly established. Zabin and Hayward (1993) also reviewed the research literature and concluded that the clinics have some impact in reducing teenage pregnancies and births. They describe a positive study in Baltimore, Maryland. Clinics were established in the early 1980s in some high schools but not in others, making it possible to compare the effectiveness of the clinics. The results were that a year after the clinics had been established, 38 percent of girls in the schools with clinics attended them for contraception advice before their first sexual intercourse, as compared with 18 percent in the schools without clinics, who sought advice elsewhere. Nine months after first sexual intercourse, 78 percent of the clinic school girls had attended the clinics, whereas 48 percent of the girls at nonclinic schools had sought contraceptive advice elsewhere. After the program had been running for 28 months, 24 percent of the 15- to 18-year-olds had become pregnant in the schools with clinics, as compared with 50 percent in the nonclinic schools. This is clearly a considerable gain, even though a 24 percent pregnancy rate is far from satisfactory and shows that school clinics do not by any means provide a total solution to the problem of teenage pregnancy.

Not all students of this issue have found such positive results for school-based clinics. One of the first U.S. cities to establish these clinics was St. Paul, Minnesota, where the clinics were set up in several high schools in 1973. Their use and effectiveness in preventing pregnancy over the next 20 years were investigated by Kirby et al. (1993). They found that about 35 percent of girls sought advice from the clinics. Initially it was claimed that birth rates were approximately halved within two to three years after the clinics had opened. However, an assessment of long-term trends of the birth rates of adolescent girls in the St. Paul high schools carried out by Kirby et al. has shown that the clinics had no discernible effect. The St. Paul clinics had two weaknesses: (1) they did not dispense contraceptives but referred girls to the hospital clinic to obtain these, and (2) they did not provide abortion services or refer students for abortions, although they did provide pregnancy tests.

A further analysis of the effectiveness of school-based clinics in the second half of the 1980s in six U.S. cities was made by Kirby and Waszak (1992). The methodology of the study was to compare the prevalence of various forms of sexual activity and pregnancy among students in the schools that had school-based clinics with those in similar schools without the clinics. The authors concluded that the school-based clinics had no effect on contraceptive usage in four of the schools. In Dallas, Texas, the girls in the clinic school actually used contraception less than in the comparison school. Only in Muskegon, Michigan, did the girls in the clinic school increase their use of contraception. Probably this was just a chance result; and taking the six cities together, the study showed that the clinics had no impact on increasing contraceptive usage. Furthermore, the clinics did not have the expected effect of reducing the number of pregnancies, because they had no effect, except for one school, on increasing contraception usage. The overall numbers of girls who became pregnant were 30 percent in the clinic schools and 24 percent in the nonclinic schools. Even in Muskegon, the one city where the presence of the clinic appeared to increase contraceptive usage, the pregnancy rate was 24 percent in the clinic school and 20 percent in the comparison school. These can only be regarded as very discouraging results. Even taken in conjunction with the more encouraging results obtained in the Baltimore study, it is difficult to be other than pessimistic about the effectiveness of the school-based clinics in reducing teenage pregnancies.

It should be noted that the six cities in the study were San Francisco, Dallas, Gary (Indiana), Muskegon, Jackson (Mississippi), and Quincy (Florida). The school enrollments ranged from 76 percent to 98 percent black, except in San Francisco where they were 30 percent black, 20 percent Hispanic, and 37 percent Filipino. There were negligible numbers of whites in any of the schools. It may be that, for a variety of reasons, minority students are more difficult to help by school-based clinics. Whether or not this is the case, we are forced to conclude that the research indicates that the effectiveness of school-based clinics is quite small, and perhaps even nonexistent. There are two principal reasons for this. The first is that attendance at the clinics is voluntary and most pregnancies and births take place among girls who have not attended the clinics. The second is that most of the clinics do not dispense contraceptives and do not solve the problem of the effort and the embarrassment occasioned by visits to physicians or pharmacies to obtain the prescriptions and the contraceptives.

School-based clinics raise the same problem as does instruction in contraception given as part of sex education lessons in schools, which is that they appear to condone sexual intercourse among teenagers, and many parents and adults are unhappy about this. Dryfoos (1990) and Zabin and Hayward (1993) have concluded from a review of research that the presence of school-based clinics does not increase sexual activities among teenagers, but not everyone may be reassured by this conclusion.

Despite the negative results of much of the research on the effectiveness of school-based clinics, it would be desirable to persevere with them. The clinics would be more effective if they actually dispensed contraceptives, rather than only giving advice and prescriptions. The concern felt by many that the clinics encourage sexual activity is a problem, but one that could be mitigated by the clinics offering general medical advice, including that on sexually transmitted diseases. We should not expect that school-based clinics are likely to have a major impact on preventing teenage pregnancies and births, but if they had a minor positive effect they would be worthwhile.

7.   PROMOTING THE USE OF CONTRACEPTION

Apart from some young teenagers, virtually everyone knows about contraception. Nevertheless there are still many people who do not use it efficiently, and it can be anticipated that this will continue. The principal impediments to the efficient use of contraception are the cost and the effort and foresight required to get them in advance of their required use. Both of these impediments need to be overcome.

So far as the cost impediment is concerned, the position in the United States is that approximately 15 percent of women use public-funded clinics that provide contraception free or at low cost. The clinics are largely situated in towns and cities, so it is difficult for those who live in rural areas to get to them. The 85 percent or so of women who do not use these subsidized clinics obtain their contraceptive services from their physicians, to whom they have to pay fees. In many cases these fees are covered by medical insurance. Only 56 percent of physicians who provide contraception accept Medicaid reimbursement. The most widely used contraceptive is the pill, which is taken by about 40 percent of U.S. women at an annual cost of about $200 at 2001 prices. This is a significant cost, particularly to poor women and teenagers. The position regarding free or subsidized contraception deteriorated in the United States from 1980 into the early 1990s. Family planning providers have received some federal funding under Title X, but the funds received declined steadily in real terms, so that by 1992 they were about one-third of the 1980 figure (Daley & Gold, 1993). Social services grants for the provision of contraception also declined in the last two decades of the twentieth century to about one-third of that in the 1970s. Medicaid funds increased slightly, but in 1990 Congress enacted legislation to reduce them. The net result of these cutbacks in expenditure has been that total funds for family planning providers by the 2000s were only around two-thirds of those received in 1980, the result of which has been inadequate provision of contraceptive services. Waiting times for appointments increased to six weeks in many clinics, and six weeks is a long time to wait for contraception.

Although the contraceptive pill is the most widely used method of contraception by U.S. women, it is not a wholly effective method because it has to be taken daily and preferably at the same time each day, which women can easily forget to do. The effect of this is that a number of women who get the pill nevertheless become pregnant. For instance, a study of teenage mothers carried out in the early 1990s by Polaneczky, Slap, and Forke (1994) found that 38 percent of those who had been given the contraceptive pill had nevertheless become pregnant within 18 months because they had failed to take it regularly. The most efficient form of contraception is Norplant, the capsule injected under the skin, which provides fully effective contraception for around five years. In the late 1990s, the average cost of Norplant in the United States was about $370 with an additional $150 to $650 for insertion, counseling, and checking. This makes it quite a costly investment for poor women. A number of those who have worked actively on the problem of unplanned pregnancies have advocated that Norplant should be provided at little or no cost to teenage school girls at school-based clinics and to women at family planning clinics (e.g., Moskowitz, Jennings, & Callahan, 1996). This is the ideal solution to overcoming both the cost barrier to the use of contraception and the problem of ineffective use of contraception.

The costs of contraception are generally lower in Europe than in the United States, and this probably contributes to the generally lower rates of unplanned pregnancies and births. In France, Sweden, and the Netherlands, contraception is either provided free or at low-cost subsidized prices. In India, all contraception is provided free in an attempt to reduce the increase in population. Condoms are handed out at no cost by barbers, who also give advice on their use (Thomas, C., 1997). This is a model that could usefully be adopted in the United States and Europe. More generally, what is required in the United States is a considerable expansion of publicly funded family planning clinics providing free or low-cost contraception.

So far as the problem of the effort of obtaining contraception is concerned, a measure that could be usefully adopted in all countries for making contraception easier to obtain and cheaper would be to make the contraceptive pill available without prescription at subsidized cost and purchasable at drugstores, pharmacies, and other retail outlets. This proposal was made a number of years ago in *The Lancet* (The pill, 1974), in which it was argued that although it may be advisable for a small minority of women not to take the pill, these cases are not identified in the routine examinations that are typically given by physicians.

8. EMERGENCY CONTRACEPTION

Despite all the measures that could be taken to increase knowledge of contraception and to provide contraceptives that are free and obtainable with a minimum of effort, it is probable that women will continue to become pregnant as a result of unanticipated sexual intercourse and that these will be disproportionately the less intelligent and the more psychopathic. A useful backup that provides a partial solution to this problem is the availability of emergency contraception, also known as postcoital contraception or, more colloquially, the "morning-after pill." As these terms imply, it aborts newly conceived embryos. One of the most used of these abortifacients is RU 486 (Mifepristone). It is highly effective if taken within three days after intercourse. Studies in England in the early 1990s on a little over a thousand women found that the treatment was 100 percent effective and had no adverse side effects (Webb, Russell, & Elstein, 1992). Other effective abortifacients include large doses of estrogen, progestin, and damazol and insertion of a copper intrauterine device (IUD).

Emergency contraception was developed in the mid-1980s. In the United States, emergency contraception was approved by the Food and Drug Administration (FDA) for prescription by physicians in the fall of 1998. It is also permitted in Britain and most of Continental Europe. Nevertheless, it has not been greatly used. The reasons for this are that not all women know of its existence and that many who have had unprotected intercourse prefer to take the risk that they have not become pregnant rather than incurring the trouble of going to a physician and getting the abortifacient pill. So far as the knowledge of emergency contraception is concerned, in the United States, a study by Schilling (1984) found that even among sexually experienced college students who had had an abortion, 88 percent had never heard of the contraception. In a later review of the lack of knowledge of emergency contraception in the United States, Gold, Schein, and Coupey (1997) found that in the mid-1990s physicians were typically prescribing emergency contraception to only four patients a year. They described the existence of emergency contraception as "the nation's best kept secret." Furthermore, a number of pharmacies in the United States do not dispense the RU 486 abortifacient pill, largely because they anticipate trouble from antiabortion activists if they were to do so.

Knowledge of emergency contraception in Britain is also poor, although it is apparently better than in the United States. A study of London women with an unplanned pregnancy found that 40 percent were unaware of the existence of emergency contraception (Burton, Savage, & Reader, 1990). Surveys carried out in Britain in the 1990s have found that somewhere between 50 percent and 80 percent of women of childbearing age were aware of the existence of emergency contraception. However, a study of 177 women seeking an abortion in England in 1992 and 1993 found that only 13 had used the emergency contraceptive pill on some previous occasion (Gooder, 1996). Nevertheless, there is little doubt that many women who become pregnant accidentally would use emergency contraception if they knew about it and if it were made easily available. A study carried out by Duncan, Harper, Ashwell, and Mant (1990), found that 70 percent of women requesting an abortion said they would have used emergency contraception if they had known of it or how to obtain it, and another study found that 93 percent would have done so (Bromham & Cartmill, 1993).

A major problem with emergency contraception is that it has to be used within 72 hours of intercourse. Women therefore have to act quickly to secure an appointment with a physician, obtain a prescription, and then get the pills from a pharmacy. It is frequently difficult or even impossible to do all of this within the three days. This problem has been overcome in some U.S. states by allowing pharmacists to prescribe and sell drugs under collaborative arrangements with physicians. In 1996, this was permitted in 22 states for drugs providing pain relief and immunization. In 1999 this facility was extended in Washington State to the provision of emergency contraceptive pills. An investigation of its effectiveness has found that in the first four months following the project's launch the hotline set up by the project received 4,934 calls and that 2,765 prescriptions were issued (Wells, Hutchings, & Gardner, 1998). The authors of the study estimate that this facility prevented about 200 pregnancies and, because about half of unintended pregnancies are terminated, about 100 abortions. Several U.S. states also allow pharmacists to prescribe and dispense pills for emergency contraception ahead of immediate use for women who wish to have them as a standby. This is also permitted in Scotland, and research has shown that women who obtain these standby pills are more likely to use them (36 percent) than those who needed a prescription to obtain them (14 percent) (Glaister & Baird, 1998).

At the beginning of the twenty-first century there is much further progress to be made in the greater provision and use of emergency contraception. Women need to be made more aware of its existence through sex education in schools and advice columns in women's magazines and by advertisement. As of January 1, 2001, the pills can be bought over the counter in drugstores and other retail outlets in Britain without the necessity of obtaining a prescription from a physician. This could usefully be in the United States, Continental Europe, and elsewhere.

9. UNRELIABILITY OF CONTRACEPTION

One of the fundamental problems for eugenics is that contraception is not wholly reliable. There is some element of human error in its use, leading to unplanned pregnancies and births, and this is inevitably greater among the less intelligent and the more psychopathic. This is a major cause of dysgenic fertility.

In the last two decades of the twentieth century the contraceptive pill was the form of contraception most widely used by women in North America and Northern Europe (Jones et al. 1986). The contraceptive pill is a fairly effective method of contraception, but by it is no means foolproof. Obtaining the pill requires effort and forward planning. This could be made easier, but no matter how easy it is made to obtain the pill, such as by allowing it to be sold over the counter without a prescription from a physician, there will always remain an element of human error in its use. It is estimated that in the United States the pill has a failure rate of 8.5 percent per annum, as compared with 1.4 percent for the IUD and 0.4 percent for sterilization (Westoff, 1988). It is not difficult to understand the reasons for these differences. The IUD and sterilization do not require women to do anything to maintain their effectiveness, whereas in the case of the pill, some women will neglect to obtain their supplies regularly or to take them consistently. The same problem is present with the condom, which has a failure rate about the same as that of the pill, due to breakage or slippage (Westoff, 1988).

At the present time there are four forms of contraception with significantly lower failure rates than the pill and the condom. These are the IUD, the implant, the postcoital abortifacient or morning-after pill, and sterilization. The IUD is about six times more error free than the pill. It encountered problems in the late 1970s and early 1980s because it can cause pelvic inflammatory disease in a number of women, leading to infertility. This made many women reluctant to use the IUD and to prefer some less reliable form of contraception. Subsequent research showed that the risks of developing pelvic inflammatory disease from the IUD are in fact quite small; it entails an increased risk of 1.8 percent among married and cohabiting women and 2.6 percent among those with several partners (Lee, Rubin, & Borucki, 1988). Nevertheless, some women who contracted pelvic inflammatory disease filed lawsuits against the manufacturers and as a result the IUD was withdrawn from sale. In 1987 the Population Council licensed an improved copper-bearing IUD, the TCU 380A (Paragard), which provided protection for eight years. A second form of IUD is Progestasert, which works by releasing the hormone progesterone and provides protection for approximately one year. Both IUDs are used on a rather limited scale and need to be prescribed more widely for women at low risk of contracting pelvic inflammatory disease.

Contraceptive implants, of which Norplant is the best known, consist of capsules inserted subdermally, generally in the upper arm (Sivin, 1988). They release a synthetic hormone, levonorgestrel, which prevents pregnancy for a period of five to eight years. Trials began in Chile in 1974, and industrial production for routine use began in Finland in 1979. By 1988 implants were approved for general use in 11 European countries and were being used by approximately 200,000 women. Clinical trials have been run on approximately 12,000 women in a further 18 countries, including the United States, where Norplant was approved in 1990. The results show that implants are a highly effective means of contraception with a failure rate of about 1 percent over the first three years. From the fourth year onwards, failure rates rise to about 3 percent a year. Failure rates are higher in heavier women and are almost entirely absent in light women. With further research these problems should be solved, and implants will be virtually 100 percent reliable. Implants do not fail through human error and need to be used on a greater scale. The disadvantage of hormonal implants is that some women experience side effects such as nausea and headaches and decline to use them.

The third highly effective form of contraception is the postcoital pill. This needs to be more widely known and used, but it has the disadvantage that women need to act within three days after unprotected sexual intercourse; and many women, especially the less intelligent and more psychopathic, will not do this. The fourth reliable form of contraception is sterilization, but this has a finality that makes it unattractive to many women who prefer to keep their options open as regards the possibility of future childbearing.

10. NEED TO OVERCOME IMPEDIMENTS TO RESEARCH FOR BETTER CONTRACEPTIVES

Because all existing methods of contraception have disadvantages of various kinds, there is a need to develop new and fully effective forms of contraception with no adverse side effects or other problems. Unfortunately, this has become difficult. The problem for research on the development of new contraceptives lies in the time required and the cost. It takes approximately

10 to 15 years to develop a new contraceptive and costs somewhere between $50 million and $100 million at current prices. This length of time and high cost are caused by the number of steps required to bring a new contraceptive product to market, which consist of the initial synthesis of new compounds, the testing for toxic effects on animals, dosage studies, scale-up for manufacturing, negotiations with regulating agencies, and clinical testing. If the clinical testing is successful, an application for commercial manufacture and marketing is made in the United States to the FDA for clearance to market the product in the United States and to the corresponding agencies in other countries. The time and cost involved in this long sequence of steps have proved a deterrent to the university researchers, pharmaceutical companies, and research institutes that could work on the development of new contraceptives. For university-based researchers, the problem is that they are constrained to work in a fairly short time frame. University researchers have to obtain research grants, do the work, and publish their results within three to four years. They are constrained to work in this way partly to advance their own career development and partly because funding agencies for academic research normally support research only on this relative short time frame. Funding agencies are reluctant to give university researchers $50 million to $100 million to try to develop a contraceptive that might come to fruition in 10 to 15 years. A further reason why this kind of research cannot for the most part be done in universities is that it requires the cooperation of large numbers of individuals to work on the great range of problems involved in the development of a new product. University researchers almost invariably work as individuals or in small groups of two or three. They have to do this to get recognition for their work, publications, and their next research grant. For these reasons university-based researchers have sometimes made the initial discoveries of new drugs and have produced a few vaccines and medical products, but they have made little contribution to the development of new drugs and other medical technology up to the point of production.

Pharmaceutical companies have been the principal sources of research and development of new drugs and other medical products. However, they have not made much contribution to the development of contraceptives. The reason for this is that the costs involved are judged to be too large in relation to the likely profits. The two principal factors deterring pharmaceutical companies from attempting to develop new contraceptive drugs are, first, that a U.S. patent does not give exclusive rights to manufacture for sufficient time to recoup the costs of development and, second, that the costs of litigation in the United States are often extremely high when members of the public can establish that they have suffered damage from use of one of the pharmaceutical company's products. For instance, in 1986 a woman was awarded $4.7 million against the Ortho Pharmaceutical Corporation for birth defects suffered by her baby allegedly as a result of using Ortho-Gynol spermicide. Similar actions against manufacturers of the IUD because of inflammatory pelvic disease led to all but one of the companies withdrawing this product from sale in the United States during the mid-1980s. As a result of these problems, pharmaceutical companies have largely abandoned research and development on new forms of contraception.

This leaves the independent biomedical research institutes as the locus for research on new contraceptives. It was one of these, the Worcester Institute, that undertook the research and development of the contraceptive pill. The initial discovery on which the pill was based was made in 1937 when Makepeace found that progesterone suppresses ovulation in rabbits. In 1951 Gregory Pincus of the Worcester Institute met the wealthy philanthropist Mrs. Stanley McCormick, and together they worked out the program of research and development for the contraceptive pill. Mrs. McCormick donated about $2 million, and additional grants were obtained from the G.D. Searle Company and the Syntex Corporation. It took nine years to bring the pill to market, which took place in 1960, at a total cost of approximately $6 million, equivalent to around $60 million in real terms in the year 2000.

Costs of this magnitude are considerable even for the large U.S. foundations and corporations. Hence in recent years, around 70 percent of the funding for research on contraception in the United States has come from the government. In the world as a whole, about 75 percent of research on contraception is carried out in the United States. It is therefore largely in the United States that the difficulties in the way of research on contraception need to be reduced. The principal changes required are the simplification of the FDA testing requirements, an increase in the patent-protection period, and a reduction in the potential financial losses pharmaceutical companies are likely to incur through litigation. The losses from litigation could be reduced by placing a limit on the amount of damages so that compensation was given for economic loss and for pain and suffering; but punitive damages could not be awarded except in the rare cases where the manufacturer could be shown to have acted recklessly. Alternatively, actions for damages against drugs for which FDA approval has been given could be prohibited, as they are in France for drugs approved by the corresponding government agency. It needs to be accepted that many useful drugs have adverse side effects on a small number of users. The size of the damages awarded in the United States acts as a deterrent to the research and development of all new drugs to the detriment of the public and needs to be curtailed by federal government legislation.

11. STERILIZATION

Many of those who decide they do not want any more children opt for sterilization as a convenient and highly effective form of contraception. In the United States in the 1990s, approximately 31 percent of married women and 17 percent of married men were sterilized (Loose, 1998). The figure for women aged 40–49 is approximately 17 percent (Giami, 1998). In Britain, Germany, France, Sweden, and several other European countries, sterilization was easy to obtain and free in the last two decades of the twentieth century (Meredith & Thomas, 1986).

In the United States sterilization is less easy to obtain because of the bureaucratic procedures that have to be complied with and physicians' fears of litigation. The costs of sterilization can be paid by Medicaid, which stipulates that two consent forms must be completed at least 30 days in advance of the procedure. Applicants are also required to have two counseling sessions. The consent forms and documentation of the counseling sessions normally have to be sent from prenatal care clinics to the hospitals where the sterilization is to be carried out and sometimes get mislaid in the transfer. The result of this is that many women who apply for sterilization do not get the operation done. A National Institute of Health study in 1990 found that of 1,200 pregnant women on Medicaid who applied to be sterilized and who filled out the consent forms, only 59 percent were actually sterilized. The report attributed the large failure rate to "bureaucratic and institutional barriers" (Loose, 1998). There is a need in the United States to make sterilization simple and easier to obtain for women who ask for it.

12.   EUGENIC IMPACT OF ABORTION

By the end of the twentieth century fairly reliable methods of contraception had been available for about 125 years, following the development of the modern condom in the 1870s. Yet many couples do not use contraceptives efficiently, and many women continue to have unplanned pregnancies and births. Many of these women who become pregnant unintentionally resort to abortion as the solution for contraceptive failure. Two studies of women requesting abortions in Britain, one by Griffiths (1990) and the other by Duncan et al. (1990), found that 32 percent and 47 percent, respectively, had not used contraception. Similar results have been found in Australia by Hudson and Hawkins (1995). Among 1,407 women requesting abortions, 41 percent had not used contraception, and the remainder had had contraception failures of various kinds, such as condom breakage or slippage, forgetting to take the pill, and so on.

There is substantial evidence that women who have unplanned pregnancies and request abortions are predominantly less the intelligent and the more psychopathic. For instance, a British study carried out by Ziebland and Scobie (1995) examined the relationship between the numbers of pregnancy terminations per 1,000 women aged 15 to 44 in 90 English regions in relation to a deprivation index obtained from the percentage of the population on welfare. The correlation between the two measures was .61, indicating that the greater the deprivation in a region, the higher the abortion rate. Since deprivation is associated with low intelligence and high psychopathic personality, the impact of abortion is to reduce the birth rate of the less intelligent and more psychopathic. For this reason, abortion needs to be made free and easily available.

The eugenic impact of abortion in the United States has been demonstrated by Steven Levitt, an economist at the University of Chicago, and John Donohue, a lawyer at Stanford University (Levitt & Donohue, 1999). They noted that following the Supreme Court decision in 1973 effectively legalizing abortion throughout the United States, the annual numbers of abortions increased from approximately 750,000 in 1972 to approximately 1.6 million in 1980. They also noted that most of this large increase in the numbers of abortions occurred among the poor, blacks, and the underclass, who produce the greatest numbers of criminals. Hence, they conclude that approximately 1 million potential criminals who would previously have been born were aborted. They estimate that this explains about half of the reduction in crime that occurred between 1991 and 1997. In further support of this thesis, they found that states with the highest abortion rates after 1973 experienced the greatest reduction in crime some 20 years later. Furthermore, five states that allowed abortions before the 1973 Supreme Court ruling permitting abortion experienced an earlier reduction in crime. This study demonstrates the considerable eugenic benefits accruing from the legalization of abortion.

13.   THE FACILITATION OF ABORTION

Abortion was legalized in North America and in most of Europe in the late 1960s and 1970s. In the United States abortion became legal nationwide in 1973 as a result of the Supreme Court's ruling in *Roe v. Wade*. The ease of obtaining abortions varies in different countries. The model for the availability of abortion on demand is Sweden. Since 1975, abortion has been available on request and free of charge in public hospitals up to the eighteenth week of pregnancy. After the eighteenth week, abortion is difficult to obtain but is sometimes possible with the permission of officials of the National Board of Health and Welfare. Most other European countries, including Austria, Britain, Denmark, Germany, Finland, France, Italy, the Netherlands, and Norway, provide abortion free or at low cost and, in practice, on demand.

In Britain abortion is permitted on medical, sociomedical, and sociopsychological grounds, and these are liberally interpreted. An abortion can be obtained when two physicians certify that continuation of the pregnancy would involve greater risk to the mother or her children than if the pregnancy were terminated. In general, abortion is obtained fairly easily and free of charge at National Health Service hospitals. In some areas, however, physicians are unsympathetic to abortion and decline to perform the operation. The position in Canada is similar, with some variations between different provinces. In the United States abortion is less easily obtained than in Europe. The principal impediments are availability and cost. Fourteen states and the District of Columbia fund free abortions for poor women on welfare, but others have to pay appreciable sums, and in the remainder of the states everyone has to pay fees. The federal government has prohibited the use of Medicaid funds for abortion except where a woman's life is endangered. The costs of abortion are prohibitive for many teenagers and some adults. In the 1980s and 1990s a number of states introduced measures to make abortion more difficult to obtain, such as by imposing waiting periods, withdrawing subsidies, and requiring minors to obtain parental consent. As a result of these restrictions, the number of hospitals and clinics providing abortion declined, and fewer abortions were carried out. The abortion rate per 100 pregnancies fell from 29.3 in 1980 to 25.9 in 1992. As a result the percentage of illegitimate births increased in 1992 (Henshaw & Van Vort, 1994).

In a general review of this issue, Ohsfeldt and Gohmann (1994) concluded that these restrictive measures have been successful in their intention of reducing the numbers of abortions. Several studies have concluded that the availability of subsidized or free abortions funded by Medicaid increases the numbers of abortions (Henshaw, Koonin, & Smith, 1991; Meier & McFarlane, 1996). It has been estimated by Blank, George, and London (1996) that the impact of reducing financial assistance for abortion by Medicaid has been to reduce the numbers of abortions among Medicaid-eligible women by 25 percent. More generally, Lichter, McLaughlin, and Ribar (1998) concluded that the reduction in the number of abortion clinics by approximately 20 percent over the decade 1980–1990 was responsible for an increase of approximately 10 percent in the proportion of single mothers.

Another impediment to the use of abortion in the United States has been that when minors request abortions a number of states require that their parents should be informed and give their consent. By 1997, 27 states had made parental certification mandatory (Haas-Wilson, 1997). This has deterred a number of teenagers from having pregnancy terminations and has been responsible for a rise in teenage pregnancies in the 1980s and 1990s (Meier & McFarlane, 1994; Joyce & Kaestner, 1996a). The impact of the parental notification laws is illustrated by the experience in Minnesota, where a parental notification law was introduced in 1981. The abortion rate for 15- to 17-year-olds in the preceding three years (1978–1980) was 18.8 per 1,000 young women. Following the new law, the abortion rate fell to 12.8 per 1,000 young women in 1982 (Rogers, J. H., 1991). The lesson to be drawn from this is that teenagers' abortions should be confidential.

The imposition of impediments of cost, availability, and time constraints on the ease of obtaining abortions inevitably has a dysgenic impact because these are going to be too great for some less intelligent and more psychopathic women to overcome. For some of these women it is easier to let the pregnancy take its course. Simms and Smith (1986) quote some of the typical explanations given by single teenagers in Britain who found they were unintentionally pregnant, investigated the possibility of trying to obtain an abortion, but failed to obtain one: "The doctor said it could be too late to get rid of it"; "Mum was against it"; "My doctor said I couldn't have one" (p. 96). It will come as no surprise that Simms and Smith found that unintended pregnancies carried to term occur most frequently among poorly educated and low socioeconomic status women who lack the intelligence and perseverance to overcome the problems of obtaining an abortion.

While the legalization of abortion that took place in most Western nations during the 1970s was a welcome development both on eugenic grounds and on the grounds of giving women "the right to choose," there are still obstacles in the way of obtaining abortions in a number of countries, particularly the United States. These need to be removed. There is a particular problem in the United States because of the strength of the anti-abortion lobby, which is likely to prove difficult to overcome.

14. CONCLUSIONS

Negative eugenics could usefully be promoted by the more effective provision of information and services to enable the less intelligent and the psychopathic to control their fertility. The most important of these is the provision of sex education in schools, which should include information on contraception, use of emergency contraception, and abortion. This should be supplemented by the provision of school-based clinics, which should dispense contraceptives, carry out pregnancy tests, and give advice on how to obtain abortions.

It should become the norm for adolescent girls to be provided with reliable contraception before they become sexually active. The preferred forms of contraception are the IUD and the implant because these are less subject to human error than the pill and the condom. In addition, emergency contraception should be made more widely known and more easily available. Abortion should be provided free and on demand as a backup for contraception failure.

The Netherlands, the country that has gone furthest toward the achievement of these objectives, has succeeded in reducing the teenage pregnancy rate to approximately one-twelfth of that of the United States. This is an encouraging model for what should be achievable in the United States, Britain, and other countries where unplanned pregnancies and births are more numerous.

In addition, there is a need to develop more reliable forms of contraception. This requires the removal of the disincentive for pharmaceutical companies and research institutes to invent these, arising from the inadequate patent protection and the high punitive damages awarded against companies to those consumers who suffer adverse side effects. There is a need in a number of the economically developed countries, and particularly in the United States, to make sterilization and abortion easier to obtain and at reduced cost.

The measures discussed in this chapter are relatively uncontroversial and would command widespread public support. For this reason they are politically feasible objectives. If all these measures could be implemented and made fully effective, the eugenic impact would be significant although not huge. Dysgenic fertility would be reduced, although it is doubtful whether it would be eliminated completely.

13

Negative Eugenics: Incentives, Coercion, and Compulsion

1. Incentives for Sterilization: The Shockley Plan
2. The Denver Dollar-a-Day Program
3. Incentives for Women on Welfare to Use Contraception
4. Payments for Sterilization in Developing Countries
5. Dysgenic Effects of Welfare
6. Curtailment of Benefits to Welfare Mothers
7. Single and Welfare Fathers
8. Sterilization of the Mentally Retarded
9. Parental Demands for Sterilization
10. Sterilization of Female Criminals
11. Sterilization of Male Criminals
12. Conclusions

In Chapter 12, we considered the first strategy of negative eugenics—the provision of information and services on contraception and abortion—and concluded that this is likely to have only a fairly small eugenic impact. We turn now to the second strategy of negative eugenics—the use of incentives, coercion, and compulsion for those with low intelligence and psychopathic personality to restrict their fertility. Such measures can be placed on a continuum of coerciveness, ranging from the offering of financial incentives for not having children to compulsory sterilization.

1. INCENTIVES FOR STERILIZATION: THE SHOCKLEY PLAN

A possible approach to the problem of reducing dysgenic fertility is to offer those with low intelligence and psychopathic personality financial incentives not to have children. A proposal of this kind was advanced in the early 1970s by William Shockley (1972) and received quite a lot of publicity because of Shockley's fame as a Nobel laureate for the discovery of the transistor.

Shockley's proposal was to offer payments for sterilization to all nontaxpayers with IQs below 100. Most people with IQs below 100, who comprise half the population, pay taxes; so perhaps about 10 percent to 20 percent of the total population would be eligible for the payments, the precise percentages depending on the tax thresholds in any particular year. In practice, those eligible for the scheme would be the unemployed and those on low earnings and with dependants, which puts them below the tax threshold. Although the scheme was designed primarily to attract those with low IQs, the stipulation that those eligible for the payments would have to be nontaxpayers would also attract a certain number of those with psychopathic personality, one characteristic of which is antipathy to work and paying taxes. The payments proposed were $1,000 for each IQ point below 100. Thus, for example, someone with an IQ of 70 would be paid $30,000.

People with very low IQs are frequently not well informed and might not come to hear of the scheme. To overcome this potential problem, an agent's fee of 10 percent would be payable to those arranging for the sterilization. Some provision would also have to be made for the age factor because older people have fewer potential childbearing years ahead of them; therefore the payments would have to be tapered for older people, but how this should be done was not elaborated.

Shockley thought that the scheme would be self-financing. He gave the example of a person with an IQ of 70 who might produce 20 children, many of whom would be unemployed and criminals, and who would incur a cost to the government of an estimated $250,000. This would be saved by the sterilization payment of $30,000. If these figures are broadly correct, there would evidently be a substantial saving to the taxpayer as well as an eugenic gain.

Shockley presented his scheme as what he called "a thinking exercise." Thought provoking as the proposal is, it could give rise to several problems. First, there is the problem of cost. Payments of $30,000 to those with IQs of 70 in 1972 would need to be adjusted about 10-fold for inflation and would amount to about $300,000 for the early twenty-first century. Those with higher IQs would be eligible for lower but still large payments. For instance, those with IQs of 85 would be eligible for payments of $150,000. There can be little doubt that with payments of this size, large numbers of people would come forward for sterilization, and this would entail enormous costs. Furthermore, in many cases married couples would both be applicants and would receive double this sum. If this scheme were put into practice and the government started paying out checks of $600,000 to married couples to be sterilized, many of them chronically unemployed and others low wage earners, public opinion would be outraged. It is doubtful whether any government would take up the scheme. However, the cost is not a major problem in principle because it is likely that many of those whom the scheme is designed to attract would be willing to be sterilized for a much lower payment, perhaps as low as a few hundred dollars.

A second problem with the plan is that many applicants would be expected to deliberately perform poorly on the intelligence test in order to obtain larger payments. Some applicants might obtain IQs of zero and hence be eligible for payments of $1 million in present-day money. It would be possible to overcome this problem by getting rid of the low intelligence criterion and by simply making the scheme available to all nontaxpayers. Among men, virtually all will have IQs below 100, and those that do not will be largely schizophrenics and psychopaths, who could usefully be included in the plan. Hence, instead of having a graded system of payments according to IQ, there could be a flat rate payment to all nontaxpayers.

Third, a scheme of this kind could encounter problems of married women who do not work or who might give up work for a tax year in order to qualify. This problem might be overcome by taking the joint income of married couples into account so that nonworking married women whose husbands were taxpayers would not qualify.

Fourth, there is the problem of those who have already completed their families and have no intention of having more children, many of whom might well be considering having themselves sterilized anyway and who would make applications for payments.

A fifth problem is that it stigmatizes the unemployed and those on low incomes. Public and media opinion in the Western democracies has become highly critical of stigmatization of this kind. For this reason alone, governments in the Western democracies could not be expected to consider introducing such a scheme.

Critical examination of the proposal shows how difficult it is to frame proposals of this kind that would be politically feasible in the Western democracies. Nevertheless, it may serve as a useful starting point for thinking about schemes of this general kind that might be politically acceptable.

2. THE DENVER DOLLAR-A-DAY PROGRAM

A politically acceptable variant of the Shockley plan was introduced in Denver in the early 1990s. The program offered payments to single teenage mothers not to become pregnant for a second time. The program was sponsored by Planned Parenthood and called the "Dollar-a-Day" program because it paid teenage mothers under the age of 16 a dollar a day not to become pregnant. To obtain these payments, the teen mothers were required to attend weekly meetings at which they received $7 so long as they were not pregnant. If they became pregnant, they were no longer paid. After a five-year trial, the program had had significant success, insofar as only 17 percent of the girls became pregnant, as compared with a normal 50 percent repeat pregnancy rate for girls who have become pregnant before age 16 (Steinbock, 1996). Despite this success, the program was condemned by the president of the Planned Parenthood Federation of America as "coercive." This condemnation seems curious coming from the president of a society whose objective is to promote the use of contraception and responsible family planning and illustrates the contemporary hypersensitivity to any scheme that might be considered eugenic. Furthermore, it is stretching the meaning of "coercive" to apply it to offering payments to young teenagers not to become pregnant. Even if the scheme is considered coercive, it can be justified on the grounds that it is in the interests of the girls concerned as well as of society that they should not become pregnant again at such a young age. Society can legitimately coerce minors to behave in their own best interests and does so, for instance, by requiring them to attend schools.

The Denver scheme is instructive because it shows that it is politically feasible to induce young girls not to become pregnant by offering quite small financial incentives. It deserves wider implementation. The failure rate of 17 percent suggests that it needs strengthening. The best way of doing this would be to make the payments conditional on the girls using some long-lasting form of contraceptive, such as the IUD or Norplant.

3. INCENTIVES FOR WOMEN ON WELFARE TO USE CONTRACEPTION

In the 1990s, the legislatures of several U.S. states introduced measures to offer financial incentives to women on welfare to use contraception. Shortly after Norplant, the longlasting contraceptive implant, was approved in the United States in 1990, bills were introduced in Kansas, Oklahoma, Louisiana, Tennessee, and Washington offering payments of $100 to $500 to welfare mothers on condition that they had Norplant implants. The argument in favor of these bills was that welfare mothers incur additional welfare costs when they have more children and it would be more cost-effective to pay these women not to have additional babies. An objection to offering payments to welfare mothers to have Norplant insertions is that because many of them use contraceptives anyway, the payments would be unnecessary public expenditure. However, Norplant is a highly effective form of contraception, and the payments proposed were so small that the costs would be minimal in relation to the gains. Unhappily all these bills were defeated or died in committee.

The most robust bill to offer financial incentives to welfare mothers to use contraception was introduced in Ohio in 1995. It proposed a payment of

$1,000 to any new welfare mother plus an increase of 150 percent of her welfare income if she agreed to be sterilized by tubal ligation. If she agreed to Norplant, she would receive a $500 payment and a 10 percent increase in her welfare income every six months until it reached the 150 percent level. In addition, each new welfare mother would have to pass a test of parenting skills, and if she refused or failed, the baby would be placed with relatives or given to foster parents. If she did not pass the test within a year, the baby would be put up for adoption. These proposals for offering financial incentives to women on welfare to use contraception have been presented as cost-saving measures and sometimes also on the grounds that the children of women on welfare are typically brought up in poverty, do not do well at school or in employment, and are at significant risk of becoming unemployed or criminal. Proposals of this kind are also attractive on eugenic grounds and are an encouraging example of schemes that are politically feasible in Western democracies.

4. PAYMENTS FOR STERILIZATION IN DEVELOPING COUNTRIES

Several economically developing countries have offered payments to both men and women to be sterilized as a means of reducing the birth rate. The first country to introduce measures of this kind was India, where several schemes offering men payment to be sterilized were introduced in the 1960s. One of these schemes has been described by Simon (1974). Payments of various sums, conditional on sterilization, were offered to approximately 8,000 men working in nine factories of the Tata organization. The results were analyzed in terms of the amount of money offered and the men's incomes. As would be expected, the higher the payments offered, the greater the proportion of men coming forward for sterilization. However, the proportion only reached 10 percent even for the highest payments for the poorest workers. The value of the highest payment was 220 rupees, representing about $27. This seems a small sum, but in India it represented more than one month's pay. With smaller incentives of the order of 20 rupees, there was a 2.6 percent takeup among poorly paid workers. These results show that financial inducements of relatively small sums are effective in inducing a small percentage of poorly paid men to be sterilized. No doubt, larger payments would attract more volunteers. It can be inferred that the scheme would tend to be eugenic because the takeup would be principally among the lowest earners.

Another country in which payments for sterilization have been offered is Peru. For several decades, European and Asian women in Peru have typically had two or three children, whereas native American women have typically had six or seven. In 1995 President Fujimori perceived this as a problem and induced the Peruvian Congress to introduce measures giving financial incentives in the form of food and clothing to rural, native American women to be sterilized. The program was backed up by giving government health workers quotas of women to be sterilized and bonus payments for every woman whose sterilization was secured (Mosher, 1998).

The implementation of schemes like these reminds us that eugenic measures that are not politically feasible in the economically developed Western nations can be introduced in less developed countries, where public opinion is less sensitive to measures of this kind.

5. DYSGENIC EFFECTS OF WELFARE

A variant of offering girls and women on welfare incentives not to become pregnant is to remove the welfare payments to unemployed women who have babies. The thinking behind this proposal is that a number of women in the underclass have babies because this enables them to live on welfare, which is preferable to working. Charles Murray (1984, 1990, 1993) argued that the increasing generosity of welfare provision in the United States during the 1960s and 1970s made having babies a rational option for women with low intelligence, no educational qualifications, and poor employment prospects. This thesis is based on the assumption that people are rational calculators who evaluate the benefits and costs of the various courses of action open to them. Single women with low intelligence and psychopathic personality do poorly at school and are normally only able to obtain poorly paid and uninteresting employment. To many, a preferable option is to have babies. These enable them to live on welfare, which provides them with an income, accommodation, and greater freedom and leisure. Murray does not maintain that all single women who have babies calculate the precise arithmetic of the costs and benefits of having babies as opposed to working; rather many of them sense from their mothers, sisters, and girl friends that living as a single mother on welfare provides a satisfactory lifestyle compared with the alternatives. The effect of this is that they do not mind becoming pregnant, do not take measures to avoid this, and do not have their pregnancies terminated.

A number of social scientists have examined Murray's thesis and found evidence supporting it. Abrahamse, Morrison, and Waite (1988) examined the attitudes of 13,061 high school girls towards becoming single mothers. They found that 23 percent of white and 41 percent of black girls said they would consider this option. These percentages were lower among those intending to go on to college than among those planning to cease education after high school (18 percent to 34 percent, respectively, among white girls), suggesting that those with the more attractive alternative of college and satisfying employment were substantially less likely to consider single parenthood. When the sample was followed up two years later, about half of those who said they would be willing to consider single motherhood had actually become single mothers. This study confirms that for many American adolescent girls becoming a single mother is not just an accident but a consciously considered option that is deliberately chosen by appreciable numbers of those who do not have attractive alternatives.

Several studies examining the thesis have found a positive association between the size of welfare payments to single women with babies in different states and the numbers of such women. Matthews, Riber, and Wilhelm (1997) found this relationship for the years 1978 to 1988. Jackson and Klerman (1994) found that the relationship held for white women but not for black, suggesting that only whites act as rational calculators and tend to have more children when welfare benefits are high. In a further study of this issue, Leibowitz, Eisen, and Chow (1986) found that pregnant single teenagers were more likely to have their baby than to have an abortion if they would be eligible for welfare benefits. The most sophisticated analysis of the impact of welfare payments for single women to have babies is that of Clarke and Strauss. They found a positive relationship among the U.S. states for the period 1980–1990 between the size of welfare benefits for single mothers in relation to the average female wage and the birth rate to single teenage women. Clarke and Strauss (1998) summarize their conclusions as follows:

> The results for both white and black teens are remarkably consistent with theory and stress the importance of economic incentives on the choice between work and welfare. A 1 percent increase in welfare benefits appears to increase illegitimacy among both white and black teens by more than 1 percent. A 1 percent increase in female wages appears to have a more modest effect of about 0.4 percent decrease in illegitimacy for white teens but does not appear to affect illegitimacy rates for black teens. (p. 840)

This study supports the rational calculator model of the childbearing of underclass females. The model works better for white teenagers than for blacks, suggesting that white teenagers are more efficient rational calculators perhaps because whites have higher intelligence levels.

A positive effect of the size of welfare payments on single motherhood has also been found in Britain. In the 1970s, 1980s, and 1990s, approximately a quarter of the population in Britain lived in state-subsidized council housing. These were predominantly the unskilled and semiskilled and the poorly educated. Most of the better educated and the middle class buy their houses, and a few live in private rented accommodation, both of which are beyond the means of low earners.

Among the unskilled, semiskilled, and poorly educated, there are more applicants for council housing than available accommodation, and consequently applicants are placed on waiting lists and can remain on these for many years. Priority in the allocation of council housing is given to pregnant women, single parents, and families with children, according to rules formalized by the Housing (Homeless Persons) Act 1977. The effect of these rules was that young women could jump ahead on the waiting list and obtain council accommodation by becoming pregnant. Rational calculation theory predicts that some of them would do so, or at least that they would not take so much care to avoid becoming pregnant as they would if having a child did not enable them to obtain free accommodation and an income. The option is not attractive to intelligent and well-educated young women who hold good jobs and would not want to live and rear a baby in council accommodation. The rational calculator model of single motherhood has been examined for Britain by Ermish (1991), who estimated that in the 1970s and 1980s, a 10 percent increase in the real value of welfare benefits for single mothers raised the illegitimacy rate by approximately 1.5 percent.

6. CURTAILMENT OF BENEFITS TO WELFARE MOTHERS

The solution to the problem of underclass women having babies deliberately as a means of living on welfare as an alternative to working is to stop the payment of welfare benefits to these women. This is the solution proposed by Murray (1984, 1990). The effect of this would be that single women having babies would have to work; depend for support on their parents or other relatives, on the babies' fathers, or on private charity; or give the babies up for adoption. This would reduce the benefits and increase the costs of babies to welfare mothers very considerably and would almost certainly lead to a commensurate reduction in the numbers.

The proposal is probably too draconian to be politically acceptable in either North America or Europe but it is nevertheless important to state as the ideal solution. In practical terms the objective should be to reduce welfare payments to single mothers with babies to the greatest extent that is politically feasible. This was done in the 1990s in several U.S. states. The first was New Jersey, which in 1993 introduced the "family cap," which provided that single women with children would not be given additional welfare payments if they had more babies. A similar measure was introduced in Arkansas in 1998. The family cap in New Jersey appeared initially to be having its intended effect of reducing out-of-wedlock births. The first evaluation study indicated that these births fell by 29 percent. There was also an increase of 3.7 percent in abortions among Aid to Families with Dependent Children (AFDC) recipients during the first eight months following the introduction of the cap (Donovan, 1995). The introduction of the cap was by no means fully effective in eliminating births to single women on welfare because approximately 6,200 babies were born to single "capped" women in the first year following the introduction of the measure (Donovan, 1998).

There are problems in assessing the effectiveness of the cap. Among these is that there is no control group sample of women not subject to the cap and that welfare recipients often do not report the birth of an additional child to their welfare agency because they have no incentive to do so. Because these women largely cease to report additional babies, their birth rate appears to fall, whereas in fact it has not fallen. It is too early yet to form a conclusion of the magnitude of the effects of reducing welfare payments to single mothers on their birth rates; but the evidence to date suggests that over the long term, once this becomes widely known, it is likely to be effective.

7. SINGLE AND WELFARE FATHERS

Hitherto we have considered measures that should be taken to reduce the numbers of children of women with low intelligence and psychopathic personality. We need also to consider the related problem of the fathers of these children. Several studies have shown that these fathers are drawn predominantly from the least socially desirable sections of the population. For instance, Marsiglio (1987) has analyzed the American National Longitudinal Survey of Youth and found that among young men who had fathered a child at the age of 17 or less, 41 percent were high school dropouts (the dropout rate among those who had not fathered a child was 14 percent). The teenage fathers also came disproportionately from lower class families: 42 percent of the teenage fathers had fathers with less than 12 years of education, as compared with 28 percent of those young men who were not fathers. In a summary of the research on this issue, Nock (1998) concludes that "unwed fathers have lower educational attainments, are more likely to drop out of school, and are more likely to express ambivalent attitudes about the value of work than are other men; they are more likely to have been charged with crimes, are more likely to be unemployed, and, if they do marry, have less stable marriages than do men who did not have premarital births" (p. 280).

One of the problems in attempting to curtail single fatherhood is that there is a machismo subculture in which young men take pride in fathering children. E. Anderson (1990) has concluded on the basis of his research in a Philadelphia ghetto that many of these men pursue women and try to get them pregnant in order to demonstrate their masculinity. The more children they can father without marrying the mothers, the higher their status. In order to achieve this, they abandon women once they are pregnant and move on to new conquests.

Thus, the research evidence has shown that unmarried fathers, like unmarried mothers, are typically the least desirable stock from which to produce children and that unmarried fatherhood, like unmarried motherhood, needs to be curtailed. This is a difficult problem because the procreation of children is generally cost free to poor men. The solution normally proposed is to impose costs on unmarried fathers in order to deter them from having children and to reduce the tax burden of the support of these children. However, it is difficult to make these measures effective. In the United States, the 1980 census showed that only 14 percent of single mothers obtained an award for maintenance against the fathers of their children, only 7 percent actually received any payments, and only 4 percent received the full payments awarded (Cutright, 1986).

One of the problems in attempting to levy maintenance payments on the fathers of illegitimate children is that many of them do not have sufficient incomes from which to make payments. Cutright (1986) estimated that in the United States 44 percent of white and 60 percent of black young males have incomes below the poverty level, against whom maintenance payments for illegitimate children could not be levied. A further problem is that in the 1980s and 1990s, surveys have shown that many boys become sexually active and become fathers while still at school and hence cannot be required to maintain the babies for which they are responsible. Surveys have shown that black boys become sexually active at an average age of about 14.5 and urban white boys at about 16 and that many of these boys do not use contraceptives (Moore & Ericson, 1985; Zabin, Smith, Hirsch, & Hardy, 1986; Zelnik & Shah, 1983). These boys have no incomes that could be distrained upon to maintain their children, so the threat of child support would not deter them from fatherhood. It has sometimes been suggested that teenage fathers could have future maintenance orders imposed on them for payments when they become earners, but it is not likely that young teenage boys will be deterred from getting girls pregnant because of a remote possibility that at some future date they might have maintenance orders levied against them.

In the Ohio scheme for offering welfare mothers payments for the use of contraception, one requirement was that the welfare mothers should identify the fathers of their children and that these would be offered the options of paying child support, of carrying out community service work, of being sterilized and receiving a payment of $1,000, or of serving two years in prison. This is a commendable scheme and would be better still if the option of carrying out community service were removed. The scheme could usefully be expanded to all men on welfare after some limited period of, say, four months. The provisions would be similar to those for welfare mothers and would require sterilization as a condition of receiving welfare. This scheme would not prevent these men from having children in the future, which could be accomplished by the removal of sperm from the testes and the use of artificial insemination; so sterilization could not be regarded as too onerous a requirement.

8. STERILIZATION OF THE MENTALLY RETARDED

One of the principal objectives of classical eugenics was the sterilization of the mentally retarded. In the first half of the twentieth century, sterilization laws were introduced in the United States, Canada, and most of Europe. These were one of the major policy achievements of the eugenics movement. The first of the sterilization laws was introduced in the state of Indiana in 1907. The law authorized the sterilization of any habitual criminal, rapist, or "idiot or imbecile," as the more severely retarded were called at the time, whom physicians assessed as "unimprovable." For sterilizations to be carried out, there had to be unanimous agreement of the hospital's or prison's physicians and two external physicians that there was no reasonable prognosis for improvement for the individuals concerned. The law was quickly put into effect, and the Indiana surgeon H. C. Sharp (1907) reported having carried out 456 vasectomies on male retardates and criminals later in the year.

Over the next decade, 16 other U.S. states introduced sterilization legislation, a number of them adding the insane to the categories on whom the operation was to be performed. The state that pursued this policy most vigorously was California, where 3,233 sterilizations were carried out between 1907 and 1921. Of these, 1,853 were men and 1,380 were women. The men were sterilized by vasectomy and the women by salpingectomy (the cutting and tying of the Fallopian tubes). By 1925 a further 12 U.S. states passed sterilization laws, bringing the total where these laws were in place to 31 (Kevles, 1985). In the 1940s the numbers of sterilizations began to decline, and by the end of the 1960s, mandatory sterilizations had virtually ceased. Many states repealed their sterilization laws, and in others the laws were not implemented.

Laws for the sterilization of the mentally retarded, criminals, and the mentally ill were introduced in Canada and throughout most of Europe in the 1920s and 1930s. Sterilization laws were introduced in the Canadian provinces of Alberta and British Columbia in 1928 with the objective of eliminating "the danger of procreation, with its attendant risk of multiplication of the evil by transmission of the disability to progeny." Approximately 400 sterilizations were carried out in the two provinces. In the 1960s sterilization came under attack from civil liberties groups, and the law authorizing sterilization was repealed in 1977 (Sterilized, 1996).

In the 1920s and 1930s, sterilization legislation was introduced in most European countries, including Austria, Denmark, Estonia, Finland, France, Germany, Norway, Sweden, and Switzerland. A sterilization law was passed in Sweden in 1926 that permitted mandatory sterilization of the mentally retarded. It is estimated that approximately 60,000 women were compulsorily sterilized between 1936 and 1974 (Glasse, 1998). A sterilization law was introduced in Denmark at about the same time, and over the next decade about a third of the mentally retarded were sterilized (Kemp, 1957). In 1933, a sterilization law was enacted in Germany. It is estimated that 300,000 sterilizations were carried out in Germany over the next 12 years, most of which were carried out in the period 1933–1939 (Muller-Hill, 1988). Britain was one of the few European countries where sterilizations were not carried out. The objective of preventing the mentally retarded and the mentally ill from having children was achieved in Britain by segregating the sexes in institutions.

From the late 1960s onward, sterilizations were progressively reduced in

Europe, as in the United States and Canada. There is, nevertheless, a good case for reviving the sterilization of the mentally retarded and criminals. It is indisputable on both empirical and theoretical grounds that many of these people transmit their characteristics to their children by both genetic and environmental processes. In the case of the mentally retarded, their children are frequently taken away from them by social service agencies, so they do not obtain the fulfillment of rearing children. If the children remain with their parents, they are reared in poor environments that contribute to the perpetuation of the social costs of mental retardation, low intelligence, and crime, by the process that has become known as "the cycle of disadvantage," but that should be more properly known as "the cycle of social pathology."

9. PARENTAL DEMANDS FOR STERILIZATION

As the sterilization of mentally retarded females authorized by physicians largely ceased from the late 1960s onwards, a number of parents of these females began to apply to the courts to have their daughters sterilized. These parents took the view that their daughters were not mentally equipped to defend themselves against predatory males, that it would be damaging to them to become pregnant, and that, if their daughters lived at home, it would be the parents themselves who would have to bring up the child, who would be likely to be retarded or have low intelligence.

One of the first actions of this kind was the Grady case, which was initiated in New Jersey in 1981. In this case, the parents of a mentally retarded adolescent girl about to enter a sheltered workshop asked their physician to arrange to have her sterilized because they feared she would become sexually involved and become pregnant. The local hospital refused to perform the operation, and the parents took the case to court, which granted their request. The New Jersey attorney general appealed this decision to the state Supreme Court, which upheld the parents' request but ruled that the sterilization of retarded individuals always required judicial approval.

Throughout the 1980s and 1990s similar actions were brought in a number of U.S. states. Generally the parents' requests were opposed by organizations asserting that mentally retarded women had the right to have children, and the cases were taken to state supreme courts. An action of this kind was brought to a conclusion in the Michigan Supreme Court in early 1998. In this case, Donna and Richard Wirsing applied to the courts for permission for the sterilization of their daughter Lora, who was born with brain damage, had an IQ of 35, and could not read, write, tell the time, or care for herself. Their application was opposed by the Michigan Protection and Advocacy Service, which argued that mentally retarded women should be permitted to have babies. The state Supreme Court upheld the parents' action (Tobin, 1998).

Similar actions were brought in Britain in the 1980s. The law on whether physicians could perform sterilizations at the request of the parents was uncertain until 1987 and 1988, when it was clarified by three cases. In the first of these, the local authority responsible for a mentally retarded minor, Miss B, applied for her to be made a ward of court and then applied to the High Court for permission to sterilize her. The High Court gave its approval, and the case went to the Court of Appeal and then to the House of Lords. Both of these upheld the approval, and the sterilization was carried out. In the second case, the Court of Appeal approved the sterilization of "Jeanette," a 17-year-old girl with a mental age of 5 and an IQ of approximately 30. The third case took place in 1988 and concerned Miss F, aged 36, who was mentally retarded and a voluntary inpatient in a hospital. She had formed a sexual relationship with another mentally retarded patient. Her mother and the doctors agreed that pregnancy would be undesirable, that neither Miss F nor her partner was able to use contraception reliably, and that sterilization was in Miss F's own best interests. Application for sterilization was made to the court and was granted. This case also went to the Court of Appeal and the House of Lords to clarify the position once and for all. The decision to sterilize Miss F was upheld and the operation carried out.

Another case of this kind was determined in Scotland in 1996. In this, the mother of a severely autistic woman, aged 32, petitioned the court to have her sterilized. "I" was a resident in a hospital and had formed a relationship with a man with Down's syndrome. The Scottish court approved the mother's petition.

Thus, over the course of the twentieth century, the wheel on the sterilization of mentally retarded women turned full circle. In the first half of the century, this was widely regarded as sensible on eugenic grounds. In the three decades following World War II, sterilizations were reduced and eventually ceased as a result of pressure from civil liberties lobbies. In the last two decades of the century, sterilization was once again permitted as a result of the requests and the legal actions of the parents of mentally retarded women. The sterilizations that took place in the last two decades of the century were not carried out on eugenic grounds, but in the best interests of the retarded young women. Nevertheless, they achieved a eugenic objective and are to be welcomed.

10. STERILIZATION OF FEMALE CRIMINALS

From the mid-1960s onward there have been more than 20 legal cases in the United States in which judges have given women convicted of child abuse the option of a term of imprisonment or probation together with temporary sterilization for a number of years by Norplant implants (Steinbock, 1996). The rationale for this sentencing policy was that the judges considered these women unfit to rear children and that they should therefore be prevented from having more, at least for a few years. Both options presented serve this purpose, but the option of having a contraceptive implant has the advantage that the woman can continue to bring up her children.

One case of this kind occurred in 1984 when Ruby Pointer was convicted of the felony of child endangerment. She had adopted for herself and her two sons a macrobiotic diet, which was seriously deficient in nutrients. The result of this was that her sons were severely malnourished, so much so that one of them had suffered permanent neurological impairment. The judge took the view that she was likely to cause serious damage to any subsequent children she might have and consequently that it would be desirable to prevent her from having any more children. He therefore sentenced her to a five-year term of probation and ordered her to undergo counseling and to refrain from becoming pregnant during this period. The requirement that she refrain from becoming pregnant was overturned on appeal. Seven years later it came to light that she had subsequently had three daughters, all of whom were severely malnourished and unable to speak. This vindicated the judge's view that she was unfit to rear children.

A second case concerned Tracy Wilder, a 17-year-old high school student who became pregnant in 1990, gave birth, and killed her child by suffocation. She was sentenced to 2 years in prison followed by 10 years of probation, conditional on her completion of high school, acceptance of psychological and contraceptive counseling, and use of contraception during the probation period. The American Civil Liberties Union and the Family Research Council (an anti-abortion group) opposed the contraception provision, but Tracy Wilder opted to accept it rather than risk an appeal that might have led to a longer term of imprisonment.

A third case is that of Darlene Johnson, a 27-year-old woman with a criminal record for check fraud, theft, disturbing the peace, battery, and burglary. In 1990 she had four children and was pregnant. In December of that year she was convicted of child abuse. She and her boyfriend had beaten the children with the buckle end of a belt so severely that the state removed the children and placed them in foster care. The judge, Howard Broadman, could have sent her to prison but offered probation conditional on her using Norplant for three years after the birth of her child. Her attorney moved for a reconsideration of the Norplant condition on the grounds that it violated her right to privacy; but Judge Broadman denied the motion, contending that the condition was justified on the grounds of reformation, rehabilitation, and public safety. The judge maintained that Johnson was likely to abuse any further children she might have and therefore that the public interest would be served by preventing her having more children. Johnson appealed; but before the appeal could be heard, she was convicted of taking cocaine and was sentenced to five years imprisonment.

A fourth illustrative case of this sentencing policy occurred in 1993 and concerned Barbara and Ronald Gross, who were convicted of attempted aggravated sexual battering of two of their children. The judge offered them the alternatives of either 5 years imprisonment or 10 years probation on the condition that Barbara Gross was sterilized. In addition to the abuse of the children, Barbara Gross had a schizophrenic disorder and was mentally retarded. At the time of the sentence she had four children and was pregnant. Of the two alternatives offered by the judge, she opted for sterilization.

In these and similar cases many people will no doubt accept that the judges were right in deciding that the women were unfit mothers and likely to cause harm to any future children and that it would be desirable to prevent further pregnancies. It is preferable for these women to be put on probation conditional on temporary sterilization than to send them to prison, which in most cases would serve little useful purpose. These judges' decisions were not made ostensibly on eugenic grounds, but they furthered the eugenic objective of preventing these women from having children, at least for a limited period. The eugenic objective should be to support these judicial sentences and to promote their use more often, together with the stipulation of longer periods of contraception and, preferably, permanent sterilization.

11. STERILIZATION OF MALE CRIMINALS

It may be considered a curious feature of the cases in which convicted females were offered probation conditional on the use of contraception for a period of years that these orders were not extended to their male partners, who are typically, although not invariably, equally culpable of the child abuse. It is equally in the social interest that men convicted of child abuse should not have children, and it would be desirable to extend the option for sterilization to them in return for more lenient sentences.

It would also be desirable to extend the sterilization option more generally to male criminals. Imprisonment has little effect either in the reforming of criminals or as a deterrent and incurs considerable social costs of maintenance in prison and reduced employability on release. A better alternative, from the point of view of reducing future criminal offending and the promotion of eugenics, would be for judges to offer convicted male criminals the alternatives of imprisonment or castration accompanied by probation. Castration consists of surgical removal of the testes, with the result that neither semen nor the male sex hormone testosterone are produced. Normally the effect is to reduce the sex drive quite considerably, although sexual performance is still possible. The effect of castration and the lowering of the sex drive is to reduce sexual and violent offending.

A number of European countries have used castration to reduce future offending by men convicted of sexual crimes. This was common throughout Europe in the Middle Ages for rape. In the twentieth century, castration for sexual offenders was introduced first in Denmark in 1929 and soon afterward in Germany, Norway, Finland, Iceland, Sweden, and Switzerland. Only in Denmark was it carried out compulsorily from 1929 to 1967. In the other countries castration has been offered to offenders as an alternative to imprisonment or for a reduction in the length of the sentence.

Research has shown that castration is effective in reducing sexual crimes among sex offenders. Some of the best evidence for this comes from Denmark. At one institution 285 sex offenders were castrated and "research showed that the incidence of relapse into new sexual crime fell to a minimum after surgical castration" (Hansen & Lykke-Olesen, 1997, p. 196). Of 21 prisoners surgically castrated between 1935 and 1970, only two committed further sex crimes, and these two had been given testosterone substitute therapy by their physicians. Of 24 prisoners who refused castration and had consequently to spend a further average of eight years in prison, 10 committed further sex crimes after their release.

A further study on this issue in Denmark concerned 738 sexual offenders castrated between 1929 and 1959. This study found that their reoffending rate was 1.8 percent, as compared with a reoffending rate of 9.7 percent among noncastrated offenders (Ortmann, 1980). This result has been confirmed in Germany, where a 1970 law allowed judges to offer convicted offenders the option of a reduced term of imprisonment conditional on castration. About 400 castrations were carried out from 1970 to 1986 (Raine, 1993). Wille and Beier (1989) followed up 99 of them for 11 years after their release from prison and found that 3 percent of them reoffended, as compared with 46 percent among a matched group of sex offenders who had not been castrated. They reviewed 10 other studies of castrated sex offenders, which all showed reoffending rates of between 0 percent and 11 percent.

In addition to reducing sex offending, there is evidence that castration reduces violent crime in general. The reason for this is that high levels of testosterone generate aggression. Reviews of the literature pointing to this conclusion have been published by Thiessen (1990), and Raine (1993). The role of testosterone in motivating aggression is probably the major reason why violent crime is much more frequently committed by males than by females and why the peak rate of crime among males, approximately between the ages of 13 and 25, coincides with the peak rates of the secretion of testosterone. Hence violent crime would almost certainly be reduced by the castration of violent criminals in addition to sexual offenders.

An alternative to castration, which may be more acceptable to public opinion, is temporary chemical sterilization. This is achieved by taking the drugs cyproterone acetate (CA) and medoxyprogesterone acetate (MPA or Depo-Provera), which reduce the output of testosterone. This was introduced in the courts in the 1960s for sexual offenders in Germany and Switzerland and subsequently in a number of court cases in the United States as an alternative to incarceration. It is effective in reducing the sexual drive (Thiessen, 1990). These effects last only as long as the drugs are taken, and when they cease to be taken, the sexual drive returns. This makes sterilization by drugs more acceptable to public opinion but less effective because sexual performance is still possible and the reduction in sexual drive only lasts as long as the drugs are taken.

The offering of castration by the courts to offenders and violent criminals as a condition of more lenient sentences has not been made on eugenic grounds, but it promotes the eugenic objective of reducing the fertility of criminals. It should be supported and promoted for more widespread use in the sentencing of criminals.

12. CONCLUSIONS

In this chapter we have considered a variety of programs for the provision of incentives, coercion, and compulsion for the promotion of negative eugenics. These schemes are of three main kinds. The first consists of offering incentives for contraception or sterilization to the target populations. Some measures of this kind have been introduced in the United States, such as the Denver Dollar-a-Day scheme, which gives teenage mothers a dollar a day so long as they do not become pregnant, and payments to single mothers on welfare to have Norplant implants. These programs have had some positive effect and deserve to be implemented more widely.

The second type of scheme imposes penalties on the target populations for having children. For single women on welfare, the penalty consists of the withdrawal of welfare payments. The political problems of implementing this measure for existing welfare mothers would be considerable. A more moderate measure would be ceasing to pay single welfare mothers additional benefits when they have additional children. Schemes of this kind are politically feasible and have been tried, and the results are promising. Probably if they were implemented on a greater scale and over a longer time period, so that they became widely known, they would have a greater impact. More schemes of this kind need to be introduced.

Consideration has been given to the imposition of penalties on the fathers of the children of single mothers. These are difficult to impose on the target populations, who are frequently untraceable, unemployed, or such low earners that payments cannot be levied on them. The problem of deterrence of these fathers is intractable and probably insoluble in the Western democracies.

The third and most robust form of negative eugenics consists of the sterilization of the mentally retarded and criminals. This was widely practiced in the United States, Canada, and Continental Europe in the middle decades of the twentieth century but largely ceased from the early 1970s. Nevertheless, as compulsory sterilization was phased out, a number of the parents of mentally retarded girls and women have requested the sterilization of their daughters, and these requests have been granted by the courts in the United States and Britain.

There has also been some progress in the sterilization of criminals. In the United States the courts have begun to offer women convicted of child abuse the option of imprisonment or the use of contraception or sterilization to prevent them from having further children, and in a number of cases the women have chosen contraception or sterilization. In a parallel development, male sexual offenders in Europe have been offered the option of castration plus reduced terms of imprisonment. Many have opted for castration, and this has been shown to be effective in reducing further offending. These options offered by the courts should be given more widely.

The measures discussed in this chapter that are politically feasible in the Western democracies at the beginning of the twenty-first century would only have quite small impacts. Nevertheless, even measures with small impacts have some value, both in themselves and in establishing a basis on which more effective measures can be developed. For these reasons, all the measures discussed in this chapter are useful contributions to the promotion of negative eugenics.

14

❦

Licenses for Parenthood

1. The LaFollette Plan
2. The Westman Plan
3. The Lykken Proposal
4. Critique of the Lykken Proposal
5. An Effective Eugenic Licensing Program
6. Conclusions

❧❧❧

As an alternative to the series of piecemeal measures of negative eugenics discussed in Chapters 12 and 13, we will now consider proposals for comprehensive parental licensing programs. The essential feature of these schemes is that they require everyone to have a parental license to have children. These licenses would be granted only to those regarded as fit to be parents, and the criteria for fitness for parenthood would include both the genetic qualities and the suitability of the couples concerned.

This was essentially the idea proposed by Plato in *The Republic*, in which only those with the best qualities would be permitted to have children or, in effect, would be granted licenses for parenthood. The idea was elaborated by Francis Galton in his eugenic utopia, Kantsaywhere, an account of which has been given in Chapter 1. In the second half of the twentieth century, parental license proposals have been formulated by Hugh LaFollette, J. C. Westman, and David Lykken. The merits and feasibility of their ideas are considered in this chapter.

1. THE LAFOLLETTE PLAN

The first of the recent proposals for licenses for parenthood was made in 1980 by the American political scientist Hugh LaFollette (1980). He began by contending that some parents are unfit to rear children, notably those who neglect their children, ill-treat them, subject them to violent physical abuse, and even kill them. He noted that research has shown that a large proportion of the children of unfit parents become criminals. He asserted that these parents are incompetent and that they impose costs on society. To mitigate these costs, LaFollette argued that the state should take steps to prevent these children from being born. To make this proposal effective, he proposed that all couples should be required to obtain a license certifying their competence in child rearing before they are permitted to have children. This was the first use of the term *license* in this context.

In justification of this proposal, LaFollette pointed out that the state already requires people to acquire a license before they are permitted to undertake a number of activities that might cause social harm if performed incompetently. He gives the example of the automobile driving license. Incompetent drivers are a potential danger to the public, so the state reasonably requires people to demonstrate their driving competence and acquire a license before they are permitted to drive on the public highway. Similarly, physicians, lawyers, and pharmacists are required to obtain licenses certifying their competence. Practicing these professions without a license is illegal. The justification for this is that society would suffer if unqualified people practiced medicine, the law, or pharmacy; and steps need to be taken to ensure that this does not happen. LaFollette argued that the same case can be made for rearing children. Here, too, incompetent parenting imposes social costs; and to prevent these, parenthood should be licensed.

In a further justification of this proposal, LaFollette noted that the state already vets prospective adoptive parents for their fitness to rear children. Why, he asks, do we not allow just anyone to adopt a child? The answer is that we recognize that some people are unfit to rear children and that these people should be screened out in assessing the suitability of couples applying to adopt a child. Because society in effect requires the licensing of prospective adoptive parents, it should extend the principle to natural parents.

As regards the practical implementation of the scheme, LaFollette (1980) proposed that all prospective parents should be assessed for their child rearing competence by a psychological examination. This would consist of a personality assessment that would be designed to identify "the violence-prone, easily frustrated, or unduly self-centered" (p. 191). These are essentially the psychopathic, although he did not use that term.

LaFollette conceded that the psychological examination for fitness for parenthood would not be foolproof. No doubt some couples would be denied the parental license who would make adequate parents, while others would be granted the license who would turn out to be unsatisfactory parents. But, he argued, this is no different from the licensing of automobile drivers, physicians, lawyers, and pharmacists. No doubt a number of those who fail their tests for an automobile license and the qualifying examinations to practice medicine, the law, and pharmacy could nevertheless drive automobiles without having accidents and work as physicians, lawyers, and pharmacists without harming the public. Conversely, some of those who pass the automobile driving test turn out to be incompetent drivers, and some people succeed in qualifying as physicians but turn out to be incompetent doctors, and so forth. These competency tests for licensing are blunt instruments, but they unquestionably identify a number of the most incompetent thus protecting the public.

LaFollette realized he would have to consider the problem of the enforcement of his parental licensing plan. He conceded that it would be difficult to prevent unlicensed couples from producing children and suggested that this would best be dealt with by taking away the children of these parents and having them adopted or fostered.

LaFollette did not advance his parental licensing scheme on eugenic grounds. He did not point out that socially pathological behavior is transmitted genetically from parents to children, as well as environmentally, by example and poor child rearing practices. He did not mention low intelligence or mental retardation as disqualifications for obtaining the parental license, nor did he have any proposals to prevent babies being born to unlicensed parents. These are all weaknesses in his scheme. Nevertheless, his proposal was a valuable contribution to challenging the contention that everyone has a right to have children, and it stated the undoubted truth that some couples are unfit to be parents and should be prevented from having children.

2. THE WESTMAN PLAN

The concept of licenses for parenthood was revived in the mid-1990s by John Westman (1994, 1996), a professor of psychiatry at the University of Wisconsin. Westman's basic premise is the same as LaFollette's, namely that some parents are incompetent at the task of rearing children. These parents "cannot control their own impulses and either neglect or tyrannize their children; they are incapable of tolerating frustration and of postponing gratification; they are unable to handle responsibility for their own lives, much less for their children's lives; their incompetence as parents can be the result of immaturity, personality defects, or mental disorders" (1994, pp. 29–30). Westman contended that these incompetent parents fail to socialize their children effectively, with the result that the children develop into antisocial personalities who are themselves unable to rear children competently. Thus, the cycle of antisocial personality is "perpetuated from generation to generation." He argued that incompetent parents are heavily concentrated in the underclass: "Incompetent parenting is a component of the psychopathological lifestyle of the underclass, consisting of crime, drug addiction, unemployment, welfare dependency, and single motherhood" (p. 30).

Westman estimates the numbers of incompetent parents from the numbers of ascertained cases of child abuse and neglect and concluded that approximately 4 percent of American parents are incompetent. He estimated that approximately 8 percent of single mothers and about 3 percent of couples are incompetent parents. To deal with this problem, Westman proposed that couples should be required to obtain a parental license before they have children. Three criteria would have to be met to obtain the parental license. First, the individual would have to be 18 years of age; secondly, an pledge would have to be given not to abuse or to neglect the child; third, a course in child rearing would have to be taken and passed. The third provision implies that the mentally retarded and possibly also the borderline retarded would not be given the license because these would be likely to perform poorly on the test taken at the end of the course.

Couples who failed to obtain the parental license but had children would have the children taken away from them and placed with adoptive parents or foster parents. Parents who obtained a license but subsequently neglected or abused their children would have their license suspended and would be placed under supervision, or their children put into foster care. The parents would be given further instruction in proper child rearing. If they failed to improve, their license would be revoked, and their children would be permanently removed on the grounds that incompetent parents inflict considerable social damage and that society has a right to protect itself against this. And he noted that society requires licenses for a range of activities that may cause harm if they are carried out incompetently. Furthermore, he contended, children have a right to competent parenting, and if this is not provided by their biological parents, society should ensure that it is provided by adoptive or foster parents.

There are two weaknesses in Westman's scheme. First, it does not contain any proposals to prevent unlicensed couples from having children. As Westman (1996) wrote, "It would not be a birth control measure" (p. 49). Thus, it fails to address the problem of the birth of babies with genetic tendencies for low intelligence, mental retardation, and psychopathic personality. It is uncertain how much the removal of these children from their parents and having them reared by adoptive parents would improve their low intelligence and psychopathic tendencies. The major problem is to reduce the birth incidence of babies of unlicensed parents. To achieve this, some form of punishment would be required, analogous to the punishment of those who drive automobiles without a driving license.

A second weakness concerns the magnitude of the problem of removing a large number of children from their parents. State agencies already remove neglected and ill-treated children from their parents and put them into foster homes. Westman evidently envisioned that this should be done more extensively because he estimated the number of children who should be removed from their parents in the United States at 3.6 million. It is questionable whether adoptive parents could be found for all these children. It would be necessary to pay foster parents to rear them, or they could be reared in institutions. This would be costly, but possibly cost-effective. Despite these weaknesses, Westman's proposal is to be welcomed as an attempt to come to grips with the problem of irresponsible, incompetent, and psychopathic couples having children.

3. THE LYKKEN PROPOSAL

The most recent proposal for parental licensing comes from David Lykken (1995), a psychology professor at the University of Minnesota. He began his discussion by making a distinction between psychopaths and sociopaths. He used the term *psychopath* to denote those who have a genetic deficiency in the capacity to become socialized and *sociopath* for those who have no genetic impairment but who have been reared by parents who have failed to socialize them. The two types are empirically indistinguishable, and probably most psychopaths result from both genetic predisposition and poor environments. Nevertheless, the distinction is useful for thinking about how society should attempt to deal with the problem of psychopathic personality.

Lykken's solution to the problem of sociopathic personality, in his sense of the word, is to remove the babies of psychopathic/sociopathic parents and have them reared in more effective socializing environments. To put this proposal into effect, Lykken proposes a system of parental licensure. As with previous proposals, couples wishing to have a baby would be required to apply for a license. To obtain this, the couple would have to be married, economically independent, and have no criminal record or debilitating illness. They would have to pass a course in parenting lasting about 10 weeks, the purpose of which would be partly to teach the psychological skills of child rearing and partly to eliminate those who were not strongly committed to becoming parents. Lykken did not list intelligence as a criterion for obtaining a parental license, but his proposal would select against those with low IQs because many of them would not be economically independent or able to pass the examination on child rearing. He did not give an estimate of what proportion of the population would be likely to be refused a license for parenthood.

Lykken recognized that if the program were to be introduced, many unlicensed babies would be born. He suggested several ways to deal with this problem. First, a number of these unlicensed babies would be unplanned, and the parents would be permitted to regularize their position by applying for retrospective licenses. No doubt a number of these unlicensed parents would fail the licensing test. In this event the baby would be taken away from the mother and put out for foster care or adoption, or it would be placed in institutions staffed by effective socializing personnel, such as former military noncommissioned officers. Second, Lykken recognized that there would be a great many unlicensed babies who would have to be taken away from their parents and that this would give rise to administrative problems and costs. For this reason, it would be desirable to reduce the number being conceived. He suggested that this might be done by requiring the mother and the father (mothers would be legally required to name the fathers of their babies) to make regular payments for the maintenance of the child until it is grown up or adopted. These financial penalties would hopefully deter unlicensed couples from having babies. Third, women who gave birth to a second unlicensed baby would have to submit to the implantation of a long-lasting contraceptive, such as Norplant.

Lykken did not advance his parental licensing plan on eugenic grounds. His primary objective is not to prevent psychopathic children from being born, but to ensure that potentially psychopathic children are reared by effective socializing agents rather than by their own ineffective parents. Nevertheless, his plan would have some eugenic impact insofar as it might deter or prevent some psychopathic couples from having children.

4. CRITIQUE OF THE LYKKEN PROPOSAL

Although Lykken's proposal is not ostensibly designed to promote eugenics, it would have a eugenic impact insofar as it is intended to curtail births to psychopathic unlicensed couples, and its effectiveness in achieving this objective is the criterion by which we need to consider it. From the eugenic point of view, there are five points to be made about the plan.

First, the sanctions imposed on unlicensed couples who have babies would probably have some deterrent effect. These babies would be taken away from their mothers, so the incentives for single women to have a baby as a means of living on welfare would be removed. Nevertheless, many single mothers do not have their babies with the primary objective of living on welfare but simply by mistake. These would not be deterred from having babies by having the babies taken away from them.

Second, the requirement that an unlicensed couple having a baby would have to pay for the baby's maintenance until it is adopted or if it is not adopted for its care in a foster home or institution until it is adult is unlikely to be an effective deterrent. Most unlicensed couples would be those with low intelligence and psychopathic personality, and it is doubtful whether these would be deterred from having babies by the prospect of having financial sanctions imposed on them in the distant future. Furthermore, maintenance provisions would be largely unenforceable because many unlicensed couples would be unemployed or have low incomes or would be impossible to trace.

Third, the provision that women who have a second unlicensed baby should be prevented from having more children by being given a long-lasting hormonal contraceptive implant would undoubtedly have some impact on the birth rate of women with low intelligence and psychopathic personality. Some of these women have a large number of children, and it would certainly be a gain if these could be reduced. However, it would be better from the perspective of making Lykken's plan workable, and from the perspective of eugenics, if women were required to use a long-lasting contraceptive implant after having their first unlicensed baby, rather than their second.

Fourth, a potential problem in the proposed plan is whether sufficient numbers of couples could be found to adopt the unlicensed babies who would be removed from their mothers. In the United States, this would probably be possible for white unlicensed babies, but it would be more difficult for blacks and Hispanics. There are many white couples who are anxious to adopt white babies but are unwilling to adopt blacks. Surveys have shown that approximately 87 percent of white couples who would like to adopt white babies would be unwilling to adopt black babies (Meezan, Katz, & Russo, 1978). In the last two decades of the twentieth century, about 40 percent of the babies put up for adoption were black; but because most white couples are unwilling to adopt black babies, adoptive homes for many of these could not be found (Posner, 1989; Courtney, 1997).

Fifth, the problem of finding adoptive families for black babies would be compounded because the percentage of black babies compulsorily removed from their mothers would be much greater than that of whites. This can be anticipated because black teenagers have much higher birth rates then white teenagers and would produce a greater proportion of unlicensed babies. The result of this would be that a larger number of black babies, and probably also of Hispanics, would be born to unlicensed women and couples and would be taken away from their mothers to be reared by foster parents or in institutions. This racial imbalance in the impact of the program would be expected to produce protests from the minority community, which would make the program difficult to implement politically. Despite these potential problems, Lykken's parental licensing proposal, like the rather similar ones of LaFollette and Westman, makes a useful contribution to the promotion of the concept of parental licensing and is a valuable thinking exercise from which to start in the formulation of an effective scheme for a eugenic society.

5. AN EFFECTIVE PARENT LICENSING PLAN

The major problem in formulating a parental licensing program is that it is difficult to devise an effective and practical way to prevent unlicensed couples from having children. There are two ways in which this could be attempted. The first is to impose punishments. This would draw on the analogy between having unlicensed babies and the unlicensed driving of automobiles or the unlicensed practice of medicine, the law, and pharmacy, where unlicensed practitioners are punished by fines and imprisonment. Neither the LaFollette nor the Westman plans provide for the punishment of unlicensed couples having babies. The Lykken plan does provide for punishment in the form of financial maintenance orders, but it is doubtful whether these would be an effective deterrent or would be enforceable. In default of financial penalties having a deterrent effect, couples who have unlicensed babies could be imprisoned. It can be anticipated that this would have some deterrent impact as the knowledge of this sanction spread. There would, however, be three problems. The first is that the least intelligent and those with the most pronounced psychopathic personality would be the least likely to be deterred. The second is the large numbers of unlicensed parents who would probably have to be sent to prison. According to Westman's estimate, about 4 percent of couples in the United States would not be fit to receive parental licenses. If a significant proportion of these were not deterred from having babies by the prospect of imprisonment, a lot of them would have to be imprisoned at considerable cost. A third problem is that unlicensed couples could not be imprisoned for life or even for long periods. When they were released, they would be likely to reoffend, as many psychopaths do for other crimes. For all these reasons, punishments to deter unlicensed couples from having babies are unlikely to be very effective.

The second way to prevent unlicensed couples from having babies would be to sterilize all children at around the age of 12. This is probably the only method for making the plan effective. When the children grow up and wish to have children, they would be required to obtain the license and could then have the sterilization reversed. There are several ways in which the sterilization and its later reversal could be carried out. At the age of 12, girls could be required to have some form of long-lasting contraception, such as an IUD (intrauterine device) or Norplant. At appropriate intervals they would be required to have their IUD checked or their Norplant capsules replaced. Women wishing to have children would have these contraceptive devices removed after obtaining their parental license. Boys would have to be sterilized by vasectomy. When they became adults and found partners with whom they wanted to have children, they would have to apply for parental licenses. If these were granted, the vasectomies would be reversed. If this failed, sperm could be taken from their testes and used to fertilize their partners.

This is probably the only practical solution to the enforcement of a parental licensing program. Although it will no doubt seem draconian, it should be noted that most couples use contraception until they wish to have children, so the measures would only make compulsory what most couples already do voluntarily. It would also dramatically reduce the number of abortions, and many would regard this as a desirable feature of the proposal. The mandatory sterilization of boys is more contentious than the requirement of long-lasting contraception for girls, but it could well come to be recognized as the best solution to the otherwise intractable problem of irresponsible men who do not care if they get girls pregnant or who take pride in fathering numerous children whom they frequently abandon. In time, the mandatory sterilization of boys might come to seem no more objectionable than inoculations against infectious diseases, which are taken for granted as desirable. Alternatively, technological developments may provide a means for sterilizing boys for a temporary period, which would be more acceptable to public opinion than vasectomy. A possible development of this kind would be the production of a virus causing temporary sterilization. Contraceptive viruses were developed in Australia in 1997 for the sterilization of rabbits, kangaroos, and other pests. Different viruses can produce sterilization for varying lengths of time. The ideal for humans would be a contraceptive virus acting for about 10 years that could be given to 12-year-old boys. When they were aged 22, they could apply for licenses for parenthood. If they failed to obtain these, they could be vasectomized. This would not preclude them from reapplying for the parental license and, if this were granted, having children by a reversal of the vasectomy or by the extraction of sperm from testes and the artificial insemination of a licensed partner.

6. CONCLUSIONS

The parental licensing plans of the kind proposed initially by Galton and revived in the closing decades of the twentieth century by LaFollette, Westman, and Lykken are broadly similar. They all adopt as their starting point the position that some couples are unfit to rear children, that these couples are principally psychopaths, and that their children would tend to become psychopaths. To prevent this evil, psychopathic couples should be prevented from having children by the operation of a parental licensing program to prevent them from having children. Galton included intelligence among his criteria for granting the parental license, but this is not stated explicitly in the proposals advanced by LaFollette, Westman, or Lykken. Galton was clearly right on this point, and the LaFollette, Westman, and Lykken proposals need to be extended to include mental retardation as an additional criterion for withholding the parental license.

The basic assumption of parental licensing plans that some couples are unfit to have children and should be prevented from having them runs counter to the United Nations Declaration on Human Rights, which asserts that everyone has the right to have unlimited numbers of children. The United Nations Declaration cannot be accepted. There are indisputably some couples who neglect, ill-treat, abuse, and even kill their children. These people are clearly not fit to be parents, and the right to parenthood cannot be extended to them. Societies recognize this by having legal provisions and social agencies to remove children who are ill-treated from their parents and to punish the parents for their child abuse.

Parental licensing proposals contend that those who are unfit to be parents can be identified and the parental license withheld from them. This contention has been disputed by John Harris, a professor of bioethics at the University of Manchester, England. Harris (1998) wrote, "I know of no reliable evidence for any criteria of adequate parenting which can be applied to potential parents rather than to actual parents who have proved their inad-

equacy in objective ways" (p. 95). This assertion that it is impossible to predict who would be an unfit parent cannot be accepted. Most people who reflect
on this question should have no difficulty in concluding that the mentally
retarded, those with long criminal records, and heroin and cocaine addicts do
not make fit parents and that it would be desirable to prevent them from
having children.

While it should be possible to establish in principle that some people are
unfit to be parents and to deny them the parental license, the problem of
preventing them from having children is formidable. The commonest solution proposed to deal with this problem is to remove the children from the
parents and have them adopted, fostered, or placed in institutions; but this is
unlikely to make a significant contribution to preventing their birth.

Galton and Lykken both proposed that punishment would be necessary to
deter unlicensed couples from having babies. These punishments would be
either financial, in the form of fines or maintenance orders for the support of
their children, or imprisonment. It is doubtful whether these punishments
would have strong deterrent effects on the production of babies by unlicensed
couples, who would be those with mental retardation, low intelligence, and
psychopathic personalities, none of whom are readily deterred by the prospect of future punishment. Financial punishment would not be effective
because those denied parental licenses would typically have low incomes and
would be difficult to trace. If the punishment were imprisonment, it would be
impractical to send large numbers to jail for long periods of time. However,
once it became widely known that to have an unlicensed baby would be
punished by a term of imprisonment, and if abortions were free and easily
obtainable, it can be reasonably anticipated that many of those who had
unlicensed pregnancies would have abortions. The most effective solution to
the problem of preventing births to unlicensed couples would be to require
all 12-year-old girls to have some form of long-lasting contraception, such as
the IUD or the contraceptive implant, which could be removed when they
had obtained their parental license. Similarly, all 12-year-old boys would have
to be sterilized by vasectomy or by some other technology, and the sterilization could be reversed when they obtained their parental license and wanted
to have children.

15

⌒

Positive Eugenics

1. Financial Incentives to Have Children
2. Responses to Financial Incentives
3. Selective Incentives for Childbearing
4. Taxation of the Childless
5. Positive Eugenics in Singapore
6. Ethical Obligations of the Elite
7. Immigration
8. A Eugenic Immigration Policy
9. Conclusions

⌒⌒⌒

Positive eugenics consists of policies to induce those with high intelligence and sound personality, the best educated, and high earners to have more children. Strategies for achieving this objective need to be targeted on the elite, who can be broadly equated with the professional class. These strategies consist of providing them with financial incentives for childbearing and of the promotion of a sense of ethical obligation to have children. A further strategy for the promotion of positive eugenics consists of the adoption of an immigration policy designed to accept immigrants with good educational qualifications and professional skills. In this chapter we consider how such policies could be formulated and implemented.

1. FINANCIAL INCENTIVES TO HAVE CHILDREN

The major means of promoting positive eugenics proposed by eugenicists has been the provision of financial incentives for those with desirable qualities to have children. In 1908 Galton advanced this proposal in a lecture to the British Eugenics Education Society. He suggested that local eugenics committees should be established, which would collect funds and award grants to the best young people in the neighborhood to encourage them to get married and have children. Galton (1908) wrote, "I look forward to local eugenic action in numerous directions, of which I will now specify one. It is the accumulation of considerable funds to start young couples of 'worthy' qualities in the married life, and to assist them and their families at critical times" (p. 646).

A number of subsequent eugenicists advanced a variety of proposals by which the state would provide financial incentives for the genetically desirable to have children, rather than the private charities suggested by Galton. Fisher (1929) proposed that tax allowances against income should be given for children on the grounds that as only the professional and middle classes paid income tax at this time, only they would benefit from the measure. He also proposed that the state should give allowances for children proportional to their fathers' incomes, such that fathers with high incomes would receive higher allowances per child than fathers with low incomes. This would provide incentives for high-earning fathers, assumed to have high intelligence and desirable personality traits, to have more children, while avoiding the provision of similar incentives to low-earning fathers (Fisher, 1932).

Later in the 1930s, Cattell (1937) proposed that couples who had produced one highly intelligent child should be offered a grant from the government to have another, on the assumption that the next child would be likely to resemble the first. A similar plan was suggested in the early 1950s by Blacker (1952), a prominent British eugenicist. He proposed that teachers and others holding responsible positions in local communities should identify couples who had produced one or two gifted children and that these "favored married couples whose reproductive lives have already begun should be encouraged, and perhaps helped, to have as many children as their natural inclinations prompt" (p. 311). He suggested that this should be done by giving state allowances for children proportional to the parents' earned incomes, on the assumption that parents with high earned incomes would have genetically desirable qualities.

Programs of this kind were introduced in Germany in 1934 and 1935; in these programs the government provided loans to couples assessed as psychologically and biologically sound, and 25 percent of these loans were written off for each baby they produced. In several cities financial grants were given for third and fourth children born to families assessed as genetically desirable. In 1936 Heinrich Himmler set up special maternity hospitals for the wives and mistresses of members of Hitler's SS to provide the best medical care during their confinement (Kopp, 1936).

2. RESPONSES TO FINANCIAL INCENTIVES

Although the provision of financial incentives to have children is an obvious strategy for positive eugenics, it is important to assess how effective such policies would be. Some studies have addressed this issue by conducting surveys in which women are asked what factors determine the number of children they intend to have. In general, studies of this kind have shown that financial considerations are a significant factor in decisions about family size and suggest that women would respond to financial incentives by having more children. In a British study of 1,458 married women carried out in 1973, two-thirds said they intended to have the number of children they could afford (Cartwright, 1976). A study carried out in Japan in 1993 found that more than half of working mothers in their twenties and thirties said they would have more children if their employers provided paid maternity leave and gave them an additional allowance for housing (Kazue, 1995).

Other studies have addressed this question by ascertaining whether fertility increases when financial incentives for childbearing are raised. Over the course of the twentieth century, a number of countries gave financial incentives to couples to have children, in the form of either child allowances or income tax reductions. Analyses of the effects of these financial incentives have generally concluded that they are positive but quite small (e.g., Glass, 1940; Schorr, 1970). The impact of financial incentives for children has been analyzed in a cross-country study by Gauthier (1991). He collected information on cash benefits and maternity leave payments for 22 Western countries for the years 1970–86 and examined how far the magnitude of the two benefits was related to fertility. The results showed that the associations were positive but low, suggesting a small but positive effect of state payments for children on fertility.

Further evidence for this conclusion comes from U.S. studies of the effects of Medicaid financial support for childbearing. Joyce and Kaestner (1996b) examined this in a study of trends in birth rates and abortion rates between 1985 and 1991. Over this period, a number of U.S. states increased Medicaid eligibility to provide financial assistance for childbirth, making many more women eligible for free health care. The result of this was that the proportion of births financed by Medicaid increased from 14.5 percent in 1985 to 32 percent in 1991. Coincident with this increase in Medicaid assistance, fertility in the United States rose from 65.4 births per 1,000 women aged 15 to 44 in 1986, to 69.6 births per 1,000 in 1991. Over the same period, the abortion rates fell from 28.0 per 1,000 women in 1985 to 25.9 per 1,000 in 1992. Further examination of this issue in the states of South Carolina, Tennessee, and Virginia for the years 1986–91 concluded that increases in Medicaid assistance for childbirth decreased the abortion rate by approximately 3.5 percent and were responsible for a corresponding increase in the birth rate (Joyce & Kaestner, 1996b; Joyce, Kaestner, & Kwan, 1998).

3. SELECTIVE INCENTIVES FOR CHILDBEARING

A major problem for positive eugenics is the designing of policies for childbearing that provide selective incentives for the more intelligent and those with stronger moral character to have children but that avoid giving these incentives to the less intelligent and the more psychopathic. Flat-rate child allowances and other forms of payment for childbearing cannot be expected to have a eugenic impact because they are more attractive to low earners, for whom these payments are larger, proportional to their incomes. For high earners flat-rate payments will inevitably be too small to have any significant impact. The ideal strategy of providing selective financial incentives for the intelligent and those with stronger moral character to have children is probably not politically feasible in democratic societies. However, a politically practical alternative would be to provide incentives for childbearing and child rearing that are proportional to incomes, such that those with high incomes benefit most. The most straightforward ways of doing this are by providing tax relief for children and for child care expenses and also by the provision of paid maternity leave. Many countries already provide incentives of this kind, although some do not. For instance, the United States provides tax credits for children and child care, but these are not given in Britain and in a number of countries of Continental Europe. Introducing and increasing these tax allowances should be politically feasible because it is difficult to object to couples being given financial help for children.

Paid maternity leave is another measure that has a selective impact on high earners. The experience of Sweden of the impact of this on the fertility of high-earning women is instructive. In the early 1980s Swedish women were given paid maternity leave, the duration of which was progressively increased until by 1986 it was for 30 months. If a woman had a second child within this 30-month period, she received full pay for a further 30 months, and similarly for subsequent children. The effect of this has been that women who are not in employment do not gain any advantage because they continue to receive the same welfare benefits; but highly paid professional women gain substantially in comparison to the loss of income they would otherwise have suffered by giving up work for five years to have two children, or seven and a half years to have three children. Swedish women in employment have responded to this financial incentive with the result that the total fertility rate in Sweden increased from 1983 onward from a low of 1.7 to 2.1 (Hoem, 1990, 1993).

This evidence from Sweden suggests three important conclusions. First, it confirms other studies showing that childbearing has what economists call price elasticity—if financial incentives to have children are made available, women will respond by having children. Second, it shows that the declining fertility of the economically developed nations to below replacement levels in the 1980s and early 1990s is a problem that can be relatively easily overcome by the provision of financial incentives to working women. Similar reversals of declining fertility following incentives for childbearing have occurred in Norway and Denmark over the period 1982–90 (Hoem, 1990). Third, it shows that it is possible for democratic societies to implement positive eugenic policies framed in such a way that they provide selective incentives for women who are intelligent and have strong moral character to have children, while at the same time avoiding giving these incentives to the less intelligent and the psychopathic.

4. TAXATION OF THE CHILDLESS

Another strategy for positive eugenics would be to increase the taxation of the childless. These taxes should be progressive, as they normally are, so that those with the highest incomes would be the most heavily taxed. A scheme of this kind was introduced in the first century A.D. in the Roman Empire by the emperor Augustus, who was concerned about the dysgenic impact of the low fertility of the patrician class. In an attempt to overcome this problem, Augustus introduced a substantial tax on childlessness, with the objective of encouraging the patrician class to have more children. Although this plan has not been advanced in the contemporary world, it would seem to have some promise as a means of inducing high earners to increase their fertility. Such a proposal would be relatively easy to introduce politically because childless high earners are few in number and would not be expected to receive much public sympathy if they were taxed more heavily. A program of this kind has much to recommend it, particularly if it were introduced in conjunction with tax allowances for children, thereby providing high earners with both a carrot and a stick to increase their fertility.

5. POSITIVE EUGENICS IN SINGAPORE

The only country where positive eugenics was explicitly pursued in the second half of the twentieth century was Singapore. Lee Kuan Yew, the prime minister from 1959 to 1990, was concerned when he noted that the Singapore census returns showed that well-educated women were having fewer children than the poorly educated, and he realized that this would have a dysgenic impact on the population. In 1987 he introduced six measures designed to correct this by encouraging women graduates and high earners to marry and to have more children.

First, a publicity campaign was launched to encourage childbearing under the rubric, "Have three, and more if you can afford it." Government spokespeople explained that the fertility decline of recent years had occurred largely among the better educated and that these needed to be encouraged to have more children. The qualification, "more if you can afford it," was based on the assumption that people with high incomes were intelligent and had desirable personality qualities and should receive special encouragement to raise their fertility. Second, tax allowances against earned income were given for all children, but only middle-class parents paid sufficient tax to benefit from these; so the effect was to give a selective incentive to the middle class to increase their numbers of children, while not providing this incentive to the working class. Third, medical fees for childbirth were made tax deductible against income for the first four children. This also gave a selective benefit to middle-class parents. Fourth, mothers with good educational qualifications were given additional tax incentives to have children. These incentives consisted of 5 percent of their income free of tax for the first child, 10 percent for the second, and 15 percent for subsequent children. Fifth, tax credits were given for the first three children to attend approved child care centers. Sixth, a special unit was set up in the civil service to bring unmarried men and women graduates together in social settings, such as dances and cruises, with the objective of promoting romance, marriage, and childbearing among the nation's elite.

The impact of these measures can be assessed by examining whether the fertility of better educated women increased following their introduction. Statistics of births to women with secondary education and above, as compared with births to women without secondary education, are shown in Table 15.1 for 1987, when the measures were introduced, and for 1990, after the measures had been in place for about three years. They suggest that the measures had a significant impact. Note that births to women with secondary education and above increased in absolute terms from 16,012 to 24,411, while those of poorly educated women remained static; and that the percentage of births to women with secondary education rose from 36.7 percent to 47.7 percent, while those to poorly educated women showed a corresponding decline (Singapore Ministry of Health, 1994). These figures are a testimony to the effectiveness of a program of positive eugenics that is based on a mix of moral exhortation and financial incentives, but without the use of coercion.

6. ETHICAL OBLIGATIONS OF THE ELITE

While it can be confidently expected that elites would respond to financial incentives to have children and to penalties for childlessness by increasing their fertility, they might not do this to the extent that would be desired. Ideally a program of positive eugenics would increase the fertility of the elite

Table. 15.1
Live Births in Singapore in 1987 and 1990, by Education of Mother

Educational Level	1987		1990	
	Number	Percent	Number	Percent
Below secondary	26,719	61.3	26,718	52.3
Secondary and above	16,012	36.7	24,411	47.7

to perhaps around four children per couple; and at the same time a complementary program of negative eugenics would reduce the fertility of those with low intelligence and psychopathic personality to zero. Even with the provision of large financial incentives, the elite might not produce this number of children. There would be three potential problems. First, rearing four children entails a considerable sacrifice of time and energy to the detriment of careers of women, and this is a sacrifice that many professional women might not be willing to make. Second, the financial incentives to be offered to these women would probably have to be very considerable for them to make these career sacrifices. Rearing children from birth through graduation from college is a costly undertaking. For the United States, Westman (1994) estimated the cost at approximately $10,000 a year. To this would have to be added the costs of income foregone by the loss or the reduction of earnings and/or payments for child care, which might amount to around a further $10,000 a year. Thus, it might be necessary to pay the elite around $20,000 a year for their third and subsequent children, on the assumption that they would have only two without these incentives. This would entail sizeable transfer payments from the remainder of the population levied through the tax system, and it is doubtful whether this would be feasible in democratic societies. Third, many elite women remain childless in their twenties and early thirties in order to devote themselves to the advancement of their careers. They intend to have children sometime in their mid-thirties, but when they reach this age they discover that suitable men with whom to share the burden of child rearing are hard to find or that they have become infertile, for women's fertility is at its highest in the late teens and twenties and falls significantly in the thirties. For one or other of these reasons, significant numbers of elite women do not have children, and probably many of these would still not have children, even with the provision of substantial financial incentives.

The problems of finding effective policies of positive eugenics for elite men to have children are less difficult because men need not sacrifice their careers to rear children, they remain fertile over a longer period of their lives, and men in their late thirties and older find it easier to find younger women willing to bear their children than women in the same age bracket do to find suitable men.

The provision of financial incentives for elites to increase their numbers of children would have to be supplemented with the attempt to promote the belief that the intelligent and those with strong moral character have an ethical obligation to have children. The view prevailing at the beginning of the twenty-first century that the intelligent and well educated are entitled to remain childless would need to be replaced by a climate of opinion in which some degree of moral stigma becomes attached to those who fail to discharge their duty to maintain and to enhance the genetic quality of the population by transmitting their genes to succeeding generations.

7. IMMIGRATION

The final strategy for the promotion of positive eugenics would consist of the acceptance of good-quality immigrants. Historically, the leading instance where immigration has had a positive impact has been the admission into the United States and Britain of Jewish refugees from Russia and Eastern Europe, principally over the years 1890–1914, and from Nazi Germany during the 1930s. Several studies have shown that these Ashkenazi Jews have mean IQs of around 115 (Herrnstein and Murray, 1944; MacDonald, 1994). They have high educational and occupational achievements consistent with their high levels of intelligence. For instance, it has been shown by Weyl (1989) that Jews in the United States are approximately five times overrepresented among the professional and intellectual elite of U.S. scientists, physicians, lawyers, writers, musicians, and businesspeople, than would be expected from their numbers in the population. At a very high level of achievement, it has been shown by H. Zuckerman (1977) that Jews, who comprise approximately 3 percent of the U.S. population, contribute 27 percent of the Nobel prizewinners. There is no doubt that Jews have made an immense contribution to the scientific, cultural, and military strength of the United States.

The contribution of Jewish immigrants to the United States is strikingly exemplified by the development of the atom bomb over the period 1939–45. The principal physicists responsible for building the atom bomb in the Manhattan Project were Albert Einstein (whose letter to President Franklin Roosevelt in 1939 was responsible for the president instigating the project), Enrico Fermi, Leo Szilard, and Klaus Fuchs, all of whom were Jewish immigrants from Central Europe; and Julius Oppenheimer, whose family were Jewish immigrants from Eastern Europe. Paradoxically, if Hitler had pursued a eugenic policy of building good relations with the Jews, the first four of these and many other gifted Jews would have remained in Europe, and Hitler could have recruited them to work on the development of a German atom bomb. When this had been built, it would have enabled him to coerce the rest of the world into submission and achieve his ambition for Germany to secure world domination. The development of the atom bomb is an instructive example of the eugenic principle that nations that possess highly intelligent manpower and apply this to the development of advanced weapons are able to defeat nations with fewer resources.

Another group of immigrants who have had a positive eugenic impact in the United States are the Chinese, the Japanese, and the Koreans. Their average intelligence level is around 105 (Lynn, 1997); and their high educational and occupational achievements, particularly in science and technology, were documented by Flynn (1991), who also argued that they have personality qualities conducive to high achievement. Among the U.S. professional and scientific elite, they are about five times overrepresented (Weyl, 1989), and they too have made significant contributions to U.S. economic, cultural, and military strength.

8. A EUGENIC IMMIGRATION POLICY

Even though Jewish and Asian immigrants as a whole have made a positive eugenic impact on the U.S. population, the best approach to immigration policy would be to select immigrants as individuals rather than by ethnic group. There is a large range of desirable and undesirable qualities within each group. The easiest way of implementing such a policy would be to admit immigrants with strong educational qualifications and professional skills, which serve as reasonably good proxies for intelligence and the absence of psychopathic personality. To some extent, selective immigration policies of this kind are operated in the United States, Canada, Australia, and New Zealand. These countries have quotas for the numbers of immigrants they accept, and some of these are reserved for those with useful skills. For instance, Australia operates a "skill stream," under which applicants for immigration are only accepted if they have vocational qualifications; and in the 1990s approximately half of immigrants were admitted under this rule (Miller, 1999).

In the United States a quota for the admission of immigrants with skills was contained in the 1952 McCarran-Walter Act. This quota was reduced in the 1965 Immigration and Nationality Act in favor of the admission of larger numbers of relatives of immigrants who had already been accepted. The effect of the change in the immigration criteria introduced in the 1965 Immigration Act appears to have been to lower the quality of immigrants. With respect to intelligence, it has been shown by Herrnstein and Murray (1994) that the children of immigrants have lower average IQs than the native-born American population. In an analysis of the National Longitudinal Study of Youth data set, Herrnstein and Murray found that the average IQ of the children of Hispanic immigrants was 81, of black immigrants 88, and of European immigrants 97.

This conclusion has been corroborated by studies of the earnings of immigrants. In the mid-1980s it was shown by Borjas (1990, 1993) that the earnings and skills of immigrants as a whole declined following the 1965 act. He showed that this deterioration in the quality of immigrants was attributable to a change in their ethnic and racial origin. Before 1965 most immigrants came from Europe and were of good quality, as defined by their educational qualifications and earnings. From the 1980s onward, most immigrants were Hispanic, black, or southeast Asian and were of poorer quality. The conclusion that recent immigrants have been of poorer quality than earlier immigrants has been endorsed by a number of subsequent investigators. The consensus emerging from these studies has been summarized by Hayfron (1998): "The results seem to indicate that, after accounting for the usual factors that determine immigrant earnings, recent cohorts have lower earnings than their native counterparts" (p. 294). Hayfron has reached the same conclusion in an analysis of the earnings of cohorts of immigrants into Norway, finding that "the quality, as measured by education and earnings, of successive waves has declined over time" (p. 301).

A reduction in the numbers of immigrants admitted on the basis of their skills also occurred in the closing decades of the twentieth century in Canada, Australia, and New Zealand. In Europe, immigrants are not generally accepted on the grounds of skills, although there is large-scale immigration of relatives, asylum seekers, and illegals. As in the United States, the impact of recent immigration into Europe appears to have been dysgenic. Immigrants in a number of European countries, including Britain, the Netherlands, Sweden, and Italy, on average have lower earnings, higher unemployment, and substantially higher crime rates than the indigenous populations (Tonry, 1997; Solivetti & D'Onofrio, 1996).

These recent trends in immigration into the United States and Europe have been a significant setback for eugenics. The eugenic objective with regard to immigration should be to reestablish the possession of skills as the major criterion for the acceptance of immigrants and the more rigorous exclusion of those lacking skills.

9. CONCLUSIONS

There are three broad strategies for the promotion of positive eugenics. These consist of the provision of financial incentives for the elite to have children, the promotion of the belief among the elite that they have an ethical obligation to have children, and the admission of good-quality immigrants, using the criterion of the possession of educational qualifications and professional skills.

In practical terms, it has to be accepted that all of these strategies would be difficult to implement in the Western democracies on a sufficient scale to have much significant eugenic impact, given the current climate of opinion. The provision of financial incentives for the elite to have children would provoke envy. Attempts to promote the ethical obligation of the elite, and especially elite women, to have children would be attacked by feminists and much of the media. The attempt to reduce the immigration of those with poor educational qualifications and skills would be attacked by the ethnic and racial groups who would be disadvantaged and by a variety of proimmigration lobbies, who are unaware or unconcerned about the negative impact on the genetic quality of the populations of much current immigration into the Western democracies. As with programs of negative eugenics, the major problem for classical positive eugenics in the Western democracies at the dawn of the twenty-first century lies, contrary to the frequent assertions of the critics, not in the identification of the traits that would be desirable to improve, nor in the genetics of what could be achieved in principle, but in the political difficulties of the introduction of measures that would have significant eugenic impacts.

16

⇆

The Ethical Principles of Classical Eugenics

1. Provision of Information and Services
2. Subsidized Information and Services
3. Incentives to Promote Positive Eugenics
4. Incentives to Promote Negative Eugenics
5. Sterilization
6. The Case of *Buck v. Bell*
7. Immigration
8. Secondary Immigration
9. The Slippery Slope
10. Conclusions

༄༅༅

It is time for us to consider the ethical principles of classical eugenics. When eugenics came under attack in the closing decades of the twentieth century, it came to be widely asserted that eugenics is ethically unacceptable. The prevailing view was captured by Raanan Gillon (1998), a professor of medicine at the University of London, when he wrote that "eugenics is widely regarded as a dirty word" that evokes "widespread revulsion" and is "morally objectionable" (p. 219). In a similar vein, Falk, Paul, and Allen (1998) wrote that by the end of the twentieth century, it came "to be taken for granted that the whole project was morally unacceptable. To label a policy 'eugenics' became *ipso facto* to condemn it" (p. 30). Or as Diane Paul (1995), a professor at the University of Massachusetts, put it more simply, "Today, eugenics is generally assumed to be bad" (p. 3).

As we have seen, there are many different policies that can promote eugenics. These can be arranged on a spectrum running from the provision of education and information to the offering of incentives and the use of coercion and compulsion. These policies present different ethical problems and need to be considered separately.

1. PROVISION OF INFORMATION AND SERVICES

The least ethically contentious form of eugenics is the provision of information about and services for contraception and abortion. As a general principle, when the state or institutions such as newspapers, advertisements, privately run charities, and so forth provide information and services, citizens are at liberty to act on the information and use the services or to disregard them. So far as eugenics is concerned, the principal information and services provided by various agencies are about contraception and abortion. Information on these matters is frequently provided in schools as part of more general sex education and may include information on the dangers of contracting AIDS and other sexually transmitted diseases. Schools in many countries provide sex education covering these topics, and there is a widespread consensus that this is acceptable and sensible. The information is not provided with any ostensible eugenic intention, but it serves a eugenic purpose because in more intelligent and responsible families, this kind of information is usually imparted by parents to their teenage children; whereas in less intelligent and less responsible families, the parents may fail to impart this information, with the result that less intelligent and more psychopathic teenagers have more unwanted and unplanned pregnancies.

Ethical objections are sometimes made to the provision of information as part of sex education in schools on the grounds that it appears to condone and/or encourage sexual activity, which should not be condoned or encouraged amongst teenagers. This may be a legitimate concern, but it is not an objection on the grounds that the provision of this education is eugenic. On balance, there is widespread consensus that unplanned teenage pregnancies are undesirable both for the girls concerned and for society and that the provision of education and information to prevent these is both ethically unobjectionable and desirable. The same goes for the provision of this information in family planning clinics. There are certainly some people who have ethical objections to contraception and abortion, but these are not sufficient grounds for prohibiting the provision of information and services about them. In a pluralist society where these things are permitted, no valid ethical objection can be made to the provision of information and services about them.

2. SUBSIDIZED INFORMATION AND SERVICES

Many countries go farther than the provision of education and information about contraception and abortion and provide free or subsidized services.

These are similar to the provision of information insofar as they are not provided with any eugenic intention but rather to promote women's health and psychological well-being. Nevertheless, they have a eugenic impact insofar as they make it easier for people to control their fertility. This free or subsidized provision of contraception and abortion is similar in principle to many other entities that governments support on the grounds that they are socially valuable and should be encouraged. Many governments provide free or subsidized education, health, sports, libraries, museums, opera, art, parks, and numerous other things on these grounds. The provision of state-subsidized contraception and abortion is no different in principle from these, and it is impossible to raise any valid ethical objections against it.

3. INCENTIVES TO PROMOTE POSITIVE EUGENICS

The state may go a little farther than the provision of information and services and offer incentives that, whether intentionally or otherwise, promote eugenic objectives. These incentives may be designed to promote either positive or negative eugenics. The ethically significant feature of the offer of incentives is that they do not compel those to whom the inducements are offered to have more or fewer children than they might otherwise have, but they do put them under varying degrees of pressure to do so.

So far as positive eugenics is concerned, a number of eugenicists have made proposals to promote eugenics by the offering of incentives. These incentives have been principally financial and consist of the offer of payments to elites to have children. As we have seen, proposals of this kind have frequently been made by eugenicists, including F. Galton, R. A. Fisher, and R. B. Cattell, and were introduced by Lee Kuan Yew in Singapore in the 1980s. These incentives may be offered by the state through the tax system or by private institutions or foundations.

It is difficult to raise valid ethical objections to the provision of these incentives. Many governments offer incentives for childbearing either by the provision of child allowances or through tax allowances. No distinction of principle can be drawn between the state provision of subsidies to a wide range of services that are considered to be in the public interest, such as education, health, the arts, sports, libraries, and the like, and the state provision of subsidies for people to have children. Nor can any objection be made to the state offering subsidies to elites. A number of state subsidies are made for the benefit of elites rather than for the general population. This is the case, for example, in the provision of subsidies for universities, opera, and the arts. Political leaders believe that it is in the public interest for these things to be subsidized; and although some people may disagree, it is impossible to make an ethical objection. Similarly, if governments believe that it would be in the public interest for elites to have more children, there can be no ethical objection in democratic societies to the offer of subsidies for them to do so.

Subsidies for elites to bear children may also be legitimately offered by private individuals or foundations. From an ethical standpoint, this is conceptually similar to the scholarships for highly intelligent young people to attend universities and to pursue postgraduate studies. These scholarships are intended for the benefit of elites. A scheme for private institutions to subsidize elites to have children was suggested by Galton (1909) when he proposed the establishment of local eugenic societies that would identify what he called the "most worthy" young couples in the area and would provide them with financial assistance to have children. A similar plan was put into effect in the United States by the Pioneer Fund in 1939 when it offered financial inducements to American Air Corps officers to have children on the grounds that these were an elite group possessing high intelligence and strong moral character and that their children would be likely to inherit these qualities and make valuable citizens (Flanagan, 1939).

It is impossible to raise ethical objections to programs of this kind. In free societies people are permitted to spend their money as they choose, so long as this does not inflict social damage, and it cannot be argued that social damage is caused by the provision of financial incentives for elites to have children. People exercise the liberty of spending their money in all sorts of eccentric ways. If they wish to spend it by offering inducements to those they regard as having genetically desirable qualities to have children, they are entitled to do so in a free society.

4. INCENTIVES TO PROMOTE NEGATIVE EUGENICS

Incentives may also be offered, either by the state or by private agencies, to promote negative eugenics. As with incentives for positive eugenics, these incentives would normally consist of financial offers to those with undesirable characteristics not to have children. Several of these plans have been described in Chapter 13, such as the proposal made by William Shockley (1972) to offer payments for sterilization to those on welfare benefits who have IQs below 100. No ethical objections can be made to programs of this kind because the individuals to whom the offers are made are free to either accept or reject them. If they accept them, they do so of their own free will and, it can be presumed, increase their utility, while if they reject them they lose nothing.

Another form of these incentive offers is making welfare payments conditional on not having more children. This proposal was also made by Shockley (1972), who proposed that women on welfare should be required to use some long-lasting contraception to ensure that they have no more children as a condition of receiving welfare payments. A parallel program could be applied to long-term unemployed men, who could be required to undergo vasectomies as a condition of continuing to receive welfare benefits. The cessation of paying welfare benefits to single mothers is a further variant of this class of proposals designed to provide financial incentives for those on welfare not to have children.

In the case of criminals, the incentive offered to not have children is not money but liberty. As we noted in Chapter 13, a number of U.S. courts have offered female criminals probation instead of terms of imprisonment on condition that they do not have children. Similar incentives have been offered in Germany to male sex offenders, who have been offered the option of a shorter term of imprisonment conditional on chemical castration for a period of years.

It is sometimes contended that offers of this general kind are coercive because those to whom the offers are made have no practical alternative to accepting them. Thus, a welfare mother who is told that she will only continue to be paid benefits on condition of having some kind of contraceptive fitted has no real choice other than to accept. A woman who agrees to this provision cannot be regarded as freely entering into a contract. The same may be said of a male criminal who is offered the choice between a long term of imprisonment and castration. Offers of this kind have sometimes been called "offers that cannot be refused," condemned as essentially coercive and therefore as unethical. In some instances, however, these offers are refused, and this has proved to be the case with a number of criminals in Germany offered the choice between imprisonment and castration. Insofar as this objection is valid, what may be called "coercive incentives" have to be justified in terms of the ethical right of the state to discourage or even compulsorily to prevent certain classes of individuals from having children. There can be no doubt that states have an ethical right to attempt to prevent people from doing things that inflict social damage, and this right is exercised through the criminal law and police. The offering of incentives not to have children is covered by this general principle.

5. STERILIZATION

One of the most ethically contentious forms of eugenics is mandatory sterilization, such as was extensively carried out on the mentally retarded and criminals in the United States, Japan, and a number of European countries in the early and middle decades of the twentieth century. From the 1970s onwards, sterilization was widely condemned as ethically unacceptable. For instance, one U.S. biologist, Lee Silver (1996), wrote that "the forced sterilizations in America were wrong because they restricted the reproductive liberties of innocent people" (p. 217). Jacques Testart (1995), a French geneticist, wrote that "negative eugenics, imposed in opposition to human liberty and dignity, has become unacceptable in the majority of democratic societies" (p. 304).

And Matt Ridley (1998), an English sociobiologist, wrote that eugenic sterilization "was oppressive and cruel because it required the full power of the state to be asserted over the rights of the individual" (p. 46).

These objections to the ethical legitimacy of sterilization fail to recognize that there are social rights as well as individual rights. While there are individual rights of freedom of expression and behavior, there are also social rights to restrict these freedoms where the exercise of them is socially damaging. This is the ethical principle on which sterilization relies. The general principle that the liberties of individuals can legitimately be curtailed in the interests of the well-being of society has long been recognized by political theorists. In the seventeenth century, Thomas Hobbes (1651) argued in his *Leviathan* that the individual's freedom to assault or rob others must be surrendered in the interests of preserving social order. In the nineteenth century this principle was restated by John Stuart Mill (1859) in his essay *On Liberty* and by Herbert Spencer (1868) as what he called the Law of Equal Freedom, which stated that "every man has the freedom to do all that he wills, provided that he infringes not the equal freedom of any other man" (p. 121). In the second half of the twentieth century, the rights both of individuals and of society were set out by Hayek (1960) in *The Constitution of Liberty*. His conclusions can be summarized in terms of three general principles. First, individuals have rights of liberty to act as they please, except where this is likely to inflict harm on others. Second, society has rights to curtail the liberties of its citizens where the exercise of these liberties may adversely affect social well-being. Third, society has rights to curtail the liberties of individuals by requiring them to behave in certain ways if this is judged likely to promote social well-being.

Thus, in terms of general principle, it is unquestionably ethically legitimate for societies to curtail individual rights when the exercise of these is judged likely to inflict social damage. This is the principle underlying the restriction of individual liberties in the Western democracies and, no doubt, in authoritarian states as well. The liberties that are curtailed can be ordered along a continuum ranging from behaviors that indisputably cause social damage to those for which there is a probability or a possibility that they may cause social damage. The likelihood of this probability or possibility of social damage and its seriousness is a matter of judgment. In cases where there is a virtual certainty that behavior causes social damage, there is a corresponding consensus that the liberties of individuals should be curtailed in the interests of preserving social well-being. For instance, no one disputes that the rights to liberty of those convicted of serious crimes should be curtailed by imprisonment in the interests of preserving the rights of other citizens to be protected from crime.

There are numerous other examples where society curtails the rights of individuals in the interests of social well-being. For instance, Western societies recognize a general right to freedom of speech, but this right does not extend to shouting "Fire!" in a crowded auditorium where it is likely to cause panic, disorder, and injury. The same principle is used in some western Western societies, such as Canada, Britain, and several countries of Continental Europe, to restrict the right to freedom of speech if this is offensive to certain racial and ethnic groups and is likely to inflame racial hatred. Some people consider that this restriction cannot be justified in what purport to be free societies. Still, if democratically elected governments impose these restrictions, it is difficult to argue that they are ethically objectionable.

It has sometimes been argued that sterilization is not justified because normally there is not a certainty but only a possibility that social problem groups will transmit their characteristics to their children. For example, Adrian Raine (1993), a psychologist at the University of California who specializes in the psychology of crime, wrote that "current research indicates a genetic *predisposition* to crime only; individuals are not born to commit crime; there can be no genetic destiny for crime as such. Eugenic solutions are not supported by the findings of genetic research" (p. 52). This argument against eugenic sterilization to curtail the reproduction of criminals cannot be accepted. Society prohibits many behaviors on the basis that there is some actuarial possibility that they may inflict social harm, but no certainty that they will do so. For instance, societies impose speed limits for driving on public roads and highways. There is no certainty that driving faster than the speed limit will cause any harm and a probability that in most cases it will not. Nevertheless, there is an enhanced possibility that it will do so, and on these grounds society legitimately prohibits it. Similarly, there is no certainty that sibling matings and parent-child matings will produce genetically impaired children, but only an elevated probability that they will do so, and this is the basis on which they are prohibited in most societies. The same actuarial argument applies to the children produced by criminals and the mentally retarded and provides the ethical legitimacy for their sterilization.

6.   THE CASE OF *BUCK V. BELL*

The leading legal case in which the issue of the ethical legitimacy of sterilization was argued in the United States is *Buck v. Bell* (1927). The background of the case was that from 1907 onward, a number of U.S. states had passed legislation for the sterilization of the mentally retarded, the mentally ill, and criminals. On several occasions, the recommendations of physicians for carrying out sterilizations were subjected to legal challenges. These challenges were based principally on the grounds that sterilization constituted an "unusual" punishment and violated the Equal Protection Clause of the Constitution, which stipulates that like persons be treated in similar manner. Courts in different states reached different decisions, some upholding the rights of institutions to sterilize their inmates and others rejecting it.

One such case where an institution made a recommendation for sterilization that was challenged in the courts concerned Carrie Buck, a 17-year-old white girl involuntarily confined in a state institution for the mentally retarded in Virginia. Carrie Buck had been intelligence tested and was found to have an IQ of 56, well below the threshold of 70 for a diagnosis of mental retardation. Carrie was the illegitimate daughter of Emma Buck, who was found to have an IQ of 50; and Carrie had an illegitimate daughter of her own, Vivian, aged six months, whom a social worker considered was also mentally retarded.

In 1924 the physicians at the state institution where Carrie Buck was confined recommended that she should be sterilized. This recommendation was challenged in the Virginia high court by a group of conservative Christians who held it was a violation of individual rights. The Virginia court upheld the decision to sterilize. In 1927, the case went to appeal to the U.S. Supreme Court, which decided by a vote of eight to one in favor of sterilization (Cynkar, 1981). The Supreme Court's decision was read by Justice Oliver Wendell Holmes. The crucial passage was as follows:

> We have seen more than once that the public welfare may call upon the best citizens for their lives. It would be strange if it could not call upon those who already sap the strength of the state for these lesser sacrifices . . . in order to prevent our being swamped with incompetence. It is better for all the world if, instead of waiting to execute degenerate offspring for crime, or to let them starve for their imbecility, society can prevent those who are manifestly unfit for continuing their kind. The principle that sustains compulsory vaccination is broad enough to cover cutting the Fallopian tubes. Three generations of imbeciles are enough. (United States Supreme Court Reports, 1927)

Following this judgment, Carrie Buck was sterilized. It is evident that in his justification for the Supreme Court's decision, Oliver Wendell Holmes was relying on the principle that there are social rights that may sometimes override the individual's rights to have children. He noted that the state requires its citizens to give their lives, in time of war, for the benefit of their country and that it requires infants to be vaccinated in the social interest of preventing the spread of contagious diseases. He pointed out that the mentally retarded impose costs on society and that any children they have are likely to be mentally retarded; and that to avert the likely social costs of these future children, the mentally retarded can legitimately be sterilized.

The case of Carrie Buck became a focus for the opposition to compulsory sterilization of the mentally retarded, which gathered momentum in the second half of the twentieth century. Among those who have used the case to mount an argument against eugenic sterilization was Stephen Jay Gould (1985). He presented five arguments against the sterilization of Carrie Buck. First, he asserts that Carrie Buck was not mentally retarded. He said that her intelligence was ascertained by the Stanford-Binet test, which he described as a "crude test, then in its infancy," and "fatally flawed as a measure of innate worth" (p. 105). In a similar vein, Kevles (1985) asserted that the Stanford-

Binet "has long been discredited as an indicator of purely general intelligence" (p. 112). These criticisms of the Stanford-Binet Test are entirely misconceived. The Stanford-Binet Test was among the best intelligence tests of its time and in updated revisions and restandardizations remained one of the leading intelligence tests throughout the twentieth century. It may be said that it was "in its infancy" in 1924, since the test had been standardized in the United States in 1916 (Terman, 1917b); but an intelligence test cannot be discredited on the grounds that it has been constructed relatively recently. Data derived from the Stanford-Binet Test has been used by Flynn (1984) for his study of the secular rise of intelligence in the United States and by Weinberg, Scarr, and Waldman (1992) in their study of the effects of the transracial adoption of black babies by white couples. It has been and remains today one of the best and most widely used intelligence tests.

Second, Gould's assertion that the test is fatally flawed as "a measure of innate worth" cannot be accepted. The test was used by Newman, Freeman, and Holzinger (1937) in their study of the resemblance of the identical twins reared apart in different families, for which they reported a correlation of .66. This means that the test does give a measure of innate intelligence to the value of .66.

Third, to support his assertion that Carrie Buck was not mentally retarded, Gould (1985) cited a letter received from one Paul Lombardo, a lawyer who visited Carrie Buck around 1980 and found her reading a newspaper and "joining a more literate friend to assist at regular bouts with the crossword puzzles. . . . [This] confirmed my impression that she was neither mentally ill nor retarded" (p. 105). Contrary to this assertion, there is no reason why Carrie Buck, with an IQ of 56, should not have been reading newspapers and assisting in crosswords. An adult with an IQ of 56 has a mental age of nine years, that is to say, has the abilities of the average nine-year-old. The average nine-year-old can read popular newspapers with ease and make sensible suggestions for the solutions to simple crossword puzzles. Thus, despite the evidence on Carrie Buck's low IQ obtained at the time, Gould rejected the results of the intelligence test and asserted on the basis of the impressions of a lawyer who observed Carrie Buck some half century later that "Carrie Buck was a woman of obviously normal intelligence" (p. 106).

Fourth, Gould asserted that Carrie Buck's daughter, Vivian, was not mentally retarded. The historical record is that Vivian spent four terms at an elementary school and died at the age of eight. The school records show that she obtained predominantly C grades and was required to repeat a term in one class. Gould said she "performed adequately, although not brilliantly, in her academic subjects" (p. 106). Her intelligence was not tested, but the fact that she was required to repeat a term indicates that she was backward. Just how backward she was cannot be determined with any degree of confidence.

Fifth, Gould asserted that Buck's mother Emma, whose IQ was ascertained at 50, was not mentally retarded. He did not argue the case but simply asserted that this IQ must be wrong. He concluded his account of the case by asserting that "there were no imbeciles, not one, among the three genera-tions of Bucks" (p. 106). Thus, Gould rejected the conclusions of the physi-cians who examined Carrie Buck in the 1920s, the intelligence test results on her and her mother, and the judgments of the lawyers and judges, including those of the Supreme Court. Gould's treatment of this case confirms the conclusions of J. P. Rushton (1997), who wrote of Gould's "career of relent-less special pleading" (p.178), and of the anthropologists Milford Wolpoff and Rachel Caspari (1997), who wrote that "Gould's essays invariably have a not-so-hidden agenda" (p. 54). Nowhere is Gould's ideological agenda more trans-parent than in his treatment of the case of Carrie Buck.

Insofar as it is possible to reconstruct the evidence regarding Carrie Buck at a distance of three-quarters of a century, the case for sterilizing her looks sound. Her IQ of 56 had been ascertained with the leading intelligence test of the time. With an IQ at this level she would not have been able to cope with life in the community and was accordingly confined to an institution for the mentally retarded. At the age of 17, she had already had one illegitimate child whose father she did not know, or at any rate would not reveal. It was likely that she would have more illegitimate children who could not have been cared for in the institution. They would have to have been placed out for adoption, so she would have derived no satisfaction from having and rais-ing them. These children would themselves be likely to be mentally retarded or borderline mentally retarded, with IQs in the 50–85 range, depending on the IQs of the fathers. They would likely have been a disappointment to their adoptive parents and would have made little contribution to society. All the evidence suggests that the decision to sterilize Carrie Buck was a sensible one. It was a case where, as Oliver Wendell Holmes and his colleagues on the U.S. Supreme Court recognized, the reproductive rights of the mentally retarded could legitimately be curtailed in the interests of the well-being of society.

7. IMMIGRATION

A further ethically controversial eugenic policy is the restriction of the admission of immigrants to those people considered desirable and the refusal to accept immigrants considered undesirable. In considering the ethical prob-lems of immigration policy, it is important to begin by distinguishing between what is called "primary immigration" and "secondary immigration." Primary immigration consists of the immigration of those whose sole reason for wish-ing to enter the prospective host country is that they would like to live there. These are called "economic migrants." Secondary immigration consists of the immigration of people who either seek asylum or wish to join relatives.

The United States is the only major country that allows primary immigra-tion. No primary immigration is permitted in Europe, Canada, Australia, or Japan. The political leaders of these countries have decided not to allow primary immigration on a variety of grounds that are not always specified but that include the views that further increases in population would have adverse effects on the environment and that immigrants are likely to impose social costs of welfare support, the education of their children, and the like. It cannot be reasonably argued that countries have an obligation to accept primary immigrants. There are about three billion people in the world living in poverty who would like to live in the affluent Western nations. It is quite impractical to admit these numbers or anything approaching them, and the refusal of most countries to accept primary immigration must be accepted as ethically legitimate.

Many countries have operated selective immigration policies, which accept some kinds of immigrants and decline others. The foremost example of selection on the grounds of national, ethnic, and racial origin is the American Immigration Act of 1924, which set quotas for the numbers of immigrants from different countries who would be admitted, based on their proportions already in the country according to the 1890 census. The effect of this was that there were large quotas for the nations of northwest Europe, from which most of the population at that time originated, and small quotas for southern and eastern Europe and from the rest of the world. It has often been asserted that the quotas of the 1924 Act were established partly on eugenic grounds on the supposition that immigrants from southern and eastern Europe were of inferior genetic stock and would dilute the quality of the U.S. population. Whether the U.S. legislators who passed the act were in fact influenced by these eugenic considerations has been disputed by Herrnstein and Murray (1994), and it is difficult to determine. The primary considerations in the minds of the legislators responsible for the 1924 American Immigration Act appear to have been their belief that immigrants from northwest Europe would be more easily assimilated than those from elsewhere and that an ethnically homogenous society has greater social harmony and cohesion than one that is ethnically diverse and divided.

Whatever the motives of the legislators, the national quotas for the admission of immigrants set in the 1924 Immigration Act have often been criticized as unethical. John F. Kennedy (1959) attacked them in his book A Nation of Immigrants, in which he argued that people of all ethnic and racial stock should be admitted as immigrants into the United States on an equal basis. More recently, the biologist Lee Silver (1997) has attacked the quotas. He wrote that "without a doubt, governmental attempts to impose eugenic policies have caused harm to individuals as well as to whole societies through restrictions on immigration" (p.153) and "restrictive immigration policies directed against particular regions of the world are still wrong because they are designed to discriminate directly against particular ethnic groups" (p. 217).

These attacks on the ethics of a national or ethnic quota system for the admission of immigrants cannot be accepted. Political leaders of countries can legitimately take the view that immigrants of their own racial or ethnic type would be more easily assimilated and make better citizens than others. For instance, the political leaders of, say, Japan might take the view that they could assimilate racially similar Koreans and Chinese and would allow some of these to settle as immigrants, but that it would be much more difficult to assimilate Caucasians, Africans, or Australian Aborigines and that consequently these should not be accepted. This would be a reasonable view; and if Japanese legislators took it, they would be discharging one of their primary responsibilities, the preservation of social harmony in Japan.

A number of countries accept immigrants with valuable vocational and professional skills but do not accept immigrants without these skills. These include Australia, Britain, Canada, and New Zealand. The United States has a relatively small quota reserved for immigrants with skills on the grounds that immigrants with skills are likely to make a positive contribution to the receiving country. Selective immigration policies of this kind benefit the receiving country and are likely to have a eugenic impact because those with valuable skills are likely to have desirable genetic qualities. It is difficult to dispute the rights of nations to select those whom they are willing to allow to immigrate according to criteria of their own choosing. If these criteria include eugenic considerations, this cannot be regarded as ethnically unacceptable.

8. SECONDARY IMMIGRATION

Secondary immigration consists of the immigration of those who wish to enter a country on grounds other than simply the desire to live there. The principal categories of secondary immigrants are the spouses and families of citizens, many but not all of whom are relatively recent immigrants, and refugees and asylum seekers. There is a widely held consensus that countries have an ethical obligation to admit the spouses of citizens. This obligation has to include the spouses' children, who need to be with their parents, up to the age of middle or late adolescence. Whether this obligation has to be extended to spouses' parents, adult children, and siblings, as is the case in a number of Western countries, is more open to question. Refugees and asylum seekers are those who have a well-founded fear of persecution in their own countries. All the Western democracies have committed themselves to accept these by signing the 1951 Geneva Convention.

In considering the ethical problems of secondary immigration, weight has to be given to the costs incurred by the accepting countries as well as to the wishes and needs of the potential immigrants. The costs to the Western democracies are of four kinds. First, there are financial costs of welfare and medical support. Second, there are population pressure costs for countries that are arguably already overcrowded. In the United States, several organizations, including the Sierra Club, Zero Population Growth (ZPG), and the Federation for American Immigration Reform (FAIR), have put out a series of publications arguing that the increases in population arising from large-scale immigration must inevitably degrade the environment by increasing urbanization, pollution of the air and water supplies, and the like. This argument has additional force in some European countries, which have substantially greater population densities than the United States. For instance, in Britain there are serious problems of overpopulation in London and the southeast, which are responsible for a high level of air pollution from automobiles, which causes a high incidence of asthma and other respiratory conditions; excessive traffic congestion; and high property values resulting from an excess of demand over supply. The population pressure in southeast Britain is such that an estimated 1.4 million new homes need to be built in the period 2000 to 2020, which will cause further degradation of the environment. Overpopulation in southeast Britain is exacerbated by an annual influx of around 100,000 immigrants, most of whom settle in that region.

Third, costs arose because many immigrants into the Western democracies are from third world countries whose populations differ racially and culturally from the indigenous populations. This inevitably causes racial and ethnic conflict and the development of legislation and a bureaucracy that attempted to contain these conflicts. This causes resentment, which the indigenous populations endeavor to defuse by affirmative action and equal opportunity policies in admission to universities and employment. These in turn generated resentment among the indigenous population, some of whom are refused admission to universities, excluded from employment, and deprived of career advancement on the grounds that they are white. These problems have been examined by Gerald Scully (1995) in a study published by the American National Center for Policy Analysis. He concluded that "countries with a common culture are more likely than culturally diverse nations to be economically prosperous and to offer their citizens more personal freedom . . . ; we find that where there are multiple cultures there is almost always conflict" (Lutton, 1999, p. 271). The same verdict is reached by LaPorto, Lopez-de-Silanes, Sohleifer, and Vishny (1999) in a paper published by the American National Bureau of Economic Research. They conclude that multicultural societies are less well governed and have poorer economic performance than ethnically and racially homogenous societies because the competing ethnic and racial groups expend too much of their energy and resources in trying to advance the interests of their own groups.

It is the recognition of the social and economic stresses of multi-ethnic and multicultural societies that gave rise in the United States to the concept of the "melting pot," the notion that immigrants of different ethnic and cultural origins could and should all be "melted" and fused into a homogenous culture. By the end of the twentieth century, it had become evident that the melting pot works for European peoples of different national origin who are nevertheless of the same race and share a common European culture and religion, but it does not work for peoples of other races and radically different

cultures and religions, many of whom have no wish or willingness to assimilate.

Fourth, the genetic cost of much of the immigration into the Western
democracies in the second half of the twentieth century has arisen from the
acceptances of large numbers of immigrants with low intelligence and a high
propensity to crime. In the United States these problems have been present
among Hispanics and blacks, as shown by Herrnstein and Murray (1994) and
in many other studies. In Britain also, immigrants from sub-Saharan Africa
have manifested the same characteristics as blacks in the United States—low
intelligence, poor educational attainment, high unemployment, high welfare
dependency, and a crime rate approximately six times greater than that of the
indigenous population (Mackintosh & Mascie-Taylor, 1984; Tonry, 1997).
These characteristics have some genetic basis, and the acceptance of these
immigrants has imposed a genetic cost on the population, both for present
and future generations.

European governments have a broad understanding of these costs of immigration and have sought to reduce the numbers of immigrants. One way of
doing this has been by the "country of first arrival" rule, which states that
asylum seekers must apply for asylum in the first safe country they reach. Other
attempts to contain the numbers of immigrants have been to return to their
own countries those asylum seekers who have no valid reasons for seeking
asylum and to reduce the welfare benefits and rights to work of asylum seekers during the period, frequently amounting to several years, during which
they have to wait for their applications to be considered.

The general ethical problem presented by these costs of immigration is
how far the European peoples of North America and Europe can be considered to have an ethical obligation to admit large numbers of immigrants who
damage the social and the economic fabric of their societies. The European
peoples of North America and Europe accepted an ethical obligation to admit refugees and asylum seekers when the numbers of these were quite small
and consisted largely of Europeans from central and eastern Europe. In the
closing decades of the twentieth century, the numbers of refugees and asylum
seekers from third world countries became so great and imposed such considerable costs that the European peoples of North America and Europe can no
longer be regarded as having an ethical obligation to accept them.

9. THE SLIPPERY SLOPE

Many of the opponents of eugenics have advanced the slippery slope argument, which states that although some eugenic measures, and perhaps even
all of those discussed in this chapter, are ethically acceptable, and perhaps in
some cases, such as the provision of education about contraception in schools,
even desirable, nevertheless once governments accept that eugenics is a legitimate objective of public policy, they set foot on the slippery slope that

inevitably leads to eugenic measures that are unethical. The ultimate end of eugenics, these critics assert, was the use of eugenic arguments by the Nazis to justify the extermination of the Jews and others in the concentration camps. For instance, Colin Clarke (1963) wrote "We have seen in Nazism where eugenics may lead. I think that it is no accident that the Nazi doctrines about sterilization were closely linked intellectually and morally to Nazi doctrine about genocide" (p. 294). H. L. Kaye (1987) wrote of "the obvious truth that eugenics has been discredited by Hitler's crimes" (p. 46); and R. L. Hayman (1990) wrote that "the eugenics movement is an anachronism, its political implications exposed by the Holocaust" (p. 1209).

These arguments cannot be accepted. First, eugenics does not require the extermination of undesirables. It is sufficient for eugenics that the mentally retarded and recidivist criminals should be sterilized. Second, eugenic considerations did not play any significant role in the Nazi program for the extermination of the Jews. Hitler did not regard the Jews as genetically inferior. No one could have reached such a conclusion in Germany in the 1920s and 1930s because it was a matter of common knowledge and observation that the Jews were exceptionally talented. Jews were prominent in business, the professions, and intellectual life. Although they constituted only approximately 1 percent of the population, Jews won 10 out of 32 Nobel Prizes awarded to German citizens between 1905 and 1931 and were thus overrepresented among this highly elite group by a factor of approximately 30 (Gordon, 1984). Anyone who asserted that the Jews were genetically inferior and hence eugenically undesirable would have forfeited all credibility, and Hitler certainly did not do so. Hitler (1943) was indisputably anti-Semitic, and this anti-Semitism was based on his views, set out in *Mein Kampf*, that the Jews had exceptionally high abilities and were consequently a threat to the German, or as he called them "Aryan" peoples, who included the British and the Scandinavians. Hitler believed that the Jews and the Aryans were the two most talented races and that they were in competition to secure world supremacy. Thus, he wrote in *Mein Kampf* that the Jews are "the mightiest counterpart to the Aryan" (p. 64). He feared that the outcome of the struggle between these two peoples might easily be "the final victory of this little nation" (p. 300). This was the reason that Hitler was determined to destroy the Jews. He believed that if he could achieve this, the Aryans would remain as the unchallenged master race. The correct understanding of Hitler's views on the Jews has been summarized by MacDonald (1998): "Hitler believed that races, including the Jews, are in a struggle for world domination, and he had a very great respect for the ability of Jews to carry on their struggle" (p. 146). The frequent assertion that Hitler exterminated the Jews on eugenic grounds is a misunderstanding of his position.

Third, among the numerous opinions expressed about the Jews in Germany in the 1930s, there were some contending that the Jews were parasitical because they worked predominantly in white-collar service occupations, such as banking, law, the theater, the media, universities, and the like. Views of this kind fueled anti-Semitism but are economically and socially illiterate and cannot justify hostility to any ethnic or racial group.

Fourth, if eugenic views had contributed to the Nazis' extermination of the Jews, gypsies, and others, this would certainly have been an ethically unacceptable misapplication of eugenics. Nevertheless, the fact that a social philosophy has been unethically applied does not imply that such a misapplication is inevitable or that the social philosophy must be rejected on this account. Numerous social philosophies that are in general commendable have, on occasion, been misapplied. For instance, Christianity consists of a generally acceptable social philosophy, and the application of Christian principles has led to many desirable outcomes, such as the abolition of slavery, the establishment of welfare provision for the destitute, and so forth. Nevertheless, Christianity has sometimes been misapplied. The Christian church has burned at the stake numerous people who disagreed with some of its tenets; and it has waged wars against unbelievers in which abhorrent brutalities have been committed, such as the Crusades in which the Christian crusaders slaughtered large numbers of women and children. These killings must be condemned. However, these deplorable episodes do not justify the total rejection of Christianity or the conclusion that an acceptance of Christianity is the beginning of a slippery slope that inevitably leads to the extermination of those who do not accept its doctrines.

Similarly, the social philosophy of socialism evolved in the Soviet Union into a tyranny with many ethically undesirable features, including the execution of large numbers of dissidents and the extermination of many millions in the gulags. Nevertheless, it would be wrong to conclude that socialism inevitably leads to such an outcome and that socialist ideals of equality, fraternity, and the like must be condemned because these principles were applied in ethically unacceptable ways in the Soviet Union. All social philosophies are capable of ethical misuse, but this does not mean that we cannot accept them because this would be to set foot on the slippery slope that could lead eventually to ethically abhorrent outcomes.

10. CONCLUSIONS

In considering the ethical basis of classical eugenics, it is useful to distinguish between four types of programs. The first of these consists of the provision of information and services about contraception and abortion that is authorized and financially supported by the state or by private agencies. There is no coercion of citizens to use this information or these services. The provision of this information and services is akin to a number of other kinds of information and services provided by governments, private agencies, and newspapers about the dangers of cigarette smoking, drug abuse, AIDS, and the like, and to subsidized education and health care. There can be no ethical objection to the provision of information and services of any of these kinds by the state or by private agencies.

The second category of eugenic program consists of incentives and inducements offered by states or private agencies to citizens. Foremost among these is the provision of financial incentives to elites to have children and to social problem groups not to have children. Also in this category is the offer of more lenient sentences to criminals, conditional on their not having children for some specified period of time. Although these incentives and inducements can be regarded as quasi-coercive in the sense that they are hard to resist, nevertheless those to whom they are offered remain free agents and are at liberty to refuse them. The offering of inducements of this kind, which are considered by governments and their agents to promote the public welfare, likewise cannot be regarded as unethical.

The third category of eugenic program entails compulsion and consists primarily of the sterilization of the mentally retarded, recidivist criminals, and psychopaths. The general ethical principle justifying the sterilization of social problem groups lies in the existence of the rights of society to curtail the individual rights of citizens where the exercise of these is likely to cause social damage. There are numerous instances in which this principle is accepted, such as the prohibition of smoking in public places and the discharge of poisons into rivers and of pollutants into the atmosphere. The sterilization of the socially undesirable is justifiable in terms of the same general principle.

The fourth category of eugenic program consists of immigration policies designed to admit immigrants with skills and characteristics likely to benefit the receiving country and of refusing to accept immigrants likely to cause social harm or to be a social burden. The affluent Western nations cannot be regarded as having an ethical obligation to accept all the many millions from the third world who would like to be accepted as immigrants. The ethical rights of nations to select those they accept as immigrants cannot be reasonably disputed.

A number of the opponents of eugenics have resorted to the slippery slope argument, which states that although a number of eugenic measures are unobjectionable in themselves, they could lead to further measures that would be unethical. This argument is unpersuasive because all sorts of measures that are acceptable might, if taken to extremes, lead to other measures that are unacceptable. For example, once society permits the practice of religion, it may be argued, it sets foot on the beginning of a slippery slope that will eventually permit unethical religions that practice human sacrifice, or once society permits the killing of animals, it is on the beginning of a slippery slope that will eventually lead to the killing of humans. It requires no more than a moment's thought to realize that virtually everything we do can be condemned on slippery slope grounds, that it could lead to ethically unacceptable actions.

This is the concluding chapter of Part III of this book, which has been devoted to the consideration of policies for the implementation of classical eugenics. It may be useful before we proceed further to take stock of the arguments. Classical eugenics has been criticized principally on the grounds that it would not work genetically and that it is unethical. We have considered these criticisms and shown that they are unfounded. The real problem with classical eugenics is that it is not possible politically in democratic countries to implement programs that would have a major eugenic impact. Nevertheless, the implementation of the numerous policies of classical eugenics discussed in preceding chapters would have some useful positive results, and it would be desirable to introduce such of them as are politically feasible.

IV

⌁

The New Eugenics

Classical eugenics attempts to improve the genetic quality of a population by altering patterns of reproduction in such a way that individuals with desirable qualities are induced to reproduce more and those with undesirable qualities to reproduce less. The "new eugenics" attempts to improve the quality of the gene pool by the use of medical technology, or what has become known as human biotechnology. *Biotechnology*, a general term for the use of technology to provide improved strains of plants and animals, is defined in the *International Dictionary of Medicine and Biology* (1986) as "the application of the biological sciences, especially genetics, to technological or industrial uses." The term "the new eugenics" as a means of improving human genetic quality by medical technology or human biotechnology was coined by Robert Sinsheimer (1969) in an article in which he predicted that medical technology would come to replace selective reproduction as the means of improving human genetic quality. Developments in medical technology, he suggested, would become the "new eugenics." Sinsheimer's article was a prescient insight into the future and what we can anticipate for the twenty-first century.

There are five interesting topics to consider about the present and future development of human biotechnology. These are, first, the technical developments that have already taken place and that can be anticipated in the future; second, the ethical issues involved in the use of human biotechnology; third, the way that human biotechnology is likely to develop in democratic Western societies; fourth, the degree to which human biotechnology may be used by authoritarian states as a means of enhancing their national power; and fifth, the impact that the use of human biotechnology for eugenic purposes by some authoritarian states is likely to have on geopolitics in the twenty-first century and beyond. We are now ready to begin our consideration of these topics.

17

～

Developments in Human Biotechnology

1. Artificial Insemination by Donor
2. Egg Donation
3. Prenatal Diagnosis of Genetic Disorders
4. Pregnancy Terminations of Defective Fetuses
5. Eugenic Impact of Prenatal Diagnosis
6. Embryo Selection
7. Genetic Engineering
8. Gene Therapy
9. Cloning
10. Is Human Biotechnology Eugenic?
11. Conclusions

～～～

There are six techniques of human biotechnology that have already been used to promote eugenics or that have the potential to promote eugenics in the future. These are artificial insemination by donor (AID); egg donation; prenatal diagnosis of fetuses with genetic diseases and disorders and termination of the pregnancies where these are identified; embryo selection, consisting of the ascertainment of the genetic characteristics of embryos grown in vitro and the selection for implantation of those with genetically desirable characteristics; cloning, consisting of the production of genetically identical copies of individuals; and genetic engineering, used here in the sense of the insertion of new genes. These are the actual and potential developments we consider in this chapter.

1. ARTIFICIAL INSEMINATION BY DONOR

AID entails taking semen from a donor and using it to fertilize a female. This technique is frequently used by animal breeders who take semen from males with desirable characteristics with the expectation that these will be transmitted to the offspring to produce genetically improved strains of livestock. Artificial insemination appears to have been first carried out on humans in the 1790s in Scotland by a physician named John Hunter. In 1884, the first recorded artificial insemination in the United States was carried out by William Pancoast at the Jefferson College of Medicine in Philadelphia, who used sperm donated by the best-looking medical student in his class (Francoeur, 1971). During the twentieth century, AID became a standard treatment for couples who wanted children but the husband was infertile; and by the end of the century many thousands of children had been conceived and born by this means.

The degree to which the use of AID has been eugenic depends on the quality of the donors and the numbers of women using it. So far as the quality of the donors is concerned, the physicians carrying out the procedure have generally understood that it is important to use donors of good quality, and they have usually done so. However, in the 1990s, it was discovered that in Britain a number of AID clinics were paying unemployed men for semen donations and that 11 percent of an estimated 9,000 donors were long-term unemployed. Some of these were long-term unemployed drug abusers who supplemented their incomes from welfare by donating semen several times a week (Brennan & Syal, 1997). The only legal restriction placed on the use of donor sperm by fertility clinics in Britain is that it must be screened for the HIV virus. This is clearly unsatisfactory, and fertility clinics should be legally required to screen sperm donors more thoroughly for health and good educational qualifications, which would serve as indirect measures of intelligence and sound personality qualities.

As regards the numbers of babies born by AID, because the procedure is confidential, no official figures are kept, so no precise statistics of the numbers of AID births are available. Nevertheless, in the United States in the early 1930s, it was estimated that around 2,000 babies a year were born through AID (Caldwell, 1934); and by the 1970s, this figure had risen to somewhere between 10,000 and 20,000 (Fiengold, 1976). In 1987 the United States government carried out a survey of the extent of the use of AID and estimated that about 33,000 AID babies were born through this means in the year 1986 (United States Office of Technology Assessment, 1987). Although this is quite a large number of babies, it represents only about 1 percent of the total annual numbers of births in the United States. In Britain it has been estimated that about 2,000 babies a year were born through AID during the 1990s, representing about 0.3 percent of all births (Galton Institute, 1999). These numbers are so low as a percentage of births that even assuming that donors of good quality have normally been used, the eugenic impact will have been negligible.

2. EGG DONATION

The counterpart of AID is egg donation, by which a fertile woman donates her egg to a woman who is infertile. Normally the egg will be fertilized in vitro with the sperm of the husband or partner, and the embryo is then implanted into the woman's uterus. Alternatively, it can be implanted in another woman who acts as a surrogate, brings the pregnancy to term, and hands over the baby after it is born. The women who donate the eggs or who act as surrogate mothers are normally paid a fee for these services. The first reported egg donation was carried out in Australia in 1984. Cynthia Cohen (1996) has described its subsequent use in a number of economically developed nations.

Egg donation can be used for eugenic purposes if the eggs are taken from women with desirable qualities that are likely to have some genetic basis. In the 1990s a number of infertile American women seeking eggs and desiring to obtain eggs of good genetic quality advertised for them in the student newspapers of Ivy League and other elite colleges, including Harvard, Yale, and Stanford. Typically, the sum offered for eggs was $5,000, but on occasions as much as $50,000 has been offered. Frequently these advertisements specify that the donors should have high SAT scores, good college grades, and certain physical characteristics, such as blue eyes. The practice of infertile women seeking eggs from students at elite universities has spread to Britain. In 1999 an advertisement for egg donors appeared in the Cambridge University magazine *Cam*.

The women placing these advertisements could no doubt obtain eggs more cheaply by advertising in the local press or by getting them from women on welfare. However, these women evidently believe that they are likely to get eggs of better genetic quality from students at elite universities. They are undoubtedly correct in this belief. They are willing to pay a premium for eggs likely to be of good genetic quality, and in doing so they are making a good investment. However, this practice has not yet become widespread; and by the beginning of the twenty-first century, elite egg donation has not been used sufficiently to have any significant eugenic impact.

An extension of egg selection appeared in October 1999, when advertisements appeared on the Internet offering eggs from models for fees of $90,000. The advertisements pointed out that the physical characteristics of models, such as beauty and slimness, are likely to be inherited. For females, they are likely to have substantial financial value for potential employment as models and for employment more generally, where the attractive are more likely to succeed than the unattractive, and to have value for self-esteem and for securing desirable husbands. This selective use of egg donation may be regarded as eugenic insofar as an increase in the numbers of beautiful women may contribute to the quality of life. However, it does not contribute to the principal objectives of eugenics, which consist of the promotion of a high civilization; the economic, scientific, and military strength of the nation state; and the genetic improvement of the human species.

3. PRENATAL DIAGNOSIS OF GENETIC DISORDERS

Prenatal diagnosis of genetic disorders and diseases consists of tests for the genetic characteristics of the fetus carried by a pregnant woman with the object of terminating the pregnancy in cases where a genetic disorder is identified. The first testing method to be developed was amniocentesis. This consists of inserting a needle through the abdomen of the pregnant woman into the amniotic fluid surrounding the fetus and the withdrawing of a fluid sample. The sample is then analyzed for the presence of defective genes and other genetic anomalies. Amniocentesis was first developed in the 1930s for the diagnosis of erythroblastosis fetalis, a blood disorder. In the mid-1950s, it was used in Denmark to ascertain the sex of the fetus and to offer termination to pregnant women at high risk of being carriers of hemophilia. Because hemophilia is an X-linked recessive gene disorder carried by females and inherited by half their sons, but only very rarely by their daughters, it is possible to avert the risk of having a child with the disorder by ensuring that it is female. In 1967 it became possible to use amniocentesis for the fetal diagnosis of Down's syndrome; and in 1968 the first abortion of a Down's fetus was reported. In 1972, Dr. J. Brock (1982), a physician in Edinburgh, discovered that amniocentesis could be used to diagnose neural tube defects, including spina bifida, the presence of which is shown by elevated levels of alpha-feto-protein in the amniotic fluid.

By 1980 amniocentesis was being used throughout virtually the whole economically developed world to test pregnant women for the presence of genetic diseases and disorders in the fetus; pregnancy terminations were offered and normally accepted in cases where a disorder was diagnosed. Initially, it was used principally to test for Down's syndrome in the fetuses carried by women in their mid-thirties and older, among whom the risk of having a Down's baby is about 1 in 75. Later the technique became extended to test for a number of genetic diseases and disorders, including single-gene diseases such as Tay-Sachs disease, Huntington's disease, cystic fibrosis, sickle-cell anemia, hemophilia, and galactosaemia, a rare disease causing liver disorder, mental retardation, and cataract; multifactorial disorders, including spina bifida and anencephaly; and other chromosomal disorders in addition to Down's syndrome.

A problem with amniocentesis is that it cannot be carried out until the beginning of the second trimester of pregnancy, that is at about 12 weeks; and it carries about a 1 percent risk of miscarriage. For this reason, it is not normally carried out unless there is a fairly high risk of the fetus having a disease or disorder, as is the case for Down's syndrome in women in their mid-thirties and older. For younger women, the risk of Down's syndrome is normally considered to be too low to incur the risk of carrying out the procedure.

In the 1980s and 1990s four further techniques for diagnosing the presence of genetic defects in the fetus were developed and became widely used. These are (1) ultrasound scan, (2) maternal serum screening, (3) fetal biopsy, and (4) chorion villus sampling. Ultrasound scan involves passing high frequency sound waves through the mother's body and analyzing the echoes reflected from the fetus. A picture of the fetus is built up, and it becomes possible to detect 80 percent to 90 percent of major structural anomalies of the brain, heart, and limbs, including anencephaly and spina bifida. The method cannot be used until 16 weeks into the pregnancy at the earliest. It does not appear to entail any risk to the mother or the fetus. By the early 1990s most pregnant women throughout the Western nations were offered an ultrasound scan, had the test carried out, and had impaired fetuses aborted.

Maternal serum screening involves taking a blood sample from the pregnant woman and analyzing it for the level of alpha-fetoprotein (AFP). A high concentration of AFP indicates a probability of anencephaly, spina bifida, Down's syndrome, and a few other less common genetic or chromosomal disorders. In these cases further diagnostic tests, such as amniocentesis, are normally carried out to ascertain whether the possibility of a defective fetus can be confirmed. Maternal serum AFP screening has become a widely used and routine procedure in prenatal care in the United States and Europe. It has the advantage that it is both cheap and risk free. It does not, however, identify all chromosomal disorders.

Fetal biopsy involves taking a sample of blood, skin, or liver from the fetus and testing it for the presence of genetic diseases. Blood samples are analyzed for the presence of hemophilia, a number of other blood diseases, and several other rare genetic diseases. The technique is carried out in the second trimester and involves a 2 percent-to-5 percent risk of fetal mortality and miscarriage.

Chorion villus sampling was first developed by Chinese physicians in the mid-1970s and began to be used fairly extensively in Western nations in the 1980s. The method involves taking a sample of chorionic tissue from the placenta and analyzing it for the presence of single-gene disorders, multifactorial disorders, and chromosomal abnormalities. The advantage of chorion villus sampling is that it can be carried out in the first trimester of pregnancy; so if a defect is detected, a termination can be carried out sooner than with other tests, thus avoiding the health risks and emotional strain of a late abortion. Chorion villus sampling entails a 1 percent-to-2 percent risk of fetal mortality, and there may also be some risk of fetal damage.

4. PREGNANCY TERMINATIONS OF DEFECTIVE FETUSES

By the late 1960s it became possible to test pregnant women for the presence of common fetal defects. This led to the use of "therapeutic abortion" to end the pregnancies when defects had been discovered. The termination of these pregnancies required changes in the law to allow therapeutic abortions to be carried out. In most Western nations, abortion was illegal up to the mid-1960s, except in special circumstances, such as where pregnancies had occurred as a result of rape. From the mid-1960s onward, abortion laws became liberalized throughout virtually the whole of the economically developed world to allow the termination of pregnancies in the case of impaired fetuses and also frequently for other reasons, such as a threat to the woman's health. In Britain, the abortion of impaired fetuses in circumstances where the woman's health was considered to be at risk was legalized in 1967. In the United States, several states legalized abortion in the early 1970s, and the legality of abortion was established nationwide in 1973 by the Supreme Court decision in *Roe v. Wade*. By the 1980s the abortion of fetuses with defects had become legalized throughout Western nations with one or two exceptions, such as the Republic of Ireland.

The percentage of pregnant women with impaired fetuses who were willing to have an abortion was typically 70 percent to 85 percent. Generally, the proportion of women opting for termination depends on the severity of the condition diagnosed. A study carried out in Canada in 1990 found that about 80 percent of women diagnosed as carrying a fetus with a serious disorder decided to have their pregnancies terminated. For Down's syndrome, 83 percent chose termination, and for the less severe disorder of Turner's syndrome, which entails some intellectual retardation and infertility in girls, 70 percent chose termination. For the less severe disorders caused by additional X or Y chromosomes that impair intelligence by only a relatively small amount, 30 percent opted for termination (Hamerton, Evans, & Stranc, 1993). Similar figures have been found in other Western countries, including Britain (Royal College of Physicians, 1989). Opinion polls have shown that 75 percent to 85 percent of the population approve of the abortion of a severely impaired fetus (Canadian Royal Commission on New Reproductive Technologies, 1993). A survey carried out in Britain in the late 1980s found that 83 percent of women aged 37 to 44 said they would have prenatal tests for Down's syndrome and cystic fibrosis and would have an abortion if medically advised to do so (Weatherall, 1991). Those who do not have the tests carried out are normally opposed in principle to all abortion. This is the position adopted by the Roman Catholic Church, although not by all Catholics.

5. EUGENIC IMPACT OF PRENATAL DIAGNOSIS

By the 1990s most pregnant women in the economically developed nations were being given ultrasound scan and maternal serum screening for the detection of fetal anomalies. The impact of these screening procedures has been mixed. The results of a survey of the effectiveness of prenatal diagnosis and pregnancy terminations of impaired fetuses on the birth incidence of babies for six of the most common genetic and congenital disorders over the years 1977–91 in Britain are summarized in Table 17.1. It will be seen that there

Table 17.1
Numbers of Babies Born with Congenital Malformations and Disorders
in England and Wales, 1977–91

	1977	1980	1983	1986	1988	1991
Anencephaly	568	342	114	52	41	22
Spina bifida	881	756	422	267	157	104
Hydrocephalus	259	222	194	138	137	102
Cardiovascular disorders	649	866	995	882	726	577
Exomphalos	170	150	176	127	147	165
Down's syndrome	425	481	497	445	428	440

Source: Office of Population, Censuses, and Surveys, 1993

were large reductions in the birth incidences of anencephaly, spina bifida, and hydrocephalus. This is because these defects are reliably detected by ultrasound scan, which is given routinely to 80 percent to 90 percent of pregnant women attending physicians or antenatal clinics. The small numbers of babies still being born with these conditions occurred because a few mothers failed to visit their medical practitioners or antenatal clinics, refused to undergo the tests, or declined to have terminations.

In contrast to the reductions of these disorders, there was little or no reduction in the birth incidence of cardiovascular disorders, exomphalos, and Down's syndrome. The cardiovascular disorders were reduced by only about 8 percent, while exomphalos, a disorder in which the intestines protrude through the abdominal wall, and Down's syndrome were not reduced at all. It is evident that these disorders were not being reliably picked up by testing with ultrasound scan and maternal serum screening. In the case of Down's syndrome, the numbers of affected fetuses identified and aborted increased over the period, but this was offset by a rise in the numbers conceived, which resulted from the increasing tendency of women to have their children at later ages when the chances of having a Down's baby are greater.

In regard to single-gene disorders, there have been some successes in reducing the birth incidence. The greatest achievement has been the reduction in the birth incidence of Tay-Sachs disease among American Ashkenazi Jews to around 3 percent of its former incidence. There have also been substantial reductions in the birth incidence of B-thalassemia, a form of anemia, among several populations in southern Europe. The overall impact of prenatal diagnosis, however, has been quite small. By the 1990s it was estimated that prenatal diagnosis had resulted in a reduction of less than 5 percent in the birth incidence of genetic diseases and disorders (Cantor, 1992).

6. EMBRYO SELECTION

Embryo selection consists of growing a number of embryos in vitro, testing them for their genetic characteristics, and selecting for implantation those with genetic characteristics regarded as desirable, while at the same time discarding those with genetic characteristics regarded as undesirable. This procedure is also known as *embryo biopsy*, which entails growing several blastocysts (embryos grown in vitro to eight cells), removing one of the eight cells, and testing it for genetic and chromosomal defects. Verlinksy, Pergament, and Strom (1990) reported the use of this procedure to screen out embryos with genes for Duchenne's muscular dystrophy and Down's syndrome, so an embryo free of these disorders could be implanted in the mother. At about the same time, another use of this technique was reported by Handyside and his colleagues at London University. They used IVF (in vitro fertilization) for two couples in which the female was a carrier for an X-linked recessive disease, which is expressed only in males. To avoid the potential birth of a boy with the X-linked disorder, the physicians tested for the sex of the embryos and implanted only females. This technique allows couples to choose the sex of their babies, whether this is to avoid having babies likely to inherit serious disorders, or simply because they prefer one sex rather than the other.

In the 1990s there was rapid progress in preimplantation diagnosis and the screening out of embryos with genetic diseases and disorders. By 1995 preimplantation diagnosis of embryos for the presence of genetic diseases and disorders was being carried out in 16 centers in various countries. Initially, this work was done to screen for genetic diseases affecting babies at birth or shortly after birth. By the late 1990s, this was extended to screen for the presence of cancer genes that would cause tumors likely to appear only in adulthood. The first use of this method for the diagnosis of cancer genes was carried out in Britain in 1996 on the embryo of a woman with familial adenomatous polyposis (FAP), a form of bowel cancer caused by a dominant gene. By the end of the twentieth century, it had become possible to screen embryos for several thousands of genetic diseases and disorders.

Preimplantation diagnosis and embryo selection is preferable to prenatal diagnosis and abortion of defective fetuses as a means of securing a healthy baby. It avoids the stress of abortion, and it greatly increases the probability of having a child free of genetic diseases. Women who use prenatal diagnosis and abortion of impaired fetuses may become pregnant again and, in the case of single-gene diseases, are at significant risk of having another fetus with the disease and having to undergo a second pregnancy termination. These stresses can be avoided by the use of preimplantation diagnosis and embryo selection. Although the procedures have not been used extensively by the beginning of the twenty-first century, they are a significant eugenic advance.

7. GENETIC ENGINEERING

The term *genetic engineering* is used here for the implantation of new genes into an organism. The new genes may be present in the species but not in the individual, or they may not be present in the species at all. In the last two decades of the twentieth century, new genes were successfully implanted into a number of plants to produce greater yields and better resistance to disease. These have become known as genetically modified foods. New genes have also been successfully inserted into a number of animals.

In the early 1980s, Wagner and Hoppe (1981) reported the successful insertion of rabbit beta-globin genes into fertilized mouse eggs. Five mice were born with the rabbit beta-globin in their hemaglobin. Two of these were mated, and five of their offspring also had the rabbit beta-globin. Thus, the rabbit genes had been incorporated into the mice genome and transmitted to the next generation.

In the mid-1980s, Ezzell (1987) inserted a new gene for susceptibility to cancer into a mouse in order to research the causes of cancer. This produced the "oncomouse," the first genetically engineered new animal for which a patent was granted.

In the 1990s, new genes were successfully implanted into a variety of animals, including mice, pigs, sheep, and cattle. These have become known as *transgenic animals*, and by the end of the twentieth century many thousands of them had been produced.

There have been several objectives of this work. One of these is to insert growth genes to produce animals, such as fish, pigs, chickens, cows, and sheep, that will grow faster. Another is to insert human genes into animals to get them to produce organs that can be used for transplants into humans. The preferred animal for this purpose is the pig. Human genes have been inserted into pigs that have the effect of coating the pigs' hearts with human proteins so the hearts are not rejected when they are transplanted into humans. The first of these pigs was born in 1992 and called Astrid; and by 1995 genetically engineered pigs' hearts had been implanted successfully into a number of primates.

A further use of genetically engineered animals is to introduce the genes for human diseases into animals, which can then be used to study the diseases. By the end of the twentieth century, human genes for cystic fibrosis, muscular dystrophy, sickle-cell anemia, Alzheimer's disease, and a number of cancers had successfully been inserted into mice. Having an animal model for a disease accelerates the rate of progress of research. Finally, transgenic sheep, goats, and cows have been produced to secrete human proteins into their milk, which can then be extracted and used for the treatment of human diseases such as emphysema.

8. GENE THERAPY

The eugenic potential of genetic engineering lies in insertion of new genes for improved health, intelligence, and personality. By the end of the twentieth century, some progress of this kind had been made with the insertion of genes for improved health. The strategy has been to attempt to treat those with genetic diseases by inserting healthy genes in the hope that these will supplant the genes for the disease. This form of gene modification has become known as *gene therapy*. There have been some successes in the use of gene therapy for the treatment of genetic diseases in mice and humans. In the 1980s Hammer, Palmiter, and Brinster (1984) reported that they had successfully used gene therapy to treat a hereditary growth disorder in mice. Later studies reported successful treatments for mice with B-thalassemia and sterility.

Experimental work on gene therapy for humans began in 1980 when Mercola and Cline (1980) introduced beta-globin genes into two patients suffering from beta-zero thalassemia, a genetic form of anemia. After two years the treatment was apparently unsuccessful. However, experimental work using gene therapy continued, and by 1992 there were some 20 medical centers in the United States, Europe, and China running gene therapy trials for the treatment of various forms of cancer, hemophilia, adenosine deaminase (ADA) deficiency, liver failure, hypercholesterolemia, and AIDS (Anderson, W. F., 1992).

From 1993 onward positive results for the use of gene therapy began to be reported and have been described by Weatherall (1991) and Kaku (1998). In 1993 gene therapy was successfully used in the treatment of a patient suffering from hypercholesterolemia, a life-threatening disorder consisting of excessively high levels of cholesterol. The basic cause of the disease is a defect in liver function, and the treatment consisted of correcting this by insertion of healthy genes into the patient's liver. In 1995 eight children in the United States and Italy were treated by gene therapy for adenosine deaminase (ADA) deficiency, a defect of the immune system, with encouraging results. In 1996 gene therapy was successfully used in the treatment of cancer at the University of Texas Medical School. The work involved the replacement of the mutated gene P-53, which is present in over half of common cancers. Although by the beginning of the twenty-first century successful treatment of genetic diseases by gene therapy has been limited, it seems probable that the techniques will be improved to the point where they can be used to treat a number of genetic diseases.

9. CLONING

Cloning is the reproduction of a genetically identical copy of a plant or animal. Plants can easily be reproduced by cloning by taking a cutting and putting it in the soil. This method of producing identical copies of plants was known in classical Greece. The word *clone* is derived from the classical Greek word *klon*, meaning "twig" and referring to the use of twigs or cuttings to grow identical copies of fruit trees.

In the second half of the twentieth century a number of animals were cloned. In the 1950s a frog was produced by cloning by King and Briggs (1956) and in 1997 the first mammal, a sheep named Dolly, was produced by cloning by Wilmut, Schnieke, McWhis, Kind, and Campbell (1997) at the Roslin Institute in Edinburgh, Scotland. Shortly afterwards Tronnson and his colleagues at Monash University in Australia produced nearly 500 identical cattle embryos by cloning. In 1998 and 1999 a number of mice were produced by cloning by Wakayama, Perry, Zuccotti, Johnson, and Yanagimachi (1998) at the University of Hawaii.

There are three ways of cloning mammals that could be adopted for humans. One is to take a cell from a mature animal, revert it to the undifferentiated embryo state before differentiation into cells for different bodily components, and implant it in the host mother. This method was used by Wilmot and his colleagues to produce the cloned sheep Dolly, which originated from a cell extracted from the mammary gland of an adult female. A second technique is to remove the nuclei of the cells of an early embryo or unfertilized egg and to substitute the nuclei of cells taken from another embryo or adult.

A third technique is to grow embryos in vitro and to split them into two or more embryos. This is what happens naturally in the production of identical twins, so one twin can be regarded as a clone of the other. It would be possible to produce a number of clones in this way. This technique was pioneered by Hall and Stillman in 1993 at the George Washington Medical Center. One of its advantages is that the genetic characteristics of one embryo can be assessed by biopsy; and if the embryo is found to be satisfactory, the other undamaged embryo can be implanted. A further advantage is that it increases the number of embryos available for implantation in the treatment of infertility. This would be the easiest method to use for the production of large numbers of cloned cattle. It would be a better way of obtaining high-quality offspring than artificial insemination using good quality males, which has the disadvantages of using large numbers of females of lower quality, thus producing offspring that are of lesser quality than the average of the two parents because of regression to the mean. The potential commercial advantages of cloning cattle are so great that research to improve the procedure will inevitably continue to be carried out until the technique is perfected. Once this has been achieved, there is little doubt that it will be technically feasible to use the technique to clone humans. Indeed it was reported that a human embryo was successfully cloned in South Korea on December 15, 1998, and was allowed to develop into four cells, after which it was destroyed (*The Times* [London], June 17, 1999).

10. IS HUMAN BIOTECHNOLOGY EUGENIC?

It is frequently asserted by the medical profession that the human biotechnologies described in this chapter are not eugenic because the physicians
carrying out these procedures are solely concerned with the well-being of their
patients and have no intention of improving the genetic quality of the population. The reason the medical profession has sought to deny that these procedures are eugenic is that by the last two decades of the twentieth century
any procedure that could be identified as eugenic was automatically condemned.

Despite the denials of the medical establishment, a number of commenta-
tors have contended that the medical applications of human biotechnology
should be recognized as eugenic because their impact is not confined to the
patients but affects the whole of society. For instance, Troy Duster (1990)
argued in his book *Backdoor to Eugenics* that these procedures are a covert
reintroduction of eugenics. Abby Lippman (1991) suggests that the denial
that these procedures are eugenic is hypocritical, writing, "Though the word
eugenics is scrupulously avoided in most biomedical reports about prenatal
diagnosis, except where it is strongly disclaimed as a motive for intervention,
this is disingenuous. Prenatal diagnosis presupposes that certain fetal conditions are intrinsically undesirable" (p. 24).

The issue of whether these biotechnologies are eugenic can be determined
either by considering the intentions of the physicians carrying out the procedures or by their effects. The criterion of the intentions of the physicians
carrying out the procedures is unsatisfactory because it is difficult to know
what their intentions are, especially because no physicians are going to admit
publicly that they are endeavoring to promote eugenics. Anonymous surveys
of physicians have shown that some of them do consider that they are practicing eugenics, while others consider that they are not. The largest study of
this kind was carried out in 1995 in 36 nations by Wertz (1998). This was a
survey of the views of 2,901 genetics professionals and physicians carrying
out prenatal diagnosis and counseling the patients on whether to terminate
fetuses with genetic diseases and disorders. The survey found that many of
these advised their patients to have terminations and agreed with the statement, "An important goal of genetic counseling is to reduce the number of
deleterious genes in the population." This is certainly a eugenic objective.
All of the 252 professionals in China agreed with it. Many also agreed with
the statement, "It is socially irresponsible knowingly to bring an infant with
a serious genetic disorder into the world in an era of prenatal diagnosis." In
the United States 55 percent of primary care physicians and 26 percent of
geneticists, as well as 44 percent of patients, agreed. Thus, appreciable numbers of professionals carrying out this procedure are in fact motivated by eugenic
considerations concerning the genetic health of the population, and not solely
by the well-being of individual patients.

Nevertheless, it is not satisfactory to use the criterion of the intentions of the physicians carrying out prenatal diagnosis and advising on the desirability of having a fetus with a genetic disorder aborted for the determination of whether this and other uses of human biotechnologies are eugenic. The adoption of this criterion means that the same procedures are sometimes eugenic and sometimes not, depending on what physicians believe are their own motives. The only sensible criterion is the effects of the procedures, and these are indisputably eugenic insofar as they contribute to the removal of genes for diseases and disorders from the population.

11. CONCLUSIONS

Although in the mid-1980s eugenics was pronounced dead by Kevles (1985), it gained a new lease of life by the development of the "new eugenics" of human biotechnology. There are six biotechnological procedures that have had an actual or potential eugenic impact. These are AID (artificial insemination by donor), egg donation, prenatal diagnosis of the genetic characteristics of fetuses and the abortion of those with genetic diseases and disorders, embryo selection, genetic engineering, and cloning. As these have been developed in the twentieth century, AID and egg donation have helped many couples but have not, as yet, had any significant eugenic effect on the population. Prenatal diagnosis has had some dramatic effects in reducing the birth incidence of the congenital malformations of anencephaly, spina bifida, and hydrocephalus and of the recessive gene disorders of Tay-Sachs disease and B-thalassemia. Embryo selection, involving the genetic diagnosis of embryos grown in vitro and the implantation of those free of genetic diseases and disorders, has not as yet had a significant eugenic impact; but its potential is considerable as a preferable alternative to prenatal diagnosis and abortion, and it could be used in the future for the selection of embryos for intelligence and personality.

Genetic engineering in the sense of implanting new genes to correct genetic diseases and disorders has also not been used on any significant scale but is likely to be developed further. Finally, cloning has not yet been carried out on humans, except for one report of a trial in South Korea. We shall consider presently how these human biotechnologies are likely to be developed in the future, but first we must consider the ethical concerns that have been widely felt over the use of these technologies.

18

⌣

Ethical Issues in Human Biotechnology

1. Artificial Insemination by Donor
2. Egg Donation
3. Prenatal Diagnosis
4. Embryo Selection
5. Selection for Sex
6. Cloning
7. Gene Therapy
8. The Case for Permitting Germ-line Gene Therapy
9. Conclusions

⌣⌣⌣

The developments in human biotechnology have raised concerns over whether these procedures are ethically acceptable and should be permitted. These issues have given birth to a new academic discipline of bioethics, with its own journals in which these questions are debated. Although human biotechnology generally, although not invariably, has a eugenic impact, the ethical problems it raises are not the same as those raised by classical eugenics. In classical eugenics the state attempts to change the fertility patterns of the population by the provision of information, incentives, or by coercion; and this raises the ethical issues of the degree to which the state can legitimately interfere in the population's reproduction. In the Western democracies, the state does not promote the use of the human biotechnologies to influence the population's fertility and promote eugenics. The biotechnologies are used by individuals to enable them to have children and by most to have children with genetically desirable qualities.

The general approach to the ethical problems raised by the human biotechnologies should be that in liberal democracies the citizens should be permitted to act as they please unless there is a reasonably strong case that by doing so they cause harm to individuals or inflict social damage. This is the general principle for the protection of individual freedoms and the curtailment of the exercise of these freedoms in liberal democracies formulated in the nineteenth century by political philosophers like John Stuart Mill (1859) and restated in the twentieth century by Friedrich Hayek (1960) and many others. In liberal democracies, the state cannot legitimately prohibit the citizens from doing things simply because many people dislike their behavior. For instance, many people dislike homosexuality, but this is not sufficient reason to prohibit it. Likewise, many people dislike the human biotechnologies, but this is not a legitimate reason to prohibit them. To justify their prohibition, it has to be shown that the human biotechnologies are likely to cause harm either to the individuals born through their use or to society as a whole. This is the general principle we shall adopt in considering the ethical problems raised by the human biotechnologies.

1. ARTIFICIAL INSEMINATION BY DONOR

Artificial insemination by donor (AID) raises two ethical issues. The first is whether it is acceptable for women to use AID at all, and the second is whether it is acceptable to select donors for what are likely to be desirable genetic qualities. When William Pancoast used AID in 1884, he realized that it would be controversial and kept it secret. It was only 25 years later that it became public knowledge. It then aroused a public outcry in which it was condemned as adultery and contrary to God's law. The young man who had been artificially conceived, however, raised no objection to the manner of his conception and could not understand the ethical problem (Francoeur, 1971).

The Roman Catholic Church condemns AID together with other reproductive technologies on the grounds that the only ethically acceptable way for women to have children is through sexual intercourse. Ratzinger and Bovone (1987) explained the Catholic position in a Vatican publication: "Artificial fertilization violates the right of the child: it deprives him of his filial relationship with his parental origins and can hinder the nurturing of his personal identity" (p. 23). Catholics have also asserted that reproductive technologies are "a threat to the stability of the family" (Bolan, 1988). However, no persuasive evidence has been produced to show that children conceived by reproductive technology are psychologically impaired or that they jeopardize the stability of their families. It is a curiosity of Catholic doctrine condemning AID that Jesus Christ and his mother, Mary, were both, according to Catholic doctrine, conceived by AID.

Some non-Catholics also regard AID as unnatural and ethically abhorrent. For instance, in 1991 two British members of Parliament, Jill Knight and Jerry Hayes, condemned the use of AID. Jill Knight asserted, "It is difficult to imagine a more irresponsible act than to assist a women to have a child in this highly unnatural way"; and Jerry Hayes stated, "I find it personally abhorrent." They both found it particularly objectionable for virgins to use AID and called for legislation to prohibit this. However, the government decided that to ban the use of AID by virgins but to permit its use by nonvirgins would be unworkable (*Washington Post*, March 12, 1991, p. 8).

While Roman Catholics and others may find AID abhorrent, in considering whether it should be prohibited, we must apply the general principle for liberal societies that people should be permitted to behave as they choose so long as this does not cause harm to individuals or social damage. It is impossible to show that any harm or social damage is caused by the use of AID and hence there would be no ethical justification for prohibiting it.

The second ethical issue in the use of AID centers on the selection of donors judged to have desirable characteristics. When women decide to have babies by AID, they and their physicians encounter the problem of selecting the donors who are to supply the semen. These donors may be, and in practice usually are, selected on the grounds that they have good qualities with respect to their health, intelligence, and personality and that these qualities are likely to have a genetic basis and will be transmitted to the baby. For this reason, women using this procedure and their physicians rarely use the semen of those with serious genetic illnesses, the mentally retarded, or criminal psychopaths, and it would be irresponsible if they were to do so.

When Robert Graham established the Repository for Germinal Choice in 1980 as a sperm bank for the storage of the semen of Nobel prizewinners, a number of critics condemned this as unethical. The bioethicist Arthur Caplan objected that it was "morally pernicious." On further consideration, however, he changed his mind and concluded, "We mold and shape our children according to environmental factors. We give them piano lessons and every other type of lesson imaginable. I'm not sure there is anything wrong with using genetics as long as it is not hurting anyone" (Bojorquez, 1994, p. 18). Caplan's second thoughts on this issue were the right ones. It is impossible to raise any principled objection to women using semen from elite sperm banks. As one of the women who used this facility observed, "First of all, we wanted a healthy baby. But we also wanted a special baby, someone who would do well, someone who would succeed. Doesn't every parent want that? Doesn't everyone want their baby to be smarter than the others?" (Bojorquez, 1994, p. 18). Or, as the Princeton biologist Lee Silver (1996) put it, "It seems only reasonable to assume that parents will want to select the best donor possible" (p. 161). It is impossible to dispute that if women want to have children who are genetically likely to have good health, high intelligence, aptitudes of particular kinds, and sound personality and if they believe that they are likely to achieve this objective by using semen from a Nobel laureate sperm bank, they are ethically entitled to do so. Furthermore, it can reasonably be argued that women and their physicians who use AID have an ethical obligation to select donors whose sperm is likely to be of good genetic quality. Most physicians are conscious of this obligation and use semen from good-quality donors. However, occasionally physicians do not act responsibly in this regard. In 1997, it was reported in the British press that some physicians carrying out AID obtained the semen from long-term unemployed men who were paid a small fee for their donations. It is well established that typically the long-term unemployed have low intelligence and poor personality qualities and are one of the least likely sections of society from which sperm of good genetic quality is likely to be obtained. In view of this, it is arguable that the physicians using the semen of the long-term unemployed were acting unethically and in violation of their professional obligation to provide proper care for their patients.

2. EGG DONATION

The same considerations apply to infertile women exercising choice in the selection of egg donors as to those who exercise choice in semen donors. In the 1990s a number of infertile American women sought what they believed would be eggs of good genetic quality by advertising for them in the student magazines of elite universities. In 1999 a British woman placed an advertisement of this kind in the Cambridge University magazine *Cam*. When this was drawn to the attention of the magazine's editor, he professed himself horrified on the grounds that "It is all faintly eugenic" and pledged that no advertisements of a similar nature would be accepted in future (*Sunday Times* [London], October 31, 1999, p. 18). This reaction is an unreasonable infringement of the civil liberties of the woman concerned and should be condemned.

In Britain, the government in 1990 sought in its Human Fertilization and Embryology Act to restrict egg donation by limiting the fee payable to egg donors to £15 plus minimal expenses. This is quite unreasonably low and invites evasion by the surreptitious payment of higher fees, which undoubtedly occurs. In a free society there can be no legitimate ethical objections to infertile women buying eggs from whomever they please and paying whatever price for them is mutually acceptable to willing buyers and sellers. Indeed, there is a good case that women seeking egg donors have an ethical responsibility to try to obtain these from women likely to be of good genetic quality.

3. PRENATAL DIAGNOSIS

Prenatal diagnosis and pregnancy terminations of fetuses with genetic diseases and disorders raises three ethical issues: first, whether the procedure is ethical in principle; second, whether it is ethical to have prenatal tests and terminations for nonserious disorders or for conditions that are not disorders at all, such as the sex of the fetus; and, third, whether it is ethical not to have the tests and terminations.

So far as the ethical principles of this procedure for serious disorders are concerned, the Roman Catholic Church and certain other religions reject its ethical legitimacy on the grounds that all abortion is ethically unacceptable. While this position may be respected, it cannot be regarded as ethically acceptable for these religions to deny the procedure to everyone, as is the case in Ireland where the Roman Catholic Church has used its power to prohibit abortion on any grounds, as a result of which many babies have been born with serious genetic diseases and disorders. The major ethical justification for prenatal diagnosis and the termination of pregnancies in cases where a genetic disorder in the fetus has been diagnosed lies in the general principle that in a free society individuals are permitted to make their own decisions concerning the conduct of their lives, unless there is a reasonable case that their behavior is likely to cause social harm. The abortion of a genetically disordered fetus cannot be regarded as likely to cause social harm and hence is ethically justified. A subsidiary justification for the procedure is that genetic diseases and disorders impose social costs of medical care and welfare provision that the rest of the population has to meet, and it is ethically legitimate to facilitate the avoidance of these costs.

The second ethical issue raised by prenatal diagnosis concerns the abortion of fetuses with nonserious disorders. Some bioethicists have argued that it is ethically permissible to abort a fetus with a serious genetic disorder, but not one with a nonserious genetic disorder or with a condition that is not a disorder at all, such as that it is female. This issue can be illustrated by the genetic condition of dwarfism, which can be regarded as not sufficiently serious to justify having a dwarf fetus aborted. Alternatively, it can be argued that dwarfs tend to suffer psychological distress and physical disabilities resulting from curvature of the spine, and pregnant women faced with the problem of whether to bear a dwarf can legitimately decide that they would prefer to have a normal baby and have the dwarf fetus terminated. In free societies where women are permitted to have abortions for reasons of convenience, the ethical legitimacy of allowing them to have dwarf fetuses terminated cannot be reasonably disputed.

The general problem in attempting to differentiate between serious genetic disorders, for which it is ethical to have an abortion, and nonserious disorders, for which it is unethical to have an abortion, is that there are more than 6,000 genetic diseases and disorders whose seriousness spans a spectrum from those that are totally disabling through less serious disorders (such as Down's syndrome) to those that are mild disorders (such as dwarfism). It would not in practice be feasible to reach any agreed consensus on a classification of these into the disorders that are sufficiently serious to allow prenatal diagnosis and termination and those that are insufficiently serious and for which this procedure would be prohibited. The only practical solution to this problem is to allow prenatal diagnosis and termination for all genetic and inherited disorders or for none.

The third ethical issue raised by prenatal diagnosis is whether it is ethical for women to refuse to have the procedure carried out. Where the genetic disorder is serious, there is a strong case that such women are acting unethically because they are bringing into the world a child whose health will be seriously impaired, causing distress to itself and to its family, and whose maintenance imposes significant costs on society. This argument can be applied to children with Down's syndrome, most of whom will never be able to look after themselves or make any positive contribution to society and who will incur medical and welfare resources that could be better directed elsewhere. To incur the risk of giving birth to such a child by refusing to have a prenatal diagnosis can reasonably be argued as unethical. This is the view taken by the American Society of Human Genetics, which posed the question of "whether or not a defective fetus should be allowed to be born," and suggested that the ethical answer is that it should not (Shaw, 1984, p. 1). This view has been endorsed by Robert Edwards, a British physician who carries out prenatal diagnosis and pregnancy terminations and who said in 1999 that "soon it will be a sin of parents to have a child that carries the heavy burden of genetic disease." He urged that all pregnant women should be tested for Down's syndrome and common genetic disorders and that it would be unethical for women to refuse to have these tests carried out and to refuse to have fetuses with serious disorders terminated (Rogers, L., 1999, p. 28).

Our conclusion is that in free societies where abortion is permitted for any reason on the general grounds of women's freedom to choose, it is ethically legitimate to allow prenatal diagnosis and pregnancy terminations for any condition. Furthermore, there is a strong case that women have an ethical obligation to have these procedures carried out and that women who refuse to allow this and subsequently bring into the world children with serious genetic disorders are behaving unethically.

4. EMBRYO SELECTION

By the end of the twentieth century it had become possible to use the technique of in vitro fertilization (IVF) to grow a number of embryos, test them for their genetic characteristics, implant those with desirable characteristics into the prospective mother, and discard those with undesirable characteristics. This procedure is carried out to prevent the birth of babies with genetic disorders; and in the future it will become possible to extend the method to selection of embryos for high intelligence, special aptitudes, personality qualities, appearance, and other characteristics. It is useful to distinguish between negative embryo selection, in which embryos with genetic diseases and disorders are discarded, and positive embryo selection, in which embryos with characteristics considered desirable (e.g., high intelligence, sound personality, attractive physical appearance), are implanted.

One ethical position on embryo selection is that it is unacceptable in all circumstances. This is the position adopted in a number of European countries and U.S. states, which in the 1990s prohibited all uses of embryo selection. These prohibitions were introduced in Austria, Germany, Norway, and Spain and in the American states of Minnesota, New Hampshire, Louisiana, and Pennsylvania. These prohibitions cannot be justified in these countries and states that allow the abortion of fetuses with genetic diseases and disorders but prohibit the rejection for implantation of embryos with these same genetic diseases and disorders. Embryo selection is more ethically acceptable than abortion because it takes place so much earlier and when most of the dozen or so embryos that are grown in vitro have to be discarded anyway. It is impossible to mount any coherent ethical case for not discarding those with genetic diseases and disorders.

One of the principal arguments against negative embryo selection is that it stigmatizes those people who have the disorders for which embryos are screened and rejected. Thus, a committee of the European Parliament (1990) has stated that the use of embryo selection "undermines our ability to accept the disabled" (p. 29). A similar view is taken by Testart (1995). This argument should apply equally strongly to the prohibition of prenatal diagnosis because both procedures are based on the assumption that it is preferable for babies to be born healthy than with genetic diseases and disorders. The argument cannot be accepted, because the gains of having healthy babies must be greater than the psychological distress that may be experienced by those suffering from the genetic diseases and disorders, distress that is caused by the knowledge that fetuses with their conditions are being aborted or embryos with them are being rejected for implantation.

A number of those who have grappled with the ethical problems of embryo selection have drawn a distinction between negative embryo selection and positive embryo selection. A position frequently adopted is that negative selection is acceptable, but positive selection is unacceptable. For instance, in 1989 the European Parliament declared that genetic techniques "must on no account be used for the scientifically and politically unacceptable purpose of 'positively improving' the population's gene pool" and called for "an absolute ban on all experiments designed to reorganize on an arbitrary basis the genetic makeup of humans" (Kevles & Hood, 1992, p. 320). Kevles and Hood commented that "the idea that genetic knowledge will soon permit the engineering of Einsteins or even the enhancement of general intelligence is simply preposterous" (p. 320). A similar distinction has been made by the British geneticist Sir Walter Bodmer and his journalist collaborator Robin McKie (Bodmer & McKie, 1994). They contend that it would be permissible to reject embryos with genetic diseases and disorders but not permissible to select embryos for intelligence or personality.

Some of those who accept the ethical legitimacy of negative selection but reject the legitimacy of positive selection for intelligence, personality, appearance, and the like, have expressed anxieties that, although negative embryo selection is acceptable, it will lead to the eventual use of positive embryo selection. Thus, in 1995 John Polkinghorne, a professor of physics at Cambridge University and a member of the Church of England General Synod, conceded the ethical legitimacy of negative embryo selection but "expressed considerable anxiety at the prospect of this being the first step towards the much broader use of embryo screening" (*Sunday Times* [London], November 5, 1995, p. 2). This is a slippery slope argument to the effect that although negative embryo selection is ethically legitimate, it might lead to unacceptable positive embryo selection. This is not a sound argument because it should be possible to permit negative embryo selection by drawing up a list of the genetic diseases and disorders that could lawfully be diagnosed and rejected, while at the same time prohibiting positive embryo selection.

An argument that has sometimes been advanced against positive embryo selection for intelligence and other qualities is that the procedure would be expensive and could only be afforded by the wealthy, and this makes it unethical. A racial dimension to this argument has been added by Dorothy Roberts (1996), a law professor at Rutgers University in New Jersey, who pointed out that whites would be able to use embryo selection more than blacks, because whites are more affluent. This, she said, would be unjust and is an argument for prohibiting both IVF and embryo selection, although she does not go as far as to positively recommend their prohibition. This view cannot be accepted. There are a number of things that can be afforded by the affluent but not by the poor, but this does not imply that these things should be prohibited. The right of citizens to spend their money as they choose is central to a free society.

The conclusion to which we are drawn is that the ethical issues of embryo selection must be determined by the same criteria as the other human biotechnologies, namely whether its use is likely to cause harm to individuals or social damage. No reasonable case can be made that allowing women to select embryos that are free of genetic disorders or that have the genetic potentialities for high intelligence, sound personality, attractive appearance, and so forth, is likely to have any harmful or socially damaging consequences. On the contrary, it is far more probable that it will produce social benefits. As and when the techniques of positive embryo selection are developed, there can be no ethical reason in free societies for prohibiting them.

5. SELECTION FOR SEX

A development in human biotechnology that has aroused some ethical concern is the selection of a child's sex. There are three techniques by which this can be done. The first consists of staining a sample of semen with dye. Sperm containing two X chromosomes, which develop into females, are heavier and absorb more of the coloring, making them brighter and distin-guishable from those with the male Y chromosome. Sperm of the desired sex are then given to the mother by artificial insemination. This procedure is legal in the United States and Britain, and clinics for carrying it out were established in the 1990s, such as the IVF Genetic Center in Fairfax, Virginia, and the London Gender Clinic. A second method for sex selection of babies is to ascertain the sex of the fetus and abort it if it is of the undesired sex, allowing the couples to try again until they conceive a child of the desired sex. Third, IVF can be used to grow several embryos, the sex of these can be ascertained, and an embryo of the desired sex can be implanted. This last procedure was made illegal in a number of European countries in the early 1990s under the general prohibition of any use of embryo selection. It was made illegal in Britain in 1997. However, it has remained legal in Italy, and a number of couples have gone to Italy to have the procedure carried out. This illustrates the ineffectiveness of prohibiting human biotechnology in particular countries and the loopholes that are inevitably present when the procedures are permitted in other countries.

It is generally believed that if sex selection is permitted, women, possibly encouraged or coerced by their husbands, would tend to select boys. This is certainly the case in China and India, where significant numbers of baby girls are killed (Lau, 1995), leading to an excess of males in the population. In India prenatal tests for the sex of embryos and the abortion of females oc-curred on such a large scale that the procedure was made illegal in 1994.

It is widely believed that sex selection is ethically unacceptable. This is the basis for its prohibition by embryo selection in several European coun-tries. The strongest argument for the prohibition of sex selection is that it would lead to an excessive number of boys and that this would be socially harmful. In particular, there would be some males who would be unable to obtain females, and these might turn to rape and become a social problem.

There are nevertheless several arguments for permitting the selection of babies for sex. In the first place, it is by no means certain that there is a widespread preference for boys in the United States and Europe or elsewhere in the economically developed world. A survey carried out in the United States in the 1970s found that approximately two-thirds of women would prefer a boy for their first child (Pebley & Westoff, 1982). However, surveys have also shown that most people would like to have two children, and whether many of these would want two boys is open to doubt. Probably most of those who had a boy as their first child would want a girl as their second. It may well be that sex selection will not lead to any significant shift in the sex ratio toward a preponderance of boys.

Second, even if sex selection does develop on a significant scale and leads to an excess of males, this could well be desirable. In such a society, males would have to compete more energetically for females, and in this competi-tion males with higher status would tend to succeed. Conversely, women would have greater choice of men and would tend to select those with higher status, as women normally do (Buss, 1999). The males with higher status who succeeded in obtaining females and who were selected by females would tend to be the more intelligent and with stronger moral character, so the effect would be eugenic and desirable.

Third, even if a significantly greater number of males were to be born, this would probably be self-correcting, as girls would acquire scarcity value and come to be desired more than boys. The preferences of couples would then be expected to switch in favor of selecting girls.

Fourth, there are instances where couples could quite legitimately wish to have a baby of a particular sex. One of these would be where couples decided to have two children and after having a child of one sex would like to have the second of the other sex. It cannot be reasonably contended that such a choice would be ethically objectionable.

The upshot of these considerations is that the likelihood of adverse consequences arising from allowing couples to choose the sex of their children is too remote a contingency for this freedom to be prohibited, that the exercise of this freedom cannot be regarded as unethical, and that couples should be permitted to exercise this choice on general libertarian grounds.

6. CLONING

The possibility that humans might be reproduced by cloning became widely condemned in the last two decades of the twentieth century. Mary Warnock (1987), a leading British moral philosopher and expert in human biotechnology, condemned cloning on the grounds that "there must be some things which, regardless of consequences, should not be done, some barriers which should not be passed. What marks these barriers out is often a sense of outrage, if something is done; a feeling that to permit some practice would be indecent or part of the collapse of civilization" (p. 8). In the United States, the bioethicist R. A. McCormick (1994) condemned the cloning of humans on the grounds that it would be contrary to "our cherished sense of the sanctity, wholeness, and individuality of human life" (p.16). Neither of these is an acceptable argument for the prohibition of cloning.

The cloning of humans was prohibited in Britain in 1990 by the Human Fertilization and Embryology Act. In the United States, the National Institutes of Health set up a Human Embryo Research panel in the early 1990s to consider the ethical issues concerning in vitro technology in general and to advise on what research should be acceptable for federal funding. The panel issued its report in September 1994. Among the recommendations for blacklisting for funding were cloning and sex selection of embryos, except to prevent genetic diseases linked to the X chromosome. However, the panel did not recommend that these should be prohibited. It recommended that it should remain legal for privately run IVF clinics to provide these services.

In 1997, following the cloning of the sheep Dolly, many eminent persons asserted that the cloning of humans should be prohibited. In October, the Council of Europe voted to outlaw the cloning of humans in 40 European nations. Ian Wilmut, the embryologist who produced the cloned sheep Dolly, said of cloning humans, "all of us would find that offensive" (Kolata, 1997, p. 6). President Bill Clinton stated in June 1997 that to fund cloning humans would be "morally reprehensible" (p. 196). Joseph Rotblat, a British Nobel Prize winner, described cloning as "a means of mass destruction" (Harris, 1998). Hiroshi Nakajima, Director General of the World Health Organization (WHO), stated, "WHO considers the use of cloning for the replication of human individuals to be ethically unacceptable as it would violate some of the basic principles which govern medically assisted procreation. These include respect for the dignity of the human being and protection of the security of human genetic material" (World Health Organization, 1997). Frederico Mayor, Director of UNESCO, issued a statement that "Human beings must not be cloned under any circumstances." A couple of days later the European Parliament (1997) passed a resolution condemning cloning in the following terms:

> In the clear conviction that the cloning of human beings . . . cannot under any circumstances be justified or tolerated by any society, because it is a serious violation of fundamental rights and is contrary to the principle of equality of human beings as it permits a eugenic and racist selection of the human race, it offends against human dignity and it requires experimentation on humans. . . . Each individual has a right to his or her own genetic identity, and human cloning is, and must continue to be, prohibited. (p. 84)

There are three ethical objections that can be made to cloning. The first is the point made by the European Parliament that "the security of genetic material would be compromised." It is difficult to discern the meaning of this phrase. It may be that what is intended is the preservation of genetic variability, but the reduction of this by producing a small number of clones would be negligible.

Second, there is the objection presented in the resolution of the European Parliament condemning cloning on the grounds that "it permits a eugenic and racist selection of the human race." A number of other critics have made the same point. This view was popularized in a novel and movie *The Boys from Brazil*, which was concerned with the attempts of the Nazi doctor Josef Mengele to make clones of Hitler, who would attempt to conquer the world. It is undoubtedly true that cloning could be used to produce replicas of individuals who might do a lot of damage. The same argument can be made against the other techniques of human biotechnologies such as embryo selection and against normal reproduction and could be used to justify the sterilization of criminal psychopaths whose children are also likely to inflict social damage. However, the cloning of undesirables is unlikely to occur on a sufficient scale in the Western democracies to justify the prohibition of all cloning and would be best dealt with by some form of parental licensing program of the kind discussed in Chapter 14.

The third ethical objection to cloning is that a cloned individual would suffer a loss of personal identity. It is contended that we all feel ourselves unique, that cloned individuals would be deprived of this sense of uniqueness, and that this would be psychologically damaging. The problem with this argument is that identical twins, who are genetically identical in precisely the same way as clones would be, do not appear to experience psychological damage from knowing that they have a genetically identical sibling. Furthermore, although a cloned individual would be genetically identical to the parent, it would not be psychologically identical because it would be reared in a different environment and have quite different experiences. A cloned individual would be much more different from its parent than one identical twin from the other, because identical twins are reared in the same environment, whereas cloned individuals would be reared in different environments.

While the ethical objections to cloning humans must be regarded as weak, there are three positive arguments in favor of allowing cloning. First, it should become possible in the future to use cloning to grow spare body parts to replace those that are defective, such as kidneys, livers, eyes, and so forth. At present, these parts are frequently replaced by transplants from donors, but unless the donors are genetically closely similar to the recipients, the donated organs are rejected. It is hard to find donors sufficiently similar to recipients. Furthermore, there are more people needing organ transplants than there are organs available. The result of this is that many people die because of the unavailability of organs. This problem could be overcome by using cloning to grow healthy versions of the defective organs and use these for transplants. This would alleviate much suffering and save many lives. Those who oppose cloning can therefore be regarded as responsible for causing a great deal of suffering and many deaths, and this is ethically problematical for them.

Second, there are significant numbers of couples who would like to have children, but the woman is infertile and cannot have children by IVF or any other form of assisted reproduction except cloning. Allowing cloning would promote the happiness of the couples concerned, and there are no persuasive arguments that it should be prohibited. Furthermore, it is to the general social advantage that such couples should be permitted to use cloning to have children because it would contribute to maintaining the size of the population in the economically developed nations, in all of which fertility is below replacement. Third, it is also possible that single people wanting children might prefer to have a clone of themselves rather than a child conceived in collaboration with an unknown partner. There is no coherent ethical case that this should be prohibited.

The conclusion to this discussion is that arguments presented against cloning are not sufficiently strong to justify its prohibition. Valid ethical objections to cloning would have to show that it caused the cloned individual psychological damage or would be likely to cause social damage. Neither of these arguments can be regarded as plausible. Furthermore, there are potential benefits to cloning, and the balance of the argument has to be that in free societies cloning should be permitted.

7. GENE THERAPY

In the 1990s several successful trials were made for the treatment of genetic diseases by gene therapy, in which healthy genes are inserted into patients to correct the deficiencies of their malfunctioning genes. In considering the ethical issues of this treatment, a distinction has to be drawn between "somatic-line" and "germ-line" therapy. Insertion of genes into somatic-line cells only affects the individuals concerned. These genes are not transmitted to their children, who are still at risk of inheriting the adverse genes. This procedure is dysgenic insofar as it facilitates the transmission of genes for genetic diseases. The genes inserted into germ cells are transmitted to children, and the insertion of healthy genes into germ cells is therefore eugenic.

Germ-line therapy has been frequently attacked as unethical, and its use on humans has been prohibited in the United States, Britain, and much of Continental Europe on the grounds of possible adverse and unknown future consequences. The case for prohibiting germ-line therapy on humans has been stated by Suzuki and Knudtson (1990). They argue first that even apparently harmful genes may be useful, and therefore that it would be a mistake to eliminate them. The classical example supporting this argument is the disease of sickle-cell anemia. This disease is caused by inheritance of two recessive genes for the condition. The inheritance of one recessive gene for the disease has no adverse effects and provides immunity from malaria. The gene therefore has advantages in the single form. This does not, however, mean that there is a sound argument for preserving the gene, especially among populations in temperate environments where there is no malaria. Even in tropical malarial environments, it would be preferable to prevent malaria by immunization and to eliminate the sickle-cell gene. Furthermore, sickle-cell anemia is a very unusual genetic disease in that a single copy of the recessive gene confers an advantage. There are no other genetic diseases known to have an advantage of this kind. So far as is known, the genes for genetic diseases are wholly disadvantageous, and it would be best to eliminate them by germ-line therapy. It might at some point in the future be found that some of these genes do have some useful properties, but if this should transpire it would be possible to reintroduce the gene and no long-term damage would be done by making a start on eliminating it.

The second argument against germ-line therapy advanced by Suzuki and Knudtson (1990) is that once this was accepted it might lead to eugenic interventions for other characteristics, such as intelligence. They write that "the history of genetics suggests that once a human characteristic—such as a particular skin color or poor performance on IQ tests—has been labeled a genetic 'defect,' we can expect voices in society to eventually call for the systematic elimination of those traits in the name of genetic hygiene" (p. 204). This is the slippery slope argument again, the general format of which is that an innovation may be beneficial but should not be allowed because it might lead to other innovations that would be harmful. This principle would prevent much and possibly all technical and scientific progress, virtually all of which has potentially harmful applications. Thus, the invention of the wheel led eventually to the development of artillery, which has caused the deaths of countless millions in warfare. Slippery slope arguments cannot be accepted as valid objections to procedures that are ethically acceptable in themselves.

The third argument advanced by Suzuki and Knudtson against germ-line therapy is that the elimination of harmful genes would reduce genetic diversity and that genetic diversity should be preserved. They argue that genetic diversity is desirable because any genes might at some future time give an advantage in combatting some unpredictable hazard. Thus, it appears that they prefer that some unfortunate people should be required to endure the suffering caused by a genetic disease on the off chance that, in the future, their defective genes might confer some sort of unknown benefit. This remote possibility cannot be accepted as an ethically valid reason for preserving these harmful genes.

Fourth, Suzuki and Knudtson (1990) conclude their discussion of this issue by asserting that "while genetic manipulation of human somatic cells may lie in the realm of personal choice, tinkering with human germ cells does not. Germ-cell therapy, without the consent of all members of society, ought to be explicitly forbidden" (p. 48). This use of the term *tinkering* for germ-line therapy betrays the emotive stance of the writers. This treatment has the wholly laudable objective of curing those with serious and life-threatening diseases and should no more be called *tinkering* than giving heart or liver transplants to patients with heart disease or renal failure. Furthermore, Suzuki and Knudtson's stipulation that every single citizen would have to consent to the use of germ-cell therapy before it can be permitted is a very stringent requirement implying that in a nation with a population of many millions, a single objector would be able to veto the legalization of the procedure. This requirement is so unreasonable that it is difficult to believe that Suzuki and Knudtson have given any serious thought to the issue.

8. THE CASE FOR PERMITTING GERM-LINE GENE THERAPY

Contrary to many objectors, there is a strong ethical case for permitting germ-line gene therapy based on two considerations: The first is that germ-line therapy corrects the genetic defect once and for all, whereas somatic-line therapy allows the defective genes to be transmitted and entails the costs, hazards, and inconvenience of medical intervention for all the descendants inheriting the disorder. A second argument for permitting germ-line therapy is that there are certain cases where this is the only way of allowing couples to have a healthy child. One of these is where both parents have the same recessive gene disorder. Because they both have two copies of the malfunctioning gene, they have no healthy genes to transmit, and all their children will have the disorder. The only way for them to have a genetically healthy child is by germ-line therapy. A further instance is women with genetic mitochondrial diseases, which are passed to all their children. In both these cases, couples are not able to have a disease-free child by prenatal diagnosis or prenatal implantation diagnosis because all their children inherit the disease. The only way of overcoming this problem would be by germ-line therapy. This would be done by repairing the genetically defective gene in the ovum before fertilization. The ovum is removed from the woman, and the defective gene is removed and replaced with a healthy gene from a donor. The ovum is then fertilized in vitro with the husband's sperm and implanted.

Although germ-line therapy has been widely condemned, there have been some dissenting voices in favor of its being permitted, including those of Daniel Koshland (1988), the editor of *Science*; and Rubenstein, Thomasma, Schon, and Zinaman (1995), physicians at the medical school of Loyola University in Chicago. These views are surely correct. Germ-line gene therapy would indisputably yield benefits in the treatment of genetic diseases, the alleviation of suffering, and the reduction of the genes for diseases and disorders. Indeed, it is the prohibition of this form of treatment that is ethically questionable.

In due course it may well become possible to apply the technique of gene therapy to increase intelligence and to correct personality disorders. This may be done by inserting existing genes for high intelligence or sound personality into those whose intelligence is low or into psychopaths or even by constructing new genes for high intelligence or sound personality. It has frequently been asserted that this would be unethical. This is the eugenic nightmare that lies at the end of the slippery slope that Suzuki and Knudtson (1990) predict is likely to transpire once germ-line gene therapy is permitted. The same view has been expressed by W. F. Anderson (1992), who has pioneered the work on gene therapy and who has written that "although the medical potential is bright, the possibility of misuse of genetic engineering technology looms large, so society must ensure that gene therapy is used only for the treatment of disease" (p. 813). These pronouncements to the effect that the use of gene therapy to improve intelligence and personality would be unethical must be rejected. Because the improvement of intelligence and personality is a desirable objective, it would be as ethically legitimate to use gene therapy to improve them as it is to use it to treat diseases.

9. CONCLUSIONS

The ethical objections that have been raised to the various forms of human biotechnology consist partly of what can be termed "yuk reactions" and partly of attempts to show that these would cause harm, either to society or to the individuals concerned. The "yuk reactions" are feelings of emotional aversion unsupported by arguments based on rational principles. A good deal of morality has always been of this kind. As the eighteenth-century Scottish philosopher David Hume (1751) observed, "Morality is generally more properly felt than judged of." The yuk arguments against the use of biotechnology to improve human genetic quality are unpersuasive because yuk reactions to various behaviors have changed quite rapidly over time. In previous historical periods, people had yuk reactions to those who professed atheism and to women who displayed their ankles, both of which were widely regarded as unethical. In the twentieth century there were widespread yuk reactions to the sale of contraceptives, to abortion, to artificial insemination by donors, and to in vitro fertilization and the like. Today, the great majority of people no longer experience yuk reactions to these things and accept them as normal and ethically legitimate. We can anticipate that it will be the same with the more recent developments in human biotechnology, like embryo selection, cloning, and germ-line gene therapy, as these are developed and people become more used to them. Yuk reactions do not form a sound basis for ethical judgments.

The only sound basis for the ethical rejection of the human biotechnologies would be that they harm the individuals concerned or cause social damage. These claims have been made, but they do not stand up to examination. No convincing arguments have been presented that any of the human biotechnologies have done any harm or are likely to do any harm in the future, either to individuals or to society.

19

The Future of Eugenics in Democratic Societies

1. Prospects for the Rehabilitation of Eugenics
2. Political Infeasibility of State Eugenic Programs
3. Artificial Insemination by Donor
4. Reduction of Genetic Diseases and Disorders
5. Development of Positive Selection for Intelligence and Personality
6. Technical Advances in Embryo Selection
7. Development of the Use of Embryo Selection
8. Ineffectiveness of Prohibition of Embryo Selection
9. The Emergence of Caste Societies
10. Conclusions

In this chapter we consider the future of eugenics in the Western democracies. We begin with a discussion of whether the classical eugenics of attempting to alter patterns of reproduction is likely to be rehabilitated, either in theory among biological and social scientists or in practice by governments. We turn next to probable developments in the human biotechnologies of prenatal diagnosis, embryo selection, and gene therapy and consider the degree to which these are likely to lead to reductions in genetic diseases and disorders and to the enhancement of intelligence and personality.

1. PROSPECTS FOR THE REHABILITATION OF EUGENICS

Although eugenics was universally rejected and dismissed in the closing decades of the twentieth century, we should anticipate that at some point in the twenty-first century eugenics will once again become accepted as a desirable and legitimate objective of public policy. This will come first among biological and social scientists, who will come to accept the truth of the core propositions of eugenics set out in the Preface to this book. By the end of the twentieth century, a wide measure of consensus had already been reached regarding the genetic contribution to many diseases, to intelligence, and to personality and regarding the importance of intelligence and personality for educational and occupational achievement. It will take longer for a consensus to develop regarding the genetic problem of the underclass and the problems of dysgenic fertility and dysgenic immigration, but in due course these will come to be recognized. The reason for anticipating this is that truth ultimately triumphs over falsehood.

A portent of the coming counterrevolution in the rehabilitation of eugenics was the publication in 1994 of *The Bell Curve*. In this, Richard Herrnstein and Charles Murray presented new evidence for the core propositions of eugenics. They demonstrated the significant contribution of low intelligence to many social problems of the underclass, including crime, long-term unemployment, welfare dependency, and teenage motherhood. They did not make explicitly eugenic proposals for dealing with these problems, but their analysis is implicitly eugenic because, as they identified the underclass as a genetic problem, it calls implicitly for a eugenic solution. Indeed, the eugenic implications of Herrnstein and Murray's analysis are among the grounds on which it has been attacked.

A further portent of the reappearance of eugenic thinking occurred in 1998 when James Watson, the co-discoverer of the structure of DNA, spoke at the Congress of Molecular Medicine and advocated the greater use of prenatal diagnosis and termination of fetuses with genetic disorders. He asserted, "The truly relevant question for most families is whether an obvious good to them will come from having a child with a major handicap. From this perspective, seeing the bright side of being handicapped is like praising the virtues of extreme poverty" (Smith, J. D., 1999, p. 132). Although somewhat opaquely expressed, this can only be construed as an endorsement of eugenics. Once eugenics comes to be accepted again by the scientific community, it will come to be accepted by informed public opinion and the media.

2. POLITICAL INFEASIBILITY OF STATE EUGENIC PROGRAMS

While we can anticipate that the intellectual case for eugenics will once again become accepted, it is much less likely that eugenics will be rehabilitated politically in the Western democracies and that Western governments will begin to introduce eugenic programs. Even in the first half of the twentieth century, when most biological and social scientists and much of informed public opinion were favorable to eugenics and Western governments introduced eugenic measures of sterilization and selective immigration, the measures introduced were not sufficiently robust to have much impact. In the Scandinavian countries, sterilizations of the mentally retarded and mentally ill ran at only about 50 a year in Finland and Norway and at about 1,000 to 2,000 a year in Denmark and Sweden (Broberg & Roll-Hansen, 1996). This is about the maximum rate that could be expected in democratic societies. Sterilizations on this limited scale would not have much effect on the genetic quality of the population, and it cannot be expected that sterilizations are likely to be reinstated on any greater scale in democratic Western societies.

By the end of the twentieth century, two social changes had occurred throughout the Western democracies that made it more difficult for governments to introduce eugenic measures than it was in the first half of the century. The first of these was that the balance between individual and social rights, which all societies have to strike, had swung strongly in favor of strengthening individual rights at the expense of social rights. Individual rights consist of the freedom of individual citizens to behave as they choose, while social rights consist of the right of societies, acting through their governments, to restrict individual rights in the interests of general social well-being. The rejection of negative eugenics in the closing decades of the twentieth century should be understood as part of this broader trend for allowing individual rights to prevail over social rights. Whereas in the first half of the century it was widely accepted that society could legitimately restrict the rights of the mentally retarded, criminals, and psychopaths to have unlimited numbers of children on the grounds that these imposed social costs, in the closing decades of the century the individual rights of these social-problem groups to propagate freely were allowed to override the social right of society to curtail their reproductive freedoms.

One of the most extreme assertions of the precedence of individual rights over social rights in the second half of the twentieth century was made by John Rawls (1971) in his book *A Theory of Justice*. In the opening pages of the book, Rawls sets out his basic premise: "Each person possesses an inviolability founded on justice that even the welfare of society as a whole cannot override. For this reason justice denies that the loss of freedom for some is made right by a greater good shared by others. It does not allow that the sacrifices imposed on a few are outweighed by the larger sum of advantages enjoyed by the many" (p. 3). The acceptance of this premise would preclude the conscription of citizens into the armed services, the imprisonment of criminals, the detention of violent schizophrenics, the withdrawal of automobile driving licenses from habitual drunkards, and the like. It is a preposterous foundation on which to build a theory of justice and of the rights and obligations of citizens, and its wide acceptance is testimony to the extent to which individual rights were accorded precedence over social rights in the United States and other Western democracies in the second half of the twentieth century. Eugenic policies implemented by the state obtain their ethical legitimacy from the contention that social rights may sometimes take prece-

dence over individual rights and in particular over individual rights of reproductive freedom. For eugenic policies to become acceptable again, there would
have to be a shift toward a greater acceptance of the legitimacy of social rights
and of the curtailment of individual rights when these inflict social damage.
At the start of the twenty-first century, there is no sign that any shift of this
kind is likely to occur.

The second social change that took place in the second half of the twentieth century that will make it more difficult to rehabilitate eugenics consists
of the growth of groups hostile to eugenics. These consist of ideologically
committed civil liberties groups and of a variety of special interest groups, all
of which have a common cause in placing the liberties of the individual above
social well-being. Two powerful special interest groups in particular can be
expected to oppose eugenic programs. The first of these consists of the administrators, social workers, medical workers, psychologists, educators, and
the like, whose careers have been built on catering for the needs of the mentally
retarded, criminals, and psychopaths and who have identified with the interests of these "clients," as they have become known. These would inevitably
oppose eugenic proposals designed to reduce the numbers of the social problem groups on whose existence their own careers depend and with whom they
have come to empathize.

A second special interest group that would be expected to oppose any
attempt to rehabilitate eugenics is the racial and ethnic minorities that would
be disproportionately affected by eugenic policies. Foremost among these are
African Americans and Hispanics in the United States and Africans in Europe, whose low average intelligence and high crime rates would make them
disproportionately subject to sterilization and restrictions on immigration. Any
proposal to introduce eugenic programs of sterilization and immigration control would inevitably be rigorously opposed by these groups and their advocates. By the closing decades of the twentieth century, it had become politically impossible in the United States for either the Republican or the
Democratic parties to reduce immigration, let alone to introduce selective
acceptance criteria, because of the voting power of the African Americans
and Hispanics, who naturally favor the admission of increasing numbers of
their own racial and ethnic groups. This problem is also present throughout
Europe where, although the ethnic and racial minorities are fewer in number,
they are still sufficiently numerous to deter political leaders from introducing
measures calculated to offend them. The same problem of adverse impact would
also be present in any attempt to introduce measures of positive eugenics,
such as the provision of financial incentives for high-earning elites to have
children. Disproportionately fewer of the ethnic and racial minorities would
qualify, except for the Asians, and on this account they would be likely to
oppose measures of this kind.

Although it has become impossible at the turn of the millennium for governments in the Western democracies to revive the classical state-mandated

eugenics of the first half of the twentieth century (and it must be considered highly improbable that any Western governments will attempt to do this in the foreseeable future), it is probable that they will continue to support educational and facilitative eugenics consisting of the provision of sex education in schools, information about contraception, and financial support for birth control clinics and abortion. This support will be provided to promote the health and well-being of women. Probably these services and facilities will gradually come to be used more efficiently as knowledge of them grows, as the morning-after pill becomes more widely known and more easily available, and as more effective forms of contraception are developed. Probably also the welfare support for underclass women to have babies by providing them with welfare incomes and housing will be reduced as understanding grows that these act as perverse incentives for childbearing among the least desirable section of the population. It is also possible that there will be some further developments in the approval of sterilization by the courts, both of the mentally retarded at the request of their parents and of criminals as a condition of more lenient sentences. These developments are likely to take place without any ostensible eugenic purpose, but they will nevertheless have a small positive eugenic impact.

3. ARTIFICIAL INSEMINATION BY DONOR

While progress in classical eugenics is likely to be, at best, modest in the Western democracies, greater advances can be anticipated in the development of the new eugenics of the human biotechnologies. The first of these to be considered is artificial insemination by donor (AID). The use of this procedure by couples where the male partner is infertile is likely to grow because male sperm counts declined and male infertility increased in the economically developed nations in the closing decades of the twentieth century. It is also probable that couples using this procedure will become more sophisticated regarding the importance of genetics and will increasingly demand that the donors are of high quality and have desirable characteristics. To meet this demand, a number of elite semen banks have been established in the United States. The first of these was Robert Graham's Repository for Germinal Choice containing the frozen semen of Nobel Prize winners and other eminent men. The Repository, which was closed down in 1999 following Graham's death, issued a catalog giving details of the achievements and physical characteristics of the donors. The semen was provided at nominal cost to women wanting to use it. By the mid-1990s, a number of other elite semen banks had been established in the United States. The Fertility Research Foundation in Manhattan, New York, stocks semen samples from scientists and Olympic athletes. Cryobank in Boston and Palo Alto, California, has samples from outstanding students from Harvard, Massachusetts Institute of Technology

(MIT), and Stanford. These foundations provide catalogs describing the characteristics of the donors, including physical appearance; skin, eye, and hair color; medical history; SAT scores; educational grades; and special talents in music, athletics, and so on. By 1999 there were catalogs on the Internet advertising semen from high-quality donors for sale at $200 a sample and providing information on the characteristics of the donors (Wood, 1999).

These elite semen banks have been ridiculed in a number of predictable quarters, such as the editorial pages of the *New York Times* (Editorial, 1982), which sneered that "if intellectual qualities were inheritable in any simple fashion, those who conceive with the help of the Nobel could count on a great deal of vanity and a dearth of plain sense. Chances are, however, they will get themselves just children." Contrary to this assertion, there is no doubt that children born by AID from Nobel Prize winners and from top-scoring students from Harvard, MIT, Stanford, and other elite universities will normally be more gifted than children produced by donors of average abilities. The impact of the use of highly intelligent donors can be quantified for intelligence by using the formula for the impact of selective breeding provided by Cavalli-Sforza and Bodmer (1971) and given in Chapter 11. For instance, if the women who use AID have an average IQ of 100 and the donors have an average IQ of 145, and the additive heritability of intelligence is .71, the children will have an average IQ of 116. This puts them in the top 16 percent of the population, a result that is a desirable outcome. However, with only about 1 percent of women in the United States conceiving by AID, the impact of this on the child population would be to raise the intelligence level by only 0.16 of an IQ point. Thus the eugenic impact of AID using elite donors, even if this were to increase significantly, is likely to be very small.

In the middle decades of the twentieth century, several eugenicists proposed that AID would have a substantial eugenic impact if large numbers of women could be induced to use it, using semen from elite donors rather than that of their husbands. This idea was advanced in Britain in the 1930s by Brewer (1937), who coined the term *eutelegenesis* (breeding from afar) for this eugenic use of AID. The scheme was also advocated at about the same time by the American eugenicist Hermann Muller (1935, 1963). Muller proposed that the semen of "excellent men" would be stored and women who wanted children would be encouraged to use it, rather than that of their husbands. The "excellent men" selected for this purpose would possess high intelligence and have socially valuable personality qualities, such as a strong moral sense, self-discipline, a strong sense of civic obligation, and energy. "How many women," Muller (1935) asked rhetorically in his book *Out of the Night*, "would be eager and proud to bear and rear a child of Lenin or Darwin?" (p. 122). The answer to the rhetorical question has turned out to be "virtually none." The proposal that men could be persuaded to have their wives or female partners inseminated by elite sperm taken from a semen bank rather than by themselves ignores a fundamental theorem of evolutionary psychology, that people are motivated to transmit their own genes. This stumbling block cannot be expected to change. We must therefore conclude that AID from elite donors is unlikely to have any significant eugenic impact in democratic societies.

4. REDUCTION OF GENETIC DISEASES AND DISORDERS

Greater eugenic progress can be anticipated in the twenty-first century through further reductions in genetic diseases and disorders. In the last three decades of the twentieth century, considerable advances were made in the prenatal diagnosis of diseases and disorders, which led to substantial reductions in the birth incidence of a number of these conditions, notably anencephaly, spina bifida, hydrocephaly, Tay-Sachs disease, and B-thalassemia. Further progress of this kind can be anticipated and is likely to lie in the development of new and more effective diagnostic procedures, in the greater use of carrier screening to identify carriers of recessive gene disorders, in the increased use of preimplantation diagnosis of embryos grown in vitro, and in the use of gene therapy.

There are a number of reasons for believing that during the twenty-first century there will be further progress in these directions in the reduction of genetic diseases and disorders. First, the existing procedures for the identification of fetuses with genetic diseases and disorders are not being utilized with anything approaching maximum effectiveness. In Britain, a study of the degree to which these procedures were being used was made in the late 1980s by the Royal College of Physicians (1989), which concluded that if all the existing procedures were used effectively, the birth incidence of genetic and congenital disorders would be approximately halved. This remains the position at the beginning of the twenty-first century in most of the economically developed world. There is good reason to believe that these procedures will come to be used more effectively in the future.

Second, it can be anticipated that in the twenty-first century there will be further advances in the screening of pregnant women for adverse genes. This is likely to be done by the compilation of registers of families carrying genes for diseases and disorders, from which physicians will be able to check whether pregnant women and their husbands and partners are at risk of being carriers. A program of this kind was begun in the 1990s in Britain with a register of all the two million or so carriers of the gene for cystic fibrosis. As these registers are compiled for increasing numbers of genetic diseases, it will become possible to diagnose and terminate many more genetically impaired fetuses.

Third, it can be anticipated that new and more effective methods will be discovered for the identification of fetuses with genetic disorders and diseases.

What these will be is necessarily unpredictable, but there is every reason to expect that further technical progress will be made in this direction.

Fourth, medical and public opinion are largely in favor of the use of prenatal diagnosis for genetic diseases and disorders and of the terminations of the pregnancies of impaired fetuses. A few physicians are unwilling to perform or to recommend the use of these procedures, and some pregnant women are unwilling to have them carried out. These are mainly Roman Catholics who are opposed on principle to all pregnancy terminations, but these are a small minority. In most Western democracies, 80 percent to 90 percent of women opt to terminate the pregnancies of impaired fetuses.

Fifth, some of this opposition to pregnancy terminations of impaired fetuses is likely to be overcome by the use of preimplantation diagnosis (PID) of embryos grown in vitro. This avoids the ethical issues and the stress of pregnancy termination. Although this procedure has not yet become widely used and is little known, it seems likely that many women identified by carrier screening as being at risk of having a genetically impaired fetus would be willing to use it.

Sixth, physicians have come under increasing legal pressure to give their pregnant women patients full advice on the availability of prenatal diagnosis for impaired fetuses. This has arisen from legal actions taken in the United States by parents of genetically impaired children who have sued their physicians for failing to advise them properly about the availability of the tests. In response to these actions, the American College of Obstetricians and Gynecologists advised its members in 1984 to inform all pregnant patients of the availability of prenatal tests and in particular of maternal serum alpha-fetoprotein (MSAFP) screening, so that they would have a defense in any medical malpractice suit following the birth of a baby with neural tube defects and other abnormalities. The growing propensity of the public to sue physicians for negligence is likely to increase the use of prenatal diagnosis for genetic diseases and disorders and of pregnancy terminations.

Seventh, further progress can be anticipated in gene therapy for the treatment of genetic diseases and disorders, and germ-line therapy is likely to be used to prevent the transmission of defective genes to children. Although germ-line therapy was prohibited in a number of Western democracies in the closing decades of the twentieth century, the unreasonableness of this prohibition will come to be understood, and germ-line therapy will come to be permitted. When the techniques of germ-line therapy are perfected and the treatment becomes legal, major advances can be anticipated in the reduction of the genes for genetic diseases and disorders.

For these reasons we can anticipate that in the twenty-first century the birth incidence of babies with genetic diseases and disorders will be considerably reduced. It is even possible that eventually virtually all babies will be born free of genetic diseases and disorders.

5. DEVELOPMENT OF POSITIVE SELECTION FOR INTELLIGENCE AND PERSONALITY

Although further progress in negative eugenics entailing selection against genetic diseases and disorders can be confidently anticipated, progress in positive eugenics for the selection of those with high intelligence and desirable personality traits is likely to take place more slowly. There are three reasons for this. First, most of the genes determining intelligence and personality have not yet been identified, although this can be anticipated early in the twenty-first century. Second, it is virtually certain that a number of genes will be involved, and this will make the technical problems of selection more difficult than for single-gene diseases and disorders. Third, there is more opposition to positive selection for intelligence and personality than to negative selection against genetic diseases and disorders, and this has resulted in positive selection being legally prohibited in a number of the Western democracies. Nevertheless, this opposition is likely to diminish over time, and advances in positive selection are likely to take place over the longer term.

It is theoretically possible that prenatal diagnosis could be used to identify fetuses with low intelligence and undesirable personality traits when the genes for these characteristics are discovered and become identifiable in the fetus and that women might opt to have their pregnancies terminated if a particularly unfavorable set of genes were identified. However, it seems improbable that this procedure would be used on any appreciable scale. This is partly because the stress of undergoing a pregnancy termination and trying again for a baby with a better set of genes would probably be too great for significant numbers of women to contemplate. The eugenic potential of prenatal diagnosis lies almost exclusively in the reduction of genetic diseases and disorders.

It is doubtful whether either cloning or genetic engineering will make any significant contribution to positive eugenics in the Western democracies at least for the foreseeable future. With regard to cloning, there can be no doubt that it will become technically feasible to clone humans. In due course it seems likely that some people will opt to have babies by cloning. In the Western democracies this development will be retarded by the prohibition of cloning, which has already been made illegal in a number of European countries. In the United States there have been a number of calls for the banning of cloning of humans, including one made by President Bill Clinton on January 9, 1998, although it has not yet been made illegal. However, the banning of cloning is unlikely to be effective because it would be virtually impossible to persuade every country in the world to prohibit it. Some countries would permit cloning, and some physicians would be willing to carry it out. In 1998 Dr. Richard Seed in the United States and Dr. Severino Antinori in Italy both announced that they planned to offer cloning services to humans.

We can envision that a small but significant percentage of people may come to reproduce themselves by cloning. For example, couples who are both infertile and wish to have a child who is genetically related to themselves might wish to use cloning, perhaps to have two children, each of which was a clone of one of the couple. There may be single women who wish to have a child but do not have a partner who might be drawn to cloning themselves. There may be homosexuals or lesbians who would like to have a child by cloning. There might be some individuals with a very high opinion of themselves, perhaps justified, who would wish to produce a clone. Thus, we can anticipate a small demand for cloning. This is likely to be met by the establishment of cloning clinics in countries where cloning is permitted. This would probably have a minor eugenic impact because those who produced children by cloning would be likely to be more intelligent and have sounder personality qualities than the average of the population, few of whom would be likely to go to the time, trouble, and expense of the procedure. However, the numbers of people likely to have themselves cloned would probably be so small that the eugenic impact would be negligible.

With regard to genetic engineering, it will probably become technically feasible in due course to construct and to insert new genes for improved health, higher intelligence, and better personality qualities. It is, however, difficult to envision these techniques being adopted in the Western democracies in the near or medium-term future. Experiments of this kind would undoubtedly be prohibited by many Western governments. These could be evaded if a demand existed, but it is doubtful whether any significant numbers of couples would want to incur the risks of having new genes inserted into their embryos or babies, with all the unknown side effects this might entail. Possibly in the long-term future, genetic engineering of this kind might be adopted in the Western democracies.

6. TECHNICAL ADVANCES IN EMBRYO SELECTION

Progress in positive eugenics in the Western democracies in the twenty-first century is likely to take place largely by the development of embryo selection. By the end of the twentieth century it had become possible to test embryos grown in vitro for the presence of genes for several thousand genetic diseases and to select only healthy ones for implantation. In the twenty-first century it will become possible to test embryos for the presence of genes affecting numerous other characteristics, including late-onset diseases and disorders; intelligence; special cognitive abilities, such as mathematical, linguistic, and musical aptitudes; personality traits; athletic abilities; height; body build; and physical appearance. It will then be possible for couples to examine the genetic printouts of a number of embryos and select for implantation the ones they regard as having the most desirable genetic characteristics.

Before this becomes possible, three problems will have to be solved. The first is that most of the genes for which couples can be expected to select have yet to be identified. But progress in the identification of these genes is proceeding rapidly, and the functions of their most common alleles should be discovered by the first two or three decades of the twenty-first century.

The second problem is that in the recent past it has been feasible to grow only a relatively small number of embryos in vitro. The standard procedure is that a woman wanting to conceive by in vitro fertilization (IVF) is stimulated hormonally to ovulate a number of eggs, typically a dozen or so. Couples presented with the genetic printout of 12 of their embryos would have only a limited choice and might have to make some difficult decisions. For instance, the embryo with the best set of genes for intelligence might also have the genes for the probable development of cancer or heart disease in middle age or the genes for undesirable personality, predisposing the embryo to grow into an intelligent criminal. Nevertheless, with 12 embryos to choose from, couples would in most cases be able to select one with a more attractive overall profile of health, physical, and psychological characteristics than the others. We can anticipate that in the twenty-first century the numbers of embryos that can be produced by IVF will increase substantially. It has already become possible to produce around 25 embryos at one time. It should become feasible to produce many more. At birth, girls have something like a million immature eggs in their ovaries, and even women in their forties have tens of thousands of viable eggs. It will become possible to fertilize hundreds of these and to provide couples with a large choice from which to select those with the most desirable characteristics. Probably what will happen typically is that the most preferred embryo will be implanted and that three or four others will be selected and frozen for possible future use.

The third problem is that at present only around half of embryos grown in vitro and implanted produce a successful pregnancy and develop into a baby. This moderate success rate would need to be increased to something approaching 100 percent. We can anticipate that the success rate will improve with further experience and experimentation with IVF. A further potential way of overcoming this problem may be to make a number of clones of the embryos selected for implantation. It would then be possible to introduce copies of the selected embryos into the potential mother on a number of occasions until a pregnancy was achieved.

When these technical problems are solved, couples will be able to select embryos with better health, higher intelligence, better personality qualities, and better physical abilities and energy than either of the parents possess individually. Even couples who are individually below average in respect of these qualities would be able to produce a few above-average embryos. For instance, with respect to intelligence, couples produce children with a wide range of IQs. The average difference between pairs of siblings is about 15 IQ points; and if couples could produce a hundred or so embryos, the difference between the highest and the lowest would typically be around 30 IQ points. This 30-IQ-point distribution would be on both sides of the mean of the two parents, so that couples would normally be able to select embryos with IQs around 15 IQ points higher than their own.

7. DEVELOPMENT OF THE USE OF EMBRYO SELECTION

The desire to have children who are healthy and intelligent and who have good personality qualities is strong; and hence when it becomes possible to extend the use of preimplantation diagnosis of embryos to identify their genetic potential for health, intelligence, and personality qualities, there is every reason to believe that many couples will want to use this new technology. This is likely to evolve in three stages.

First, embryo selection will increasingly be used by couples in which both are carriers of recessive gene disorders or in which one has a dominant gene disorder to screen out embryos that have inherited the adverse genes. During the embryo selection procedure, the genes for intelligence and personality will also be recorded, and couples will be able to select embryos for these characteristics, in addition to those that are free of genetic diseases.

Second, the number of embryos grown will be increased, perhaps eventually to a hundred or even more. This will increase the choice of embryos and will enable couples to select embryos with high intelligence and sound personality traits in addition to good health. When this procedure becomes widespread, it will become evident that embryo-selected children are virtually always superior to naturally conceived children with respect to their health, intelligence, and personality. This knowledge will spread by word of mouth and through the media.

Third, couples who can conceive by normal intercourse will choose to have children by embryo selection as a means of ensuring genetically better children than those conceived normally. At first, embryo selection will be used by only a few intelligent, well-informed, farsighted, and affluent couples; but over the course of time increasing numbers of couples will use it, as its advantages become increasingly apparent and understood. Embryo selection will be like other new and expensive consumer products. For example, the telephone and the automobile in the early decades of the twentieth century were initially affordable only by the wealthy; but eventually, as costs were reduced through competition and technical advances, and as societies became more affluent, they came to be enjoyed by the majority.

One of the initial constraints on the spontaneous growth of embryo selection will be the cost. In the United States in the mid-1990s, it costs anywhere between $40,000 and $200,000 to have a successful IVF implantation. The additional cost of carrying out preimplantation genetic diagnosis of these embryos would not be great, nor would the culture of greater numbers of embryos; so the expense of embryo selection would not be appreciably greater than that of IVF. In Britain the cost of IVF is much lower, amounting to only around $3,000 to $4,000. This reflects the much higher costs of medical services in the United States than in Britain and a number of other countries. It can be anticipated that as embryo selection becomes a well-established procedure, many American couples will travel abroad to clinics in lower cost countries to have it carried out. International competition will force the costs of embryo selection down.

Even if embryo selection remains fairly expensive, the desire to have successful children is strong, and it can be predicted that many couples will be willing to pay the costs to ensure that their children have a greater chance of a healthy and successful life than if they were conceived naturally. Parents spend large sums trying to secure advantages for their children, such as by paying for a quality education. Once it becomes understood that a child's chances of success are greatly enhanced by embryo selection for health, high intelligence, and sound personality qualities, couples will realize that it is more cost-effective to pay for an embryo-selected child than to pay for a quality education for a normally conceived child.

At the beginning of the twenty-first century, there is a fairly high level of public knowledge that genetic factors are important in the determination of a child's characteristics, and there is also a willingness to pay to ensure good genetic quality. This is illustrated by the growth of elite semen banks and by the growing practice in the United States of infertile women seeking eggs from high-quality donors by advertising in Ivy League college magazines and newspapers. In these ways and by these means, it can be anticipated that embryo selection will come to be used by increasing numbers of couples in the Western democracies, purely through private initiative and without any interventions from governments.

8.   INEFFECTIVENESS OF PROHIBITION OF EMBRYO SELECTION

The major constraint on the development of the use of embryo selection during the twenty-first century will be that in many countries it will be prohibited. Embryo selection was made illegal in the 1990s in a number of European countries and in several U.S. states. It may be that more countries and U.S. states will make embryo selection illegal. Nevertheless, although these prohibitions will retard the use of embryo selection, it can be confidently predicted that they will not be effective in preventing its use. Even if embryo selection were to be prohibited throughout the Western democracies, there would inevitably be some countries where it was permitted, either for eugenic reasons or for its financial advantages. Countries like Singapore, where the governments are sympathetic to eugenics, would allow embryo selection to promote their eugenic objectives. Other countries would be keen to exploit the financial advantages of allowing embryo selection clinics to be established. The financial gains would be generated from couples visiting the countries to use the clinics and from levying taxes on the physicians, biologists, and their support staff working in the clinics. There are dozens of such small countries around the world, and it is inevitable that some of them would permit and actively promote the establishment of embryo selection clinics as useful sources of revenue. These countries may become known as embryo selection havens, analogous to the present status of many of them as tax havens.

The major countries prohibiting embryo selection might even make it illegal for their citizens to use the services provided in the embryo selection havens, just as some countries prohibit their citizens from putting their money in offshore tax havens. These prohibitions would be easily evaded. It would only take a day or two for couples to travel to an embryo selection haven and have the procedure carried out. Two or three weeks later the woman would discover that she was pregnant. These women could not be prosecuted for committing a criminal offense because it would be impossible to prove that the conception had taken place by embryo selection. Once the existence of these facilities became well known, they would inevitably be used by significant numbers of couples. The situation would be similar to that concerning abortion from the 1960s onward in Europe, when abortion was prohibited in some countries but permitted in others. Pregnant women desiring an abortion in countries where it was illegal, such as Spain and Ireland, made a brief visit to abortion clinics in countries where it was permitted, such as Britain and Sweden. The prohibition of embryo selection will be no more effective than the prohibition of abortion and of other goods and services for which people have a strong need, such as the prohibition of the production and sale of alcohol in the United States in the 1920s and a number of drugs today. Although they have some effect in reducing the use of the prohibited products, none of these prohibitions has been very effective. It will be the same with embryo selection.

Eventually, governments of countries that have prohibited embryo selection will come to realize that this is unreasonable and will allow it. There are two reasons why this should be anticipated. First, embryo selection is a preferable method for preventing the birth of babies with genetic diseases and disorders than is the alternative of prenatal diagnosis and pregnancy termination. Prenatal diagnosis involves the trauma of carrying the fetus for a number of weeks, of termination, and of the option of trying for a further pregnancy in the hope, but not the certainty, of conceiving a disease-free child. Embryo selection avoids these problems, and this is a powerful argument for allowing it to screen out embryos with genetic diseases and disorders. Once this is permitted, it would be virtually impossible to draw up a list of the thousands of genes for which selection is allowed and those for which it is illegal, and the procedure will be permitted without restrictions.

Second, there are powerful civil liberties arguments that if couples wish to have children who are healthy, and intelligent and with good personality qualities, there are no valid social reasons for preventing them and that they are entitled to exercise these choices and should be allowed to use embryo selection to do so. In due course the ethical objections to embryo selection will be seen to be groundless. This will remove the ethical legitimacy of the prohibitions, and the laws criminalizing its use will be repealed. For all these reasons, the eventual adoption of embryo selection by substantial proportions of the population should be regarded as inevitable in the Western democracies. This will be the major avenue for the advancement of eugenics in the twenty-first century.

9. THE EMERGENCE OF CASTE SOCIETIES

We should anticipate that in the Western democracies, embryo selection will come to be used initially by a minority of couples; but in due course, as knowledge of the benefits of this procedure spreads, it will be adopted by a majority. Nevertheless, we cannot envision that a time will come when all babies will be conceived in this way. Unplanned births resulting from normal sexual intercourse will continue to occur. In the 1990s around 30 percent of births in the United States and Britain were unplanned. This figure may be reduced in the twenty-first century through the more efficient use of contraception, the development of more effective contraception, increasing use of the morning-after pill, and greater use of abortion to terminate unplanned pregnancies. However, there are still likely to remain some unplanned births. Possibly, embryo selection will eventually be adopted by 80 percent to 90 percent of the population and will stabilize at this level. The remaining 10 percent to 20 percent of babies will continue to be conceived by sexual intercourse. These will be born largely to couples with low intelligence and psychopathic personality who conceive by accident and do not have their unplanned pregnancies terminated.

When this point is reached, the two populations will begin to diverge genetically. A gulf will open up between the embryo-selected children and the "unplanned," as those conceived by sexual intercourse may come to be known. If, as seems probable, the parents of the unplanned come from the bottom 10 percent to 20 percent of the population for intelligence, their mean IQ would be around 80 and the mean IQs of their children around 84. The remaining 80 percent to 90 percent of the population who had their children by embryo selection would have a mean IQ of about 110. By using embryo selection they could have children with IQs about 15 points higher than their own, giving their children a mean IQ of around 125. Thus, in the first generation there would be a difference of around 40 IQ points between the average IQ of the embryo-selected and that of the unplanned. This gap would increase by around 15 IQ points in each subsequent generation because the embryo-selected would continue to have children whose IQs would be around 15 IQ points higher than their own, while the IQs of the unplanned would remain the same. Thus, in the second generation the intelligence gap between the embryo-selected and the unplanned would increase from around 40 IQ points to around 55 IQ points. This would give the embryo-selected children a huge advantage in schools, colleges, occupations, and incomes. The embryo-selected children would also be selected for sound personality traits, and this would give them an additional advantage in their education, careers, and socioeconomic status.

This will lead to the emergence of a caste society containing two genetically differentiated castes—the embryo-selected and the unplanned. The two castes will live in different areas, attend different schools and colleges, and hardly meet socially. They will normally marry and mate only within their own group, transmitting their differentiated sets of genes to their children. Virtually all the professional, white-collar, and skilled jobs would be performed by the embryo-selected. Some of the unplanned would work in unskilled and undemanding jobs, but many of them would be unemployed and unemployable. These would be a genetic underclass. They would largely be the descendants of the old underclass that emerged in the second half of the twentieth century, but the gap between them and the rest of society would be wider and more strongly genetically based. They would be a social problem, just as their parents and grandparents of the old underclass were; and they would live in a ghettoized underworld of chronic unemployment, crime, drug addiction, single motherhood, and welfare dependency. Eventually, despite strong ideological opposition, it would come to be understood that the underclass of the unplanned was primarily a genetic problem and would require genetic interventions.

How the Western democracies will handle this problem is difficult to predict. Probably for some decades they would strengthen the custodial state by incarcerating increasing numbers of the underclass in prisons, as has been done in recent decades in the United States and Europe. Possibly in due course some Western governments might require the genetic underclass to have their children by embryo selection, which would gradually raise their intelligence levels, improve their personality qualities, and make it possible to integrate them into mainstream society. Alternatively, they might prohibit them from having children by sterilizing them or by the introduction of some form of parental licensing scheme of the kind discussed in Chapter 14. The political problems of implementing policies of this kind would be considerable. Nevertheless, as it became increasingly understood that the underclass was largely a genetic problem, the search for eugenic solutions would become increasingly probable.

10. CONCLUSIONS

There is likely to be a renewal of support for eugenics in the Western democracies among biological and social scientists as consensus is reached on the contribution of genetic factors to health, intelligence, and personality; to the social and economic importance of high intelligence and sound personality; to the presence of dysgenic processes in contemporary societies and the need to counter these; and to the desirability of improving the genetic human resources of society as a means of enhancing national economic, scientific, and military strength. There is, however, little prospect for a number of decades to come of any Western governments introducing eugenic measures that will have any significant impact. The most that can be envisioned in the near and medium-term future is that the development and use of more efficient forms of contraception and the reduction of welfare incentives for underclass women to have babies will reduce dysgenic fertility. There is no prospect of any revival of the classical eugenic programs of the sterilization of social-problem groups on any significant scale, of the introduction of eugenic criteria for the acceptance of immigrants, or of the introduction of incentives for elites to have more children. Much of public opinion and the media is likely to remain hostile to any policy initiatives of this kind for the foreseeable future. Even if political leaders become sympathetic to eugenic measures as serving the national interest, they would be deterred from introducing them because of the opposition of powerful special interest groups, consisting of the bureaucracies that have grown up in welfare states to look after groups that would be the target of eugenic programs, and the growth of ethnic and racial minorities that would be adversely affected by eugenic programs and would be bound to mount strong resistance to their introduction. The growth of these special interest groups opposed to eugenics would make it more difficult to introduce eugenic measures in the twenty-first century than it was in the first half of the twentieth century and will probably make the rehabilitation of classical eugenic programs impossible.

Nevertheless, we should anticipate that the new eugenics of human biotechnology will continue to make progress. This will come first in the reduction of genetic diseases and disorders with the more efficient use of prenatal diagnosis of impaired fetuses, the further development of carrier screening of couples at risk of having a child with a genetic disorder, and of the preimplantation diagnosis of embryos grown in vitro. Already by the end of the twentieth century, these procedures had achieved substantial reductions in the birth incidence of a number of genetic diseases and disorders, and further progress in this direction can be expected. We should also anticipate that progress in the reduction of genetic diseases and disorders will be made through improvements in the techniques of gene therapy. Germ-line gene therapy is likely to become permitted because of its advantage in reducing the adverse genes in future generations.

Progress in the use of human biotechnology for the improvement of intelligence and personality will come more slowly but is likely over the longer term. This will take place through embryo selection, which will enable couples to examine the genetic characteristics of a number of embryos and to select the most desirable for implantation. The desire of couples to have children who are healthy and intelligent and have good personality qualities is strong; and when embryo selection clinics are able to offer couples the possibility of having children with the genetic profile for these characteristics, there can be little doubt that couples will make use of these facilities. The governments of some Western democracies have prohibited the use of embryo selection. This will retard its uptake, but the prohibitions will be evaded by couples who use embryo selection in countries where it is permitted. In due course the prohibitions on the use of embryo selection will be removed, as it becomes understood that they are easily evaded and have no ethical justification or socially legitimate basis. As embryo selection becomes legalized, it will be used by increasing numbers of the population, eventually leading to caste societies divided into those with high intelligence and sound personality conceived by embryo selection and those with low intelligence and poor personality qualities conceived naturally (the "unplanned"). The unplanned will be a social problem and will increasingly come to be seen as a genetic underclass to which Western democracies may eventually seek to find eugenic solutions.

20

⤸

The Future of Eugenics in Authoritarian States

1. Human Quality and National Strength
2. Political Leaders' Understanding of Eugenics
3. Attitudes toward Eugenics Outside the Western Democracies
4. Preconditions for the Emergence of Eugenic States
5. Classical Eugenics
6. Human Biotechnology
7. Mandatory Embryo Selection
8. The Cloning of Elites
9. Development of the Power of the Eugenic State
10. Conclusions

⤸⤸⤸

In Chapter 19 we concluded that eugenics will make progress in democratic societies through the use of human biotechnologies by couples wanting children with genetically desirable characteristics but that it is unlikely to be revived as an instrument of state policy. In this chapter we discuss the possibility that eugenics might be adopted by authoritarian states as a means for enhancing their national strength. We discuss how far this is probable, how such states might be expected to implement such a program, and how far it would be likely to succeed.

1. HUMAN QUALITY AND NATIONAL STRENGTH

Political leaders normally seek to promote their nations' economic, scientific, and military strength. We should anticipate that in the twenty-first century, the political leaders of some authoritarian states will realize that eugenics could be used as a means for the promotion of national strength and will embark on a eugenic program as a means of advancing this objective. Political leaders normally understand that human quality or, as economists call it, "human capital," is an important determinant of national strength, and they usually seek to improve the human quality of their populations with respect to health, cognitive skills, and moral character by a variety of environmental interventions. Thus, with regard to health, political leaders typically promote the health of their populations by providing subsidized or free antenatal and postnatal care for mothers, by ensuring that children have adequate nutrition by the provision of free school meals and food stamps, by providing subsidized vaccination and health care, by the prohibition of drugs damaging to health, and so forth.

Political leaders endeavor to promote the cognitive abilities of their populations by making education compulsory for children and adolescents; by requiring schools to teach socially and vocationally important skills, such as reading, writing, math, and science; by monitoring schools to ensure that these skills are taught effectively; and by encouraging adolescents to continue their education in college by providing scholarships, loans, and subsidies. Political leaders also endeavor to strengthen the moral character of their populations. In some countries this is done by requiring the teaching of moral education in schools, as is the case in Japan and which I have described in *Educational Achievement in Japan* (Lynn, 1988); or by encouraging or requiring the teaching of religion, a component of which is the attempt to inculcate moral values; and everywhere by using the criminal law to punish morally unacceptable behavior on the implicit assumption that this will improve the moral values of the population.

2. POLITICAL LEADERS' UNDERSTANDING OF EUGENICS

While political leaders have normally been concerned to promote the health, cognitive skills, and moral character of their populations by environmental measures, a number of political leaders have understood that it would also be possible to promote these by genetic interventions, that is, by the use of eugenics. As noted in Chapter 2, Theodore Roosevelt, president of the United States from 1901 to 1909, was well aware of the importance of the genetic quality of the population. He stated that he was "deeply concerned about the threat to the quality of the human stock of America" (Pearson, R., 1996), and he "strove to persuade healthy American men and women to have more children" (p. 51). In Britain, Arthur Balfour, prime minister from 1902 to 1905, believed in the value of eugenics as a means of enhancing national strength and spoke on this theme in his inaugural address at the first International Eugenics Congress in 1912. Another British prime minister sympathetic to eugenics was Winston Churchill, prime minister from 1940 to 1948 and from 1951 to 1956 and a member of the British Eugenics Society and sponsor of the 1912 International Eugenics Congress. The political leader who held the strongest eugenic views was, of course, Adolf Hitler; but eugenic ideas and policies were regarded favorably by political leaders throughout most of the economically developed world in the early and middle decades of the twentieth century and were widely implemented in the sterilization programs and the selective immigration policies of that period.

As late as 1974, Sir Keith Joseph, a prominent British politician, made a speech on the dysgenic effect of the high rate of teenage pregnancies among the least educated adolescent girls and argued that it would be desirable on eugenic grounds to reduce this rate. By this time the media had become unsympathetic to eugenics, and observations that would have been regarded as common sense half a century earlier were greeted by a storm of protest. By the end of the twentieth century, political leaders in the Western democracies had become aware that it is politically hazardous to express ideas of this kind. Nevertheless, a number of political leaders in the Western democracies have been aware of the potential of eugenics for enhancing national power, and no doubt many of them are still aware of this, even though they realize that it has become politically impossible to express these ideas publicly or to implement eugenic programs on any significant scale.

3. ATTITUDES TOWARD EUGENICS OUTSIDE THE WESTERN DEMOCRACIES

While in the last decades of the twentieth century eugenics became widely rejected in the Western democracies, in much of the rest of the world eugenics was viewed favorably. We noted in Chapter 2 the robust program of positive eugenics that was introduced in the late 1970s in Singapore by Lee Kyan Yew, the prime minister between 1959 and 1990, which consisted of a range of incentives for university graduates to have children. Favorable attitudes to eugenics in a number of countries outside the Western democracies were found in a survey of the views held by geneticists and physicians in 36 countries carried out from 1994 to 1996 by Wertz (1998). For instance, geneticists and physicians were asked to evaluate the statement, "An important goal of genetic counseling is to reduce the number of deleterious genes in the population." The most common form of genetic counseling consists of advice on whether to terminate a genetically impaired fetus, and to do this in order to reduce the number of deleterious genes in the population serves a eugenic purpose in addition to the promotion of the well-being of the prospective mother. Fewer than one-third of the geneticists and physicians in the larger Western democracies supported this eugenic objective of genetic counseling, but in a number of countries this objective was endorsed by the majority of geneticists and physicians, including China (100%), India (87%), Turkey (73%), Peru (71%), Poland (66%), Russia (58%), Cuba (57%), and Mexico

(52%). Even in a few of the smaller Western democracies, a majority of geneticists and physicians approved of the eugenic use of genetic counseling, including Spain (67%) and Greece (58%).

In another question in this survey, geneticists and physicians were asked whether eugenics is "the major goal of genetics." This statement was endorsed by all the geneticists and physicians in China and by majorities in India, the Czech Republic, Hungary, Thailand, Russia, and Greece. Asked whether they would give "pessimistically slanted information" designed to persuade pregnant women diagnosed as carrying a fetus with Down's syndrome to terminate the pregnancy, a majority of geneticists and physicians said they would in several countries, including China (96%), Cuba (93%), Russia (89%), Greece (87%), Thailand (76%), Hungary (75%), Belgium (71%), France (66%), and Sweden (58%). The only countries where fewer than 20 percent of geneticists said they would do this were Canada (9%), the United States (13%), and Britain (14%). Broadly similar percentages were obtained for a similarly worded question about spina bifida, cystic fibrosis, and sickle-cell anemia, and somewhat lower but still appreciable percentages were obtained for achondroplasia (dwarfism) and Klinefelter's syndrome.

This survey shows that while eugenics has become widely rejected in the United States, Canada, and Britain, there is more support for it in the rest of the world, including several European countries and Israel, and much stronger support in countries outside the Western democracies. In much of the world, the majority of geneticists and physicians favor eugenics and could be expected to support and willingly implement eugenic programs initiated by governments, just as many geneticists and physicians did in the Western democracies in the first half of the twentieth century.

The most favorable attitudes toward eugenics were found among geneticists and physicians in China, where the political leaders also favor eugenics and have already begun to implement eugenic programs. In 1988 the Chinese government introduced a program for the compulsory sterilization of all the mentally retarded in the province of Gansu and for mandatory abortions for mentally retarded women who become pregnant (Tyler, 1993). In 1993 the Chinese government passed the Eugenics and Health Protection Law, which prohibited the marriage of people with mental illness, venereal diseases, and hepatitis with the express intention of preventing them from having children. A further Act of 1994 authorized the compulsory sterilization of the mentally retarded and of those with serious genetic diseases and disorders, the mandatory prenatal testing of pregnant women, and compulsory abortion in cases where a disorder is identified (Seabrook, 1994; Beardsley, 1997). These laws applied to the entire country of China.

These measures of negative eugenics were complemented in China in 1999 with a program of positive eugenics consisting of the establishment of a state-run elite-semen bank. This bank stores donations obtained solely from senior university professors for the use of married women whose husbands are infertile. There is no shortage of donors, indicating the willingness of the Chinese academic elite to contribute to the eugenic program (In China, 1999).

4. PRECONDITIONS FOR THE EMERGENCE OF EUGENIC STATES

A number of favorable preconditions for the emergence of a eugenic state are present at the beginning of the twenty-first century. We have already noted four of these: the motivation of political leaders to enhance their national strength, the awareness of some political leaders that this could be achieved by eugenics, the existence of geneticists and physicians sympathetic to eugenics and willing to implement eugenic programs, and the power of the leaders of authoritarian states to suppress civil liberties and special interest groups that might object to eugenic policies. To these should be added four additional preconditions for the emergence of a eugenic state.

First, there is the demise of socialism and communism. Throughout nearly the whole of the twentieth century, political leaders in many authoritarian states believed that some form of socialism or communism offered the best means for developing their national strength and, at the same time, preserving their own positions of power. Towards the end of the twentieth century, socialism and communism became increasingly discredited, and their credibility was finally destroyed in 1989 with the disintegration of the Soviet Union. Now that it has become evident that communism was, in the phrase of one of its famous disciples Arthur Koestler (1950), *The God That Failed*, there is a vacuum in the agendas of political leaders of authoritarian states who are concerned to increase their national power. This is likely to be filled by eugenics. Indeed, it may be no accident that the political leaders of China began to take up eugenics after they realized that communism as a form of economic organization had failed.

Second, the scientific validity of the propositions on which eugenics is based has become increasingly accepted. At the same time, those who have maintained that there are no such things as good or bad genes, that intelligence and moral character have no genetic basis, and that eugenic programs would not work, and so forth, have become increasingly rejected. The ideological motives of those who have maintained these positions are becoming increasingly understood and their arguments discredited.

Third, blueprints for a eugenic state have been present from the time of Plato, and many attempts have been made to implement these using classical eugenics, notably the sterilization programs that were widespread in the United States and Europe in the first two-thirds of the twentieth century. These precedents for eugenic policies are well known and are waiting to be rehabilitated by political leaders of authoritarian states.

Fourth, biotechnology has reached the stage of development at which it could be used for eugenic purposes. This has indeed already started in China with mandatory prenatal diagnosis and termination of fetuses with genetic disorders. A program for the cloning of elites could be started immediately, and embryo selection for intelligence and personality is likely to be feasible within a few years.

Once technologies that serve national purposes become feasible, some country or countries will inevitably use them. It can only be a matter of time before an authoritarian state embarks on a program of using the human bio-technologies for eugenic purposes.

5. CLASSICAL EUGENICS

Authoritarian states intent on implementing eugenics are likely to intro-duce both the classical eugenics of selective reproduction and the new eu-genics of the human biotechnologies. In regard to classical eugenics, they would be expected to introduce programs for the sterilization of the mentally re-tarded and of criminals and psychopaths. These may well be carried out on a more robust scale than was attempted in the Western democracies in the early and middle decades of the twentieth century. They might target all the mentally retarded, comprising about 3 percent of the population, and all psychopaths, comprising about another 3.5 percent of the population and predominantly males. They would also be expected to introduce some form of parental licensing scheme of the kind discussed in Chapter 14. To make this effective and to prevent the birth of unlicensed babies, we can envision that all 12-year-old girls would be required to be fitted with contraception. There have already been calls for this to be done in some of the Western democracies as a means of preventing teenage pregnancies by irresponsible young women. For instance, in Britain, John Guillebaud, medical director of the Margaret Pyke Family Planning Center in London, made this proposal in February 1990: "In the future, and as a social policy, when you have an area with a huge rate of teenage pregnancies, you could go into a school, obviously with the consent of the parents, and fit this device so that everybody would start out not being able to have a baby" (Murray, 1999, p. 2). The same scheme is implicit in the American psychologist David Lykken's (1995) proposal for licenses for parenthood, because the only feasible way of enforcing the pro-posal would be to make it compulsory for young girls to be fitted with long-term contraception, which could be removed once the license for parenthood had been obtained. Now that proposals of this kind are being made in the Western democracies, we can anticipate that they are likely to be introduced in authoritarian eugenic states.

The second major measure of negative eugenics that authoritarian states would be expected to introduce would be the strict control of immigration. All immigrants, including asylum seekers, would be assessed for their health, intelligence, vocational skills, and personality, and only those who were healthy and psychologically desirable would be admitted. Illegal immigration would be prevented by issuing all citizens with identity cards, which they would be required to produce on request by the police, immigration control officers, employers, and welfare officials. Illegal immigrants without the required documentation would be deported and possibly punished to deter attempts to re-enter the country and as a deterrent to others. However, probably in practice, dysgenic immigration would not be much of a problem for authoritarian states because few would wish to emigrate to them.

Authoritarian eugenic states would also be expected to implement programs of classical positive eugenics, designed to encourage their elites to have more children. There are two strategies by which it is likely that this would be attempted. These are by the provision of financial incentives and by promoting the concept that elites have an ethical obligation to society to have children. The financial incentives for having and rearing children could be made very considerable because authoritarian states need not be concerned with the notions of social justice that exist in democracies. Authoritarian states can, and typically do, have large differentials of wealth and incomes because they do not need to seek the votes of the poor. They could easily afford to give large financial incentives to their elites to have children, and the elites would use these financial subventions to buy child care of various kinds. It is probable that the elite would respond to incentives of this kind because most people like children and would have more if they could easily afford to pay others to help take care of them.

In addition to the provision of financial incentives for child rearing, eugenic authoritarian states would also be expected to promote the idea that their elites have an ethical obligation to have children. One way they might do this would be by requiring university students, taken for practical purposes as the elite, to take courses in eugenics, such as were given in a number of U.S. universities in the 1920s and 1930s, and based on the model of the courses in Marxism-Leninism that were compulsory for students in the Soviet Union and Eastern Europe during the years of communist rule. These courses would be designed to imbue the young elite with a sense of the eugenic mission of the state and of their own obligation to contribute to this by having several children. It would not be an unrealistic objective to induce the elite to have an average of four children each, and this would approximately double their numbers in the next generation, subject to a small regression effect. A robust implementation of a program of classical eugenics of these kinds by an authoritarian state would have a significant impact on the genetic quality of the population.

6. HUMAN BIOTECHNOLOGY

Authoritarian states embarking on eugenic programs must be expected to use the human biotechnologies as these are developed, subject only to the constraint of not seriously alienating large proportions of their populations.

This provision would rule out the mandatory use of AID (artificial insemination by donor), which could theoretically be used to impregnate all women with the sperm of high-quality donors. This application of the traditional method of livestock breeding would achieve dramatic results but at a cost of the disaffection of many women who would prefer to choose the father of their children and, even more, of men who would prefer to rear their own biological children and, in general, make better fathers for them (e.g., Pinker, 1997). For this reason authoritarian eugenic states would probably not attempt to introduce the mandatory use of AID.

Authoritarian eugenic states would be expected to extend the use of prenatal diagnosis of genetic diseases and disorders by making diagnosis more efficient and by making the abortion of disordered fetuses compulsory. Prenatal diagnosis would be made more efficient by screening the entire population for the presence of genes for genetic diseases and disorders. This would make it possible to identify carriers of adverse recessive genes and to test all pregnant women for the presence of fetal genetic diseases and disorders. A rigorously implemented program of this kind should make it possible to identify most fetuses carrying genetic diseases and disorders, as has already been achieved in the Western democracies for neural tube defects, spina bifida, hydrocephalus, Tay-Sachs disease, and B-thalassemia. Termination of pregnancies where fetal disorders have been identified would be made compulsory.

This program would produce substantial savings of the medical and welfare costs of maintaining those with genetic diseases and disorders. It is estimated that in the Western democracies about a quarter of hospital beds are occupied by patients with genetic diseases and disorders (Fletcher, 1988). Medical services in the Western democracies typically consume around 8 percent of gross domestic product (GDP), so the elimination of genetic diseases and disorders would involve a saving of around 2 percent of GDP. This would release financial and human resources for use elsewhere and would make a significant, although not a huge, contribution to the enhancement of national economic strength. In the future it may become possible to carry out prenatal diagnosis for the presence of genes for low intelligence and psychopathic personality. A eugenic authoritarian state could and no doubt would require the termination of fetuses with these genes.

7. MANDATORY EMBRYO SELECTION

In the medium-term future, authoritarian eugenic states would be likely to use embryo selection as the principal means for advancing their eugenic agendas. Embryo selection for health, intelligence, and sound character would be made compulsory. Procreation by the haphazard means of sexual intercourse would be made illegal. This would be enforced by requiring all 12-year-old girls to have some form of long-lasting contraception. When women wished to have children, they and their partners would be assessed for their fitness under a parental licensing scheme of the kind considered in Chapter 14. Those who obtained the license would be permitted to have children by embryo selection.

When this program was in place, it would be possible to allow almost the whole population to have children and, at the same time, to secure major eugenic advances. This would be possible because couples would produce embryos with a great variety of genotypes, and the best two or three out of a hundred or so would have considerably better genotypes for desirable qualities than either of the two parents. The requirement that all licensed couples should have their children by embryo selection would have a major impact on the genetic quality of the population. Genetic diseases and disorders would all be identified in the embryos, and these defective embryos would be discarded, with the result that genetic diseases and disorders would be entirely eliminated.

It can be anticipated that the impact on intelligence of the use of embryo selection by the whole licensed population would be to raise the average level of intelligence of the population by around 15 IQ points in one generation. The reason for this is that if couples produced a hundred embryos, there would be a range of some 30 IQ points in the embryos' potential IQs. A few of these would be expected to have IQs about 15 IQ points higher than the average of their parents. These would be selected for implantation, so the effect would be to increase the intelligence level of the child generation by around 15 IQ points. This gain would be repeated in each succeeding generation until all the alleles for low and average IQs had been eliminated. The intelligence level of the population would be expected to stabilize at its theoretical maximum of around 200 after six or seven generations. During this period of about 200 years, the effect of the rise in the population's intelligence level would be monitored for any adverse side effects. If any of these appeared, they would be corrected.

With regard to personality traits, eugenic states would not allow the implantation of embryos with the genes for psychopathic personality. The effect of this would be a very considerable reduction in crime and social irresponsibility in the next generation. Aside from this, eugenic states would be expected to proceed with caution in regard to selection for personality because their psychological advisers would counsel that a range of different personality types would be desirable to fulfill different social functions.

Once the eugenic state had made embryo selection mandatory, it would be faced with the problem of whether couples would be allowed a free choice of which embryos to select or whether the selection would be regulated. It would be preferable to allow couples freedom of choice on the grounds of allowing the population the greatest possible personal liberties in the interests of ensuring their acceptance of the oligarchic constitution. Perhaps the government would provide genetic counselors whom couples could consult if they needed help making a decision. Probably allowing couples freedom of choice in the selection of embryos would work sufficiently well for a eugenic state because virtually all couples would want children who were healthy, and intelligent and of sound personality. Couples would be expected to differ in the embryos they selected with regard to special aptitudes such as mathematical, scientific, and verbal abilities and also with regard to personality characteristics. Probably most couples would select embryos with the potential to develop characteristics similar their own. This would produce a desirable mix of different aptitudes and personality traits. It might be that too many couples would select embryos with the genotypic potential to become outstanding in sports, pop music, films, the media, and other glamorous occupations. The eugenic state would have little use for these genotypes, except perhaps for a small number of those with outstanding sporting potential for international prestige purposes. If too many couples selected embryos of this kind, the state could set up a supervisory procedure to restrict the choices exercised and could require couples to select embryos likely to contribute to the national interests of the state.

Eugenic states would need to resolve the issue of whether all couples would be permitted to have children by embryo selection or whether some would be prohibited from having children under a parental licensing scheme. So far as the genetic qualities of the children are concerned, nearly all couples could be allowed to have children because they would be capable of producing embryos with significantly better qualities than they themselves possessed. There would nevertheless be some couples who would be unlikely to produce embryos of acceptable quality. For instance, in a few cases couples both of whose IQs were around 60 would probably not be able to provide any embryos with IQs above 80, and these would not be acceptable. Because of the element of uncertainty in the characteristics of the embryos, the best way for the eugenic state to handle this problem would be to require the physicians carrying out embryo selection to exercise discretion regarding whether any of the embryos were acceptable.

While nearly all couples should be able to produce some acceptable embryos, the eugenic state might well consider that some couples would not make suitable parents. These would be the mentally retarded, the psychotic, and psychopaths, who do not make good parents because of their cognitive and personality deficiencies. These couples would be denied licenses for parenthood under a parental licensing scheme. It can be envisioned that about 3 percent of the population who are mentally retarded and 1 percent of women and 6 percent of men who have psychopathic personalities would not be granted parental licenses on the grounds that they would not make suitable parents.

It should be feasible for a eugenic state to make it mandatory to obtain parental licenses to have children by embryo selection without the alienation of any significant section of the population. The requirement to obtain a parental license to have children would come to seem no more unreasonable than the requirement that people should obtain a driving license before they are permitted to drive automobiles on the public highways. A great majority of the population can be expected to accept that the mentally retarded, the psychotic, and psychopaths should be denied parental licenses. The idea prevailing in the Western democracies that these people should be allowed to have unlimited numbers of children, many of whom inherit their parents' characteristics and have to be taken away from their parents because these are unfit to rear them, would come to be viewed as an absurdity.

Nor does there seem to be any reason why the population should object to having their children by embryo selection. This would be regarded as an extension of existing prenatal screening procedures for identifying fetuses with genetic diseases and disorders and, in effect, selecting those that are healthy. The selection of embryos for high intelligence and sound personality would be regarded as sensible and would be welcomed by most couples. Thus, a eugenic state imposing these requirements would be expected to retain the goodwill of a great majority of the population. This is an important consideration even for an authoritarian state.

There can be little doubt that authoritarian eugenic states would secure considerable advantages over the Western democracies as a result of making embryo selection mandatory. It was argued in Chapter19 that as embryo selection for intelligence and personality become feasible, it would come to be used voluntarily in the Western democracies by significant numbers of couples who want to have children with desirable genotypes. However, it would probably take three or four generations before embryo selection came to be used by a majority of the populations in the Western democracies. Authoritarian eugenic states making embryo selection compulsory would gain a large and immediate comparative advantage.

8. THE CLONING OF ELITES

Cloning could be used to considerable effect by authoritarian eugenic states as a means of reproducing their scientific, military, and political elites. As compared with embryo selection, one advantage of cloning is that a eugenic state could attempt to reproduce its elites immediately, with a fair chance of success. The cloning of elites would not require any knowledge of the genes and alleles responsible for intelligence or the complex set of personality traits necessary for creativity and high-level achievement, such as would be required before embryo selection could be used for eugenic purposes. All that cloning requires is the perfection of the techniques, and now that a number of mammals have been cloned, this should be relatively straightforward for humans.

It has sometimes been objected that cloning has encountered a large number of failures before a success is achieved and that this rules out the cloning of humans. This objection is not persuasive because the eugenic state could afford a number of failures. In any case, technical progress in cloning is likely to reduce the failure rate.

Eugenic states would be expected to confine cloning to the reproduction of quite small numbers of their scientific, military, and political elites. We cannot envision the scenario set out by Aldous Huxley (1932) in *Brave New World* in which the whole population was reproduced by cloning. For most of the population, embryo selection would be a preferable eugenic technique because it would enable nearly the whole population to have their own biological children with more desirable genetic combinations than those of either of their parents. This would retain the goodwill of the population more effectively than cloning the best individuals in each occupation and having them reared by adoptive parents.

For the quite small numbers of its elites that eugenic states could be expected to reproduce by cloning, it would be necessary to find couples willing to rear the clones. In some cases, these would be the couples from one of whom the clone was taken. These couples would give the clones the advantages of their own knowledge and experience. Most of the clones would be reared by adoptive parents who would be selected as likely to provide good rearing environments. Suitable women would have to be found who were willing to have the cloned embryos implanted, to carry them to term, and to rear them. This should not be a problem for the eugenic state. Only a few thousand children would be likely to be required for an elite cloning program. Sufficient numbers of women willing to bear and rear clones could probably be recruited as volunteers from enthusiasts for the eugenic program. If this proved not to be the case, they could be sufficiently well paid for this service to produce the required number of women.

The cloning of elites would give eugenic states a large advantage over the Western democracies in the development of national economic, scientific, and military power. Important scientific advances are typically made by small numbers of highly gifted individuals who have hitherto appeared as a result of very unusual combinations of genes and favorable environmental conditions. Authoritarian eugenic states could produce hundreds or thousands of replicas of these highly gifted individuals by cloning and could have them reared in the most favorable family and educational environments. The cloning of political and military elites would make it possible for power to be transmitted from capable elites to their clones and would solve the succession problem that has so frequently led to the downfall of oligarchies. The cloning of elites would give authoritarian eugenic states considerable advantages because, as noted in Chapter 19, it cannot be envisioned that elites would be cloned in the Western democracies.

9. DEVELOPMENT OF THE POWER OF THE EUGENIC STATE

An authoritarian state that embarked on a eugenic program using both classical eugenics and the biotechnologies would be able to secure significant improvements in the genetic quality of its population over a time frame of about 50 years. About 10 years might be required to perfect the techniques of embryo selection and cloning to produce large numbers of children with high intelligence and the personality qualities required for creative scientific and technological achievements. In another 20 years these children would be adult, and in a further 20 years they would constitute about half the working population. At this point, about 50 years after the start of the eugenic program, they would be having a major impact on the economic and scientific productivity and military capability of the state.

The political leaders of the authoritarian eugenic states would understand that they would need to provide an optimum environment to develop to the full the genetic potential of their population. The improvement of the environmental conditions for the realization of genetic potential has always been a subsidiary aim of eugenics, for which the term *euphenics* was coined. The principal way that eugenic states would promote euphenics would probably be by identifying children with exceptionally high intelligence and aptitude in mathematics and science at the age of 12 or so and giving them an accelerated scientific education. This could possibly be accomplished in high schools attached to elite universities, where the children would be taught by the ablest professors, on the model of the university-affiliated high schools in Japan, which I have described elsewhere (Lynn, 1988), or in universities for scientifically precocious teenagers, on the model of the Academic City University in Russia.

It is known that children with strong mathematical and scientific abilities can be given an accelerated education in adolescence and that they are capable of doing good university-level work and graduating by the age of around 17 (Benbow, 1992). Given such an accelerated education, many of them should be able to do good quality research in their late teens and to obtain their Ph.D.s by the age of 21 or 22. The eugenic state would establish research institutions at which a number of the best of these postgraduates would work on scientific research and development projects. Scientists frequently do their best work in their twenties and thirties. Newton was 26 when he first conceived of the idea of gravitation; Einstein was the same age when he published his paper on the special theory of relativity; Francis Crick was 37 and James Watson was 25 when they published their research on the double helix structure of DNA. Hence, within 15 to 20 years after they came to adulthood, the new generation of gifted young scientists would be expected to be making important contributions to the scientific, engineering, and technological base of the economy. Other members of this new highly gifted generation would enter the corporate sector and would devote their abilities to the marketing of the goods and services provided by the scientific and techno-logical elites.

An idea of the impact of this new highly gifted generation on the strength of the economy of the eugenic authoritarian state can be gained from consid-ering the rapid development of the East Asian economies of Japan, South Korea, Taiwan, Hong Kong, and Singapore in the second half of the twenti-eth century. These populations have an average intelligence advantage of approximately 5 IQ points over the European peoples of North American and Europe, which I have documented (Lynn, 1997). This 5-point advantage has enabled them to achieve about three times the rate of economic growth of the Western democracies during the second half of the twentieth century and to reach approximately the same living standards by the end of the cen-tury. China has the same intelligence advantage but has been held back eco-nomically by communism. Still, after the introduction of a market economy in the last two decades of the twentieth century, it has achieved the same high rates of economic growth as the other East Asian nations. Economists such as David Landes (1998) have attempted to find the explanation for the rapid economic development of the East Asian nations and concluded that it may lie in the psychological characteristics of the populations, although without realizing that their high levels of intelligence have given them a crucial advantage. The gains made by the East Asian nations, with their 5-IQ-point advantage, provide an indication of what a eugenic state could achieve by using cloning to produce large numbers of replicas of its scientific, military, and political elites and by using embryo selection to raise the intelligence level of the population by around 15 IQ points in the first generation, with further increases in subsequent generations, together with improvements in the personality qualities responsible for high-level achievement and efficient work performance. An authoritarian state that embarked on a eugenic pro-gram of the kind described in this chapter would obtain formidable economic, scientific, and military advantages over other states, including the Western democracies.

10. CONCLUSIONS

In the twenty-first century, one or more authoritarian states are likely to develop eugenic programs as a means of improving their economic, scientific, and military strength. This is probable because virtually all governments have the enhancement of their national strength as one of their principal objec-tives. Political leaders understand that the quality of their populations is a significant factor in national strength, and they normally seek to enhance the quality of their populations by measures designed to improve their health, education, cognitive abilities, and moral standards. At some point in the future the political leaders of authoritarian states are likely to realize that it would be possible to use both classical eugenics and the new eugenics of the human biotechnologies to improve the quality of their populations genetically. We should anticipate that it is only a matter of time before some authoritarian states embark on eugenic programs.

In general terms, the situation is similar to that in the 1930s when nuclear physics had developed to the stage at which it became feasible to construct an atom bomb. Once this point had been reached, it became inevitable that some country or countries would embark on a research program to make the bomb and would then use it, or threaten to use it, to gain a military advantage. In the event, it was the United States that developed the bomb and used it in 1945 to force Japan to surrender. For the next four years, until the Soviet Union developed the bomb, the United States was the only country to possess the atom bomb and could have used it to take control of the world. It was inevitable that it made no attempt to take advantage of this opportunity because in a democracy the internal opposition to an endeavor of this kind is too strong for it to be politically feasible. Authoritarian states are not constrained by internal opposition, which can easily be suppressed. An authoritarian eugenic state that used its genetically enhanced population to develop a decisive military advantage would be likely to use it to establish world domination.

Whether or not such a development occurs, we can be confident that the eugenic use of biotechnology will make progress in the twenty-first century. This progress will be driven by the wishes of couples to have children with desirable genetic characteristics and the increasing ability of biotechnology to satisfy these wishes. It is also likely to be driven by the needs of states to increase their economic, scientific, and military strength and by the realization of political leaders of authoritarian states that this could be advanced by eugenics. Once technologies are developed that serve individual and national needs, it becomes inevitable that some individuals and states will use them. The development of biotechnologies to improve the human genome will not be different in this regard from the development of numerous other technologies throughout the course of human history.

21

❧

The Evolution of the Eugenic World State

1. Dysgenic Processes in Developing Nations
2. The United States
3. Europe
4. East Asia
5. The Emergence of Chinese Global Supremacy
6. Administration of the World State
7. Implementation of a World Eugenic Program
8. Development of Genetic Engineering
9. Conclusions

❧❧❧

Chapters 19 and 20 were concerned with the likely development of eugenics in democratic societies and authoritarian states during the twenty-first century. In this chapter we consider the probable impact of eugenic and dysgenic processes in different parts of the world and how these will affect the future balance of power between nations. Both eugenic and dysgenic processes will be present, but the balance between them will differ in different countries and regions.

The dysgenic processes will consist of dysgenic fertility and dysgenic immigration. In some countries these dysgenic forces will predominate, causing a deterioration in the quality of the populations and a consequent decline in national strength. Eugenic forces will also be present. Some of these will evolve spontaneously in democratic societies. It is also likely that some authoritarian states will adopt a eugenic program, using both classical eugenics and human biotechnology, to improve the genetic quality of the population and to increase their national strength. Eugenics will become an instrument in the struggle between nations for global supremacy. How this is likely to work out is discussed in this final chapter.

1. DYSGENIC PROCESSES IN DEVELOPING NATIONS

In the developing nations in Latin America, the near and middle East, and North Africa, there was severe dysgenic fertility in the second half of the twentieth century, which I have documented in *Dysgenics* (Lynn, 1996). Throughout these nations women with secondary education were typically having two or three children, women with primary education were having four or five children, while women with no education were having six or seven children. Only in sub-Saharan Africa was there very little dysgenic fertility because women with all levels of education were having six or seven children. Further surveys published in the journal *Studies in Family Planning* have shown a continuation of dysgenic fertility in Latin America and the onset of dysgenic fertility in sub-Saharan Africa. Figures illustrating these dysgenic trends are shown in Table 21.1. The first two rows show strong recent dysgenic fertility in Brazil and the Dominican Republic. The third row shows the fertility of women in relation to educational level for 21 sub-Saharan countries, averaged from surveys carried out over the years 1988–92 (Kirk & Pillet, 1998). Women with secondary education were having only about two-thirds of the number of children of those with no education.

This dysgenic fertility is likely to continue in the twenty-first century. In sub-Saharan Africa, present average IQs obtained from a number of countries are around 70 (Lynn, 1997). The onset of dysgenic fertility will drive average IQs lower, with seriously adverse consequences for the economies and the quality of life in these countries.

Throughout Latin America, with the exception of Argentina and Uruguay, dysgenic fertility indexed by women's educational level is associated with racial differences in fertility. In these countries there is a racial social hierarchy in which Europeans are at the top, blacks and native American Indians

Table 21.1
Total Fertility Rates of Women Aged 15–49 of Three Educational Levels in Brazil, the Dominican Republic, and Sub-Saharan Africa

Country	Education		
	None	Primary	Secondary
Brazil, 1996	5.0	3.3	1.8
Dominican Republic, 1996	5.0	3.9	2.2
Sub-Saharan Africa, 1986–92	6.6	6.0	4.3

are at the bottom, and those of mixed race are in the middle (e.g., Valle Silva, 1988). Women with low fertility are largely well-educated whites, while women with high fertility are largely poorly educated blacks, American Indians, or of mixed race. Hence Europeans will decline as a proportion of the population of these countries, while the proportions of poorly educated blacks, American Indians, and mixed-race mulattos and mestizos will increase. As the European elites of these countries decline as a proportion of the population, the efficiency of the economies will inevitably deteriorate.

The continuation of severe dysgenic fertility combined with the population explosion will produce a serious deterioration of economic and social conditions throughout the developing world. Inevitably, increasing numbers of people will seek to escape by migrating to the affluent Western democracies. These people will mostly enter as refugees, asylum seekers, and illegal immigrants. The Western democracies will find it impossible to contain their numbers and will be progressively weakened by this dysgenic immigration and the social strains of multiracial societies.

2. THE UNITED STATES

The United States will experience both dysgenic fertility and dysgenic immigration for a number of decades into the twenty-first century. The most recent evidence shows that dysgenic fertility is still present in the United States (Loehlin, 1997; Lynn, 1998, 1999b). This is likely to persist as significant numbers of well-educated and intelligent women opt to remain childless in order to further their careers and to preserve their affluent lifestyles, while poorly educated and less intelligent women continue to have children either because of their inefficient use of contraception or deliberately in order to live on welfare as a preferable alternative to working. There are no signs that a spontaneous solution to this problem is likely to emerge, and we should anticipate that this dysgenic fertility will continue for the foreseeable future.

More serious and intractable will be the problem of dysgenic immigration. This began on a significant scale with the 1965 Immigration Act, which led, by the 1980s, to the admission into the United States of about one million immigrants a year, consisting largely of Hispanics and also of Asians and blacks. This immigration will continue and the numbers of Hispanics and blacks will also increase as a result of their greater fertility as compared with whites. In 1992 the American Current Population Survey showed that the number of children of Hispanic women aged 35 to 44 was 2.47 and of black women 2.23, as compared with 1.89 for whites. These fertility differences are likely to continue for the indefinite future.

As Hispanics and blacks become an increasing proportion of the U.S. population, there will be three predictable consequences. First, because Hispanics and blacks have lower intelligence levels than whites at approximately 92 and 85 IQ points, respectively, the intelligence level of the population will fall, causing economic productivity to decline and generating a number of social problems associated with low intelligence.

Second, an Hispanic-led coalition of nonwhites will become the dominant political force. The United States will then detach itself from its alliance with Western Europe. This alliance, formalized in the North Atlantic Treaty Organization (NATO) in the second half of the twentieth century, was based on the ethnic affinity and common cultural heritage of the peoples of Western Europe and the dominant group of ethnic Western Europeans in the United States. As whites cease to be the dominant U.S. group and are replaced by a coalition of nonwhites, the ethnic and cultural basis of the U.S.-European alliance will disappear. An increasingly Hispanic United States will ally itself in global politics with Hispanic Latin America.

Third, the increasingly multiracial nature of the U.S. population will generate enormous internal strains on social cohesion. The major racial and ethnic groups will continue to perform at different levels in education and earnings, with whites and Asians performing best, Hispanics performing intermediately, and blacks performing worst. These differences will continue to generate resentment among Hispanics and blacks, who will lobby to obtain compensation for them by affirmative action and set-aside quotas, reserving business contracts for themselves. The different racial groups and their advocates will also strive to secure increased immigration quotas and amnesties for illegal immigrants of their own peoples. Crime rates will escalate because Hispanics and blacks have much higher rates of crime than whites and Asians. For instance, in 1996, incarceration rates calculated by the U.S. Department of Justice per 100,000 population were 193 for whites, 688 for Hispanics, and 1,571 for blacks, while Asian crime rates were somewhat lower than those of whites (Taylor & Whitney, 1999; see also, Rushton, 1995). To escape black and Hispanic crime, there will be increasing white flight and also "Asian flight" from the black-Hispanic cities to white and Asian communities in suburbs and satellite towns where whites and Asians will increasingly come to live in fortified estates. The legal system will break down as judges and juries increasingly return perverse verdicts favoring their own racial and ethnic groups, as has already occurred in parts of New York in what has become known as a Bronx jury.

Some people have predicted that as the quality of life for whites and Asians deteriorates, interracial conflicts will become so severe that they will lead to civil wars between different racial groups and the eventual breakup of the United States into racially homogenous independent states. For instance, Michael Clough of the Council on Foreign Relations has written, "The state could be set for a series of economic and cultural wars pitting regions of the country against each other" and "America is destined to become a country of distinct, relatively independent regions, each with its own politicocultural economies, metropolitan centers, governing elites, and global interests" (Masters, 1999, p. 4). The models for this scenario are the breakup of the former

Soviet Union in 1991 and of Yugoslavia in 1998 into culturally and ethnically independent states. A conceivable outcome of this racial strife is that the largely white northern and midwestern states will secede from the Union, while the Hispanic majorities, which will appear in the southern states, will opt to join Mexico or to form an independent Hispanic nation. A more likely scenario is that the United States will continue as one country, become increasingly Hispanicized, and come to resemble the Hispanic republics of Latin America. As this happens, the United States will experience growing lawlessness, political anarchy, racial conflict, and huge disparities in wealth between rich and poor. Possibly when Europeans lose their political power, they will seek to regain this by overthrowing democratic institutions and replacing them by military rule, as has happened periodically throughout Latin America. However the details of the decline of the United States work out, it will forfeit its position as the leading world economic, scientific, and military power and eventually cease to be a major force in global politics.

3. EUROPE

Europe is likely to continue to experience some dysgenic fertility and dysgenic immigration in the twenty-first century. Europe does not permit primary immigration (immigrants who have no other reason for immigration than the wish to live in more affluent countries than their own), and there are no lobbies for permitting primary immigration. Nevertheless, there is substantial immigration of asylum seekers and illegal entrants—secondary immigration. In Britain there were approximately 0.3 million blacks and South Asians in 1961, as recorded in the census of that year. In the 1991 census the numbers had grown to approximately 3 million. This increase is likely to continue because of the young age structure of these populations and through further immigration. The 1991 British census found that non-European immigrants comprised about 5 percent of the population; but among children from 0 to 9 years of age, they comprised 9.3 percent of the population (Coleman, 1995). The number of non-Europeans is also likely to grow through the continued immigration of asylum seekers and illegal entrants. It is even possible that the 10-fold increase in the numbers of immigrants in Britain over the years 1961 to 1991 will continue at the same rate, producing a nonwhite majority sometime in the second half of the twenty-first century.

Similar increases in the numbers of immigrants have occurred in other Western European countries. A large number of these are from Turkey, North Africa, and the Near East. Many of them are Muslims, whose first loyalty is to their Islamic faith, and these will not be assimilable, at least for many generations. A second major group consists of Africans from sub-Saharan Africa and the Caribbean who have entered Europe from former colonies or as refugees from the African civil wars. Immigrants from sub-Saharan Africa manifest the same low intelligence, poor educational attainment, high rates of unemployment, welfare dependency, and crime as blacks in the United States. In Britain, blacks have an average IQ of approximately 88 (Mackintosh & Mascie-Taylor, 1984), a rate of unemployment approximately twice as great as that of whites (Blackaby, Drinkwater, Leslie, & Murphy, 1997), and a crime rate approximately six times greater than that of whites (Smith, D.J., 1997). In France, blacks have an unemployment rate approximately 50 percent higher than that of whites (Model, Fisher, & Silberman, 1999) and a crime rate approximately eight times greater than that of whites (Tournier, 1997). In Sweden, blacks have a crime rate approximately two and a half times greater than that of whites (Martens, 1997). In the Netherlands, immigrants from Surinam, Turkey, and North Africa have average IQs of 89, 88, and 84, respectively (Te Nijenhuis, 1997). These immigrants have caused social problems and racial conflicts of the kind experienced in the United States and white flight from black inner-city ghettos. These social problems and costs will increase as their numbers grow through relatively high fertility and the further immigration of asylum seekers and illegal entrants.

It will be impossible for European nations to make any significant corrections to these dysgenic processes because of the opposition of special interest groups and a predominantly liberal media. Thus, as the twenty-first century unfolds, Europe will be weakened by dysgenic fertility and by dysgenic immigration, but this will take place more slowly than in the United States. In the middle decades of the century, Europe will therefore be left as the principal power base of the European peoples. Europe will be weakened militarily by the loss of the United States as an ally but will be able to develop its own military capability to replace this. Europe is likely to be strengthened by its evolution into some form of federal state and by the incorporation of the nations of eastern and southeastern Europe, and possibly even of Russia. For these reasons, Europe will be a formidable global power for the foreseeable future.

4. EAST ASIA

Dysgenic processes were quite weak in the nations of East Asia in the closing decades of the twentieth century. Dysgenic fertility had ceased in Japan by the 1980s and was minimized in other East Asian countries (Lynn, 1996). There has been virtually no immigration, dysgenic or otherwise, into East Asian nations, except for Japan, where there has been some immigration from other Asian countries; but the amount of this has been too small to have any significant dysgenic impact and is likely to remain so.

The intelligence levels in the East Asian nations are high, with an average IQ of around 105, the evidence for which I have given in Lynn (1997; in press). These high intelligence levels have been a major factor in the rapid economic development of Japan, South Korea, Taiwan, Hong Kong, Singapore, and China in the second half of the twentieth century, during which they achieved rates of economic growth about three times greater than the Western democracies. These high rates of economic growth can be projected forward into the twenty-first century with the result that these countries will become increasingly powerful. The peoples and political leaders of the nations of East Asia are potentially or actually sympathetic to eugenics. Eugenic programs were introduced in Singapore and China during the last years of the twentieth century. In the twenty-first century, more ambitious and sophisticated eugenic programs are likely to be adopted in these and possibly in other East Asian countries.

There are five reasons for anticipating a development of this kind. First, the political leaders and the peoples of these countries do not share the high priority accorded to individual rights at the expense of social rights that developed during the second half of the twentieth century in Western nations and that has been principally responsible for the rejection of eugenics. Throughout East Asia there is a greater acceptance of the legitimacy of social rights, which provide the political and ethical legitimacy for eugenics. Second, this value system is expressed in the favorable attitudes toward eugenics among geneticists and physicians in China as found in the survey carried out by Wertz (1998), which showed that Chinese geneticists and physicians recommend pregnancy termination on explicitly eugenic grounds to women carrying fetuses with genetic disorders. Third, the political rulers of Singapore and China had already introduced eugenic programs in the last two decades of the twentieth century, suggesting a willingness to implement further eugenic measures. Fourth, the peoples of East Asia have the high levels of intelligence and scientific expertise necessary to develop and implement the potentially eugenic human biotechnologies of embryo selection, cloning, and genetic engineering. Fifth, the political leaders of at least some of these countries are likely to have the political will to implement serious eugenic programs. This is suggested by the draconian one-child policy introduced in China in 1979, which stipulated that couples were only permitted to have one child. This edict was enforced by the compulsory fitting of IUDs, compulsory abortion, and, as a further deterrent, the imposition of heavy fines amounting to approximately half of annual earnings for couples having a second child. At the same time, couples complying with the policy were given rewards in the form of cash payments and better housing, food rations, and child health care (Short & Fengying, 1998).

By the early 1990s these policies had reduced the total fertility rate in China to 1.9. A state that succeeds in imposing population policies of this kind should not have any difficulty in introducing programs of both classical eugenics and the new eugenics of the human biotechnologies. The political leaders of more authoritarian East Asian states are likely to have both the motivation and the means to introduce robust eugenic programs, and we should anticipate that some of them will do so.

Because of its sheer size, China will inevitably emerge as the most powerful of the East Asian nations. There is every reason to expect that the rapid rate of economic growth achieved by China in the closing decades of the twentieth century will continue, with the result that by the middle decades of the twenty-first century China will achieve parity with Europe in economic, scientific, and military strength. As China grows in power during the twenty-first century and the strength of the United States declines, China and Europe will evolve as the two foremost world powers. A struggle for global supremacy will develop between them, resembling the arms race between the United States and the Soviet Union in the second half of the twentieth century. In this contest Europe will be at a long-term disadvantage because of the difficulties of achieving an agreed military strategy among the 25 or so nations of which the European Union, or federal state, will consist and because of the progressive loss of social cohesion resulting from continued immigration and population growth of non-European peoples. China will have the advantages of a racially homogenous nation state and culture and of the high intelligence level of its population. In addition, Europe will not be able to introduce eugenic programs to enhance its population quality, while China is likely to develop further the implementation of eugenic programs introduced in the closing years of the twentieth century. China's use of eugenics and particularly the potential use of the human biotechnologies of embryo selection, cloning, and genetic engineering are likely to give her a decisive advantage in this struggle for global supremacy, giving her ultimate victory and emergence as the world superpower.

5. THE EMERGENCE OF CHINESE GLOBAL SUPREMACY

As China gains supremacy over Europe in economic, scientific, and military strength sometime in the second half of the twenty-first century, China can be expected to use its power to take control of the world and establish a world state. There are two reasons why this development should be expected. First, the political leaders of dictatorships and oligarchies have normally attempted to increase the size of the territories they control. History records a succession of political leaders who have devoted themselves to this objective, including Alexander the Great; a series of Roman generals and emperors who colonized most of the known world; Genghis Khan; and the British, French, and Dutch oligarchies who, among them, colonized much of the world between the sixteenth and nineteenth centuries. In the twentieth century, Hitler aimed to conquer the world, and later in the century the leaders of the Soviet Union had the same aim (Schweller, 1998). It is sometimes argued that the Chinese are an exception to this general principle, as if the Chinese lacked the gene for territorial expansion. This is improbable. In the second half of the twentieth century China annexed Tibet and fought a frontier war with India in 1962. China seeks to take over Taiwan. There is no reason to suppose that the future political leaders of China will be any different in their territorial ambitions from those of other oligarchies.

Second, during the twenty-first century there are likely to be increasing numbers of unstable states that will develop nuclear and biological weapons capable of inflicting considerable damage and with unpredictable consequences for the whole of humanity. At the present time Iraq and North Korea present the greatest threat of this kind, but others are likely to emerge. With technological advances and the spread of information, this threat will grow. The Chinese leaders are likely to form the view that it would be in their best interests for China to take control of the world and use its power to disarm these rogue states. They will see this as the best way of preserving themselves and the rest of humanity from the dangers arising from the use of these weapons.

Once China has developed a superior military capability, it will probably not be necessary to use this to establish world domination. The mere threat of its use should be sufficient to coerce the rest of the world into submission. If, however, some stubborn states refused to be coerced, it would become necessary to use some of these weapons on those countries to demonstrate their effectiveness and to enforce submission, in the same way as the United States dropped the atom bomb on Hiroshima and Nagasaki to force the surrender of Japan in 1945. One or two examples of this kind should be sufficient to coerce the world into acceptance of Chinese authority.

6.   ADMINISTRATION OF THE WORLD STATE

Once China has secured global supremacy, it would be expected to establish a world state and to administer it in much the same way as previous colonial powers have ruled their empires. It would appoint its own ethnic nationals as governors and senior support staff of its provinces and would recruit middle- and lower-ranking military commanders, police officers, administrators, and the like, from the nationals of the subject populations. This is the way the Romans, British, French, and numerous other former colonial powers have administered their empires, and it would be an obvious model to adopt. Alternatively, China might find sufficient numbers of compliant nationals of its subject peoples to run the provinces under direction from Beijing on the model of the Soviet empire and the German rule of some of occupied Europe during World War II. The political rulers of the Chinese world state would have a number of expert advisers, among whom their historical advisers would alert them to the tendency of empires to break up after several centuries and of the need to ensure that this was not allowed to happen.

The Chinese world state would not permit the manufacture or possession of weapons, except by its own peoples or by others under strict supervision. These weapons would be used to supply the military detachments that it would maintain throughout its colonies to suppress insurrections that would be likely to erupt from time to time. Apart from these minor confrontations, there would be world peace. This will bring to an end the long period of warfare between independent nation states and will be one of the benefits of the world state.

The Chinese oligarchy would be expected to retain its autocratic character. It would realize that it would be impossible to run a world state as a democracy. If a democratic constitution were established, with countries given independence and voting powers along the lines of the United Nations, the oligarchy controlling the world state would find itself outvoted. It would be deprived of its authority, and the independent countries would form coalitions to promote their own self-interests. The oligarchy would see no reason to allow this to happen and to forego the advantages gained from having secured world power. It would view democracy as an experiment that was tried by Europeans for a century or so and failed. It would learn this lesson of history and would not regard the democratic experiment as worth repeating.

Once the world state is established, it will come to be seen as the final step in the progressive aggregation of independent states into larger units, such as occurred with the unification of Germany and Italy in the nineteenth century, with the formation of the European Union in the twentieth century, and with attempts to establish a world authority in the shape of the League of Nations and the United Nations in the first and second halves of the twentieth century. The establishment of the world state will come to be seen as the inevitable culmination of these historical processes. It will come to be seen as equally inevitable that the peoples who finally achieved world domination were those with the highest intelligence levels and that the long struggle for world supremacy between the Oriental and the European peoples would eventually be won by the Orientals.

7.   IMPLEMENTATION OF A WORLD EUGENIC PROGRAM

Once China has established the world state, it will be concerned with raising the prosperity of its subject populations, just as other colonial powers have been. One of its first measures to promote this objective will be to introduce worldwide eugenic programs. These will include programs of both positive and negative eugenics. With regard to negative eugenics, one of its first objectives will be to reverse the dysgenic fertility that appeared in Europe, the United States, and the rest of the economically developed world in the middle and later decades of the nineteenth century and persisted into the twentieth century and that developed later in most of the remainder of the world. It can be expected that in its European and North American provinces, the Chinese will introduce the same eugenic measures that had been pioneered in China, consisting of both the classical eugenics of parental licensing and the new eugenics of the mandatory use of embryo selection for conception.

The Chinese may well also introduce the cloning of the elites of the European peoples. The Chinese will be aware that while they and other Oriental peoples have a higher average intelligence, the European peoples have a greater capacity for creative achievement, probably arising from a higher level of psychopathic personality, enabling them more easily to challenge existing ways of thinking and to produce creative innovations. This will be part of human genetic diversity that the Chinese will be keen to preserve and foster. They will regard the European peoples rather in the same way as the Romans regarded the Greeks after they had incorporated them into the Roman empire. Although the Romans had conquered the Greeks by their military superiority, they respected the Greeks for having developed a higher level of civilization than they themselves had been able to achieve. The Chinese will view their European subject peoples in a similar manner.

The economically developing world will present more of a problem because of the large numbers of people and the huge explosion of those segments of the population with the low intelligence and weak moral character caused by several generations of dysgenic fertility. By the time China assumes control of the world, these trends will have had a devastating impact on the economic and social life of these already impoverished nations and also on North America and Europe as large numbers of refugees and economic migrants continue to enter the Western democracies.

The world state would not be able to find sufficient numbers of geneticists and physicians in developing countries required to implement a program of embryo selection to reverse the adverse effect of several generations of dysgenic fertility. It would be expected to deal with this problem by introducing a robust program of classical eugenics. To reduce the population numbers it would probably introduce the one-child policy that was implemented in China in the 1980s and 1990s. This would be supplemented by a rigorous system of licenses for parenthood in which elites were permitted and given financial incentives to have several children in order to reverse the impact of dysgenic fertility. Later, as these problems were brought under control and as living standards improved, it would be expected to introduce the medical facilities to provide embryo selection; and over the course of several decades, these measures would produce significant improvements in the genetic quality of third world populations and would begin to produce improvements in third world economies.

8. DEVELOPMENT OF GENETIC ENGINEERING

Over the longer term, the world state can be expected to set up research and development programs of genetic engineering for the construction and insertion of new genes. These would build on the techniques of gene therapy that were pioneered in the United States and Britain in the last two decades of the twentieth century for the treatment of genetic diseases and disorders and that consisted of the insertion of healthy genes to take over the function of defective genes.

The next stage of this research program will entail the construction and insertion of new genes, not present in the human genome, for improved health, intelligence, and personality. The development of this technique should be feasible in principle because it would adopt the methods used successfully in the late twentieth century for the production of a number of genetically modified foods and "transgenic" animals. The functions of some of these new genes can already be surmised. With regard to health, new genes might be constructed for the deferment of aging, enabling people to use accumulated knowledge, experience, and skills over an extended life span. With regard to intelligence, new genes might be constructed for larger brain size. It is well established that brain size is associated with intelligence (Rushton, 1995). There is little doubt that the relationship between brain size and intelligence is a causal one and that humans with larger brains would have increased cognitive abilities.

It is difficult to predict what other kinds of new genes might be devised for the improvement of the human genome. Nevertheless, it is impossible that humans can have reached their genetic optimum and are incapable of further improvement. Just what new genes could be constructed and inserted into the human genome will be a research problem for the biologists and geneticists of the world state. It is possible that hundreds or even thousands of new genes for greater capacities could be constructed, which would enable humans to solve problems well beyond their present capabilities. It is likely that in due course these will make it possible to colonize planets in other solar systems in preparation for the time when the earth becomes uninhabitable. This will be the culmination of the ability of humans to use their intelligence to adapt themselves to new environments and will be the final achievement of eugenics.

9. CONCLUSIONS

Eugenic and dysgenic forces will play a significant role in the development of the balance of power between nations during the course of the twenty-first century. In the developing countries of Latin America, North Africa, the Near and Middle East, and South Asia, the severe dysgenic fertility present in the second half of the twentieth century will continue. Dysgenic fertility will develop in sub-Saharan Africa. The resulting genetic deterioration of intelligence and personality will cause serious economic, political, and social problems throughout the economically developing world.

In the United States there is likely to be a continuation of dysgenic fertility and, more serious, large-scale dysgenic immigration that will produce a Hispanic-black majority in the second half of the century. This may lead to the breakup of the United States along racial lines, with the secession of some northern and midwestern states with large white majorities to form an independent, largely white state and of southern states with Hispanic majorities to form another independent state or to join up with Mexico. More probably, the United States will remain a single nation in which deteriorating population quality and racial conflict will progressively weaken its position as a leading world power. In the second half of the twenty-first century, the United States will come increasingly to resemble the Latin American republics, in which Europeans form a minority of the population, and will cease to be a major player in world politics. Europe also will continue to experience dysgenic fertility and dysgenic immigration, but their negative impact will be less severe there than in the United States. As the power of the United States declines during the second half of the twenty-first century, Europe will become the major power center of the European peoples.

Eugenics will make progress in the United States and Europe through medical advances in prenatal diagnosis and pregnancy terminations of fetuses with genetic diseases and, in the foreseeable future, through couples using embryo selection to have children with good health, high intelligence, and sound moral character. Eugenics will be advanced through the private initiatives of couples concerned to have children who are healthy, intelligent, and of sound personality. Eugenic programs will not be introduced by states on any significant scale because democratic political structures and the opposition of special interest groups and the media opposed to eugenics will make the implementation of state eugenics impossible.

The nations of East Asia are likely to develop their economic, scientific, technological, and military strength during the twenty-first century by virtue of the high intelligence levels of their populations and the absence of any serious dysgenic processes. These countries have not allowed the growth of an underclass with high dysgenic fertility, and they have not permitted dysgenic immigration. China will continue its rapid economic development and will emerge as a new superpower in the early middle decades of the twenty-first century. Chinese economic, scientific, and military strength is likely to be increased by the further development of the eugenic programs introduced in the 1980s and 1990s and particularly by the introduction of the new eugenics of embryo selection and the cloning of elites. As the power of the United States declines, China and Europe will emerge as the two superpowers. A global conflict will develop between them in which Europe will become progressively weakened by dysgenic forces and China progressively strengthened by eugenic programs.

This conflict will eventually be won by China, which will use its power to assume control of the world and to establish a world state. This event will become known as "the end of history." Once China has established a world state, it can be expected to administer this on the same lines as former colonial empires by appointing Chinese governors and senior military and administrative support staff in charge of the provinces of its world empire or by allowing nationals of its subject peoples to administer the provinces under Chinese supervision. The establishment of a Chinese world state will inevitably not be welcomed by the peoples of the rest of the world, who will become colonized populations governed by an oligarchy based in Beijing. There will be no democracy, and a number of freedoms will be curtailed, including freedom to publish seditious material and to have unlimited numbers of children. There will, however, be certain compensating benefits. There will be no more wars between independent nation states with the attendant dangers of the use of nuclear weapons and biological warfare. It will be possible to deal with the problems of dysgenic fertility, global warming, deforestation, the population explosion in the developing world, the AIDS epidemic, and similar global problems that cannot be tackled effectively in a world of independent nation states. Among the world state's first objectives will be the reversal of dysgenic processes and the introduction of eugenic programs throughout the world. Over the longer term the world state will set up research and development programs for the use of genetic engineering to improve the human genome and to produce a new human species able to solve hitherto unsolvable problems and to colonize new planets. This will be the ultimate achievement of Galton's vision of using eugenics to replace natural selection with consciously designed human selection.

This scenario for the twenty-first century, in which China assumes world domination and establishes a world eugenic state, may well be considered an unattractive future. But this is not really the point. Rather, it should be regarded as the inevitable result of Francis Galton's (1909) prediction made in the first decade of the twentieth century, that "the nation which first subjects itself to a rational eugenical discipline is bound to inherit the earth" (p. 34).

References

Abrahamse, A. F., Morrison, P. A., & Waite, L. J. (1988). *Beyond stereotypes: Who becomes a single teenage mother?* Santa Monica, CA: Rand.

Accardo, P. J., & Whitman, B. Y. (1990). Children of mentally retarded parents. *American Journal of Diseases of Childhood, 144,* 69–70.

American Psychiatric Association. (1994). *Diagnostic and statistical manual of mental disorders.* Washington, DC: Author.

Anderson, E. (1990). *Streetwise: Race, clan and change in an urban community.* Chicago: University of Chicago Press.

Anderson, V. E. (1974). Genetics and intelligence. In J. Wortis (Ed.), *Mental retardation and developmental disabilities: An annual review.* New York: Brunner-Mazel.

Anderson, W. F. (1992). Human gene therapy. *Science, 256,* 808–13.

Andreason, N. C. (1987). Creativity and psychiatric illness. *Psychiatric Annals, 8,* 113–19.

Auletta, K. (1982). *The underclass.* New York: Random House.

Ayer, A. J. (1984). *More of my life.* London: Methuen.

Bachu, A. (1991). *Fertility of American Women: June 1990.* Washington, DC: U.S. Government Printing Office.

Baird, P. A., Anderson, T. W., Newcombe, H. B., & Lowry, R. B. (1988). Genetic disorders in children and young adults: A population study. *American Journal of Human Genetics, 42,* 677–93.

Baroody, A. J. (1988). Number comparison learning by children classified as mentally retarded. *American Journal of Mental Retardation, 92,* 461–71.

Barrick, M. R., & Mount, M. K. (1991). The big five personality dimensions and job performance: A meta-analysis. *Personnel Psychology, 44,* 1–26.

Bauer, G. (1994 August 15). Illegitimate rhetoric. *National Review,* 59–60.

Beardsley, T. (1997). China syndrome: China's eugenics law makes trouble for science and business. *Scientific American, 276*(3), 33–34.

Beauchamp, T. L., & Walters, L. (1994). Introduction to reproductive technologies and surrogate parenting arrangements. In T. L. Beauchamp and L. Walters (Eds.), *Contemporary issues in bioethics*. Belmont, CA: Wadsworth.

Benbow, C. P. (1992). Academic achievement in mathematics and science of students between ages 13 and 23: Are there differences among students in the top one percent of mathematical ability? *Journal of Educational Psychology, 84*, 51–61.

Berger, J. (1999, January 10). Yale gene pool seen as route to better baby. *New York Times*, p. 15.

Billig, M. (1998). A dead idea that will not lie down. *Searchlight, 277*, 8–13.

Biology and Society. (1987). Sterilisation in mental handicap. *Biology and Society, 4*, 55.

Birch, H. G., Richardson, S. A., Baird, D., Horobin, G., & Illsley, R. (1970). *Mental subnormality in the community*. Baltimore, MD: Williams and Wilkins.

Blackaby, D., Drinkwater, S., Leslie, D., & Murphy, P. (1997). A picture of male and female unemployment among Britain's ethnic minorities. *Scottish Journal of Political Economy, 44*, 182–97.

Blacker, C. P. (1952). *Eugenics: Galton and after*. Cambridge, MA: Harvard University Press.

Bland, R. C., Orn, H., & Newman, S. C. (1988). Lifetime prevalence of psychiatric disorders in Edmonton. *Acta Psychiatrica Scandinavia, 77*, 24–32.

Blank, R. H. (1995). International symposium on critically ill newborns. *Journal of Legal Medicine, 16*, 183–88.

Blank, R. M., George, C. C., & London, R. (1996). State abortion rates: The impact of policies, providers, politics, demographics and economic environment. *Journal of Health Economics, 15*, 513–53.

Blum, R. W., & Resnick, M. D. (1982). Adolescent decision making: Contraception, abortion, motherhood. *Pediatric Annals, 11*, 797–805.

Bodmer, W. F., & McKie, R. (1994). *The book of man*. London: Little Brown.

Bojorquez, J. (1994, December 19). In his image. *Sacramento Bee*, p. 18.

Bok, G. (1983). *Racism and sexism in Nazi Germany: Motherhood, sterilization, and the state*. Signs, 8, 413–32.

Bolan, W. F. (1988). Statement of New Jersey Catholic Conference in connection with public hearing on surrogate mothering. Newark, NJ: Commission on Legal and Ethical Problems in the Delivery of Health Care.

Borjas, G. J. (1985). Assimilation, changes in cohort quality, and the earnings of immigrants. *Journal of Labor Economics, 3*, 463–89.

Borjas, G. J. (1987). Self-selection and the earnings of immigrants. *American Economic Review, 77*, 531–53.

Borjas, G. J. (1990). *Friends or strangers*. New York: Basic Books.

Borjas, G. J. (1993). The intergenerational mobility of immigrants. *Journal of Labor Economics, 11*, 113–35.

Bouchard, T. J. (1993). The genetic architecture of human intelligence. In P. A. Vernon (Ed.), *Biological approaches to the study of human intelligence*. Norwood, NJ: Ablex.

Brazzell, J. F., & Acock, A. C. (1988). Influence of attitudes, significant other, and aspirations on how adolescents intend to resolve a premarital pregnancy. *Journal of Marriage and the Family, 50*, 413–29.

Brebner, J. (1998). Happiness and personality. *Personality and Individual Differences, 25*, 279–96.

Brennan, P. A., Mednick, S. A., & Jacobsen, B. (1996). Assessing the role of genetics in crime using adoption cohorts. In G. R. Bock and J. H. Goode (Eds.). *Genetics of criminal and antisocial behavior*. Chichester, UK: Wiley.

Brennan, Z., & Syal, R. (1997 August 17). Sperm donor payments to end. *Sunday Times* (London), p. 7.

Brewer, M. (1937). *Eugenics and politics*. London: Eugenics Society.

Brickley, M., Browning, L., & Campbell, K. (1982). Vocational histories of sheltered workshop employees placed in projects with industry and competitive jobs. *Mental Retardation, 20*, 52–57.

Brigham, C. C. (1923). *A study of American intelligence*. Princeton, NJ: Princeton University Press.

Brimelow, P. (1995). *Alien Nation*. New York: Random House.

Brinchmann, B. S. (1999). When the home becomes a prison: Living with a severely disabled child. *Nursing Ethics, 6*, 137–43.

Broadhurst, P. L. (1975). The Maudsley reactive and non-reactive strains of rates: A survey. *Behavior Genetics, 5*, 299–319.

Broberg, G., & Roll-Hansen, N. (1996). *Eugenics and the welfare state*. East Lansing: Michigan State University Press.

Brock, D. I. (1982). *Early diagnosis of fetal defects*. Edinburgh: Churchill Livingstone.

Brody, N. (1992). *Intelligence*. San Diego, CA: Academic Press.

Broman, S. J., Nichols, P. L., Shaughnessy, P., & Kennedy, W. (1987). *Retardation in young children*. Hillsdale, NJ: Erlbaum.

Bromham, D. R., & Cartmill, R.S.V. (1993). Are current sources of contraception advice adequate to meet changes in contraceptive practice? A study of patients requesting termination of pregnancy. *British Journal of Family Planning, 19*, 179–83.

Brooner, R. K., Bigelow, G. E., Strain, E., & Schmidt, C. W. (1990). Intravenous drug abusers with antisocial personality disorder: Increased HIV risk disorder. *Drug and Alcohol Dependence, 26*, 39–44.

Brown, B. S., & Courtless, T. F. (1967). *The mentally retarded offender*. Washington, DC: President's Commission on Law Enforcement and Administration of Justice.

Bruce, E. J., Schultz, C. L., & Smyrnios, K. X. (1996). A longitudinal study of the grief of mothers and fathers of children with intellectual disability. *British Journal of Medical Psychology, 69,* 33–45.

Bruininks, R. H., Hill, B. K., & Morreau, L. E. (1988). Prevalence and implications of maladaptive behaviors and dual diagnosis in residential and other service programs. In J. H. Stark, F. J. Menolascino, M. H. Albarelli, and V. C. Gray (Eds.), *Mental retardation and mental health.* New York: Springer-Verlag.

Brunner, H. G., Nelen, M. R., Breakfield, X. O., Ropers, H. H., & van Oost, B. A. (1993). Abnormal behaviour associated with a point mutation in the structural gene for monamine oxidase A. *Science, 262,* 578–80.

Brunner, H. G., Nelen, M. R., & van Zandvoort, P. (1993). X-linked borderline mental retardation with prominent behavioral disturbance: Phenotype, genetic localisation, and evidence for disturbed monoamine metabolism. *American Journal of Human Genetics, 52,* 1032–1039.

Budiansky, S. (1997). *The nature of horses.* London: Weidenfeld and Nicholson.

Burack, J. A., Hodapp, R. M., & Zigler, E. (1997). *Handbook of mental retardation and development.* Cambridge: Cambridge University Press.

Burleigh, M. (1994). Psychiatry, German society, and the Nazi euthanasia program. *Social History of Medicine, 7,* 213–28.

Burt, C. L. (1915). The general and specific factors underlying the primary emotions. *British Association Annual Reports, 84,* 694–96.

Burt, C. L. (1937). *The backward child.* London: University of London Press.

Burt, C. L. (1952). *Intelligence and fertility.* London: Eugenics Society.

Burt, C. L. (1966). The genetic determination of differences in intelligence: A study of monozygotic twins reared together and apart. *British Journal of Psychology, 57,* 137–53.

Burton, R., Savage, W., & Reader, F. (1990). The morning after pill. *British Journal of Family Planning, 15,* 119–21.

Buss, D. M. (1999). *Evolutionary psychology.* Boston: Allyn and Bacon.

Caldwell, J. H. (1934, March). Babies by scientific selection. *Scientific American, 150,* 124–25.

Campenella, T. (1613). *Civitas solis.* Rome.

Canadian Royal Commission on New Reproductive Technologies. (1993). *Proceed with care.* Ottawa: H. M. S. O.

Cantor, C. (1992). The challenges to technology and information. In D. J. Kevles and L. Hood (Eds.), *The code of codes.* Cambridge, MA: Harvard University Press.

Card, J. J., Petersen, J. L., & Greeno, C. G. (1992). Adolescent pregnancy prevention programme: Design, monitoring, and evaluation. In B. C. Miller, J. C. Card., R. L. Paikoff, & J. L. Petersen (Eds.), *Preventing adolescent pregnancy*. London: Sage.

Carrell, A. (1935). *L'homme, cet inconnu*. Paris: Larousse.

Cartwright, A. (1976). *How many children?* London: Routledge and Kegan Paul.

Cattell, R. B. (1937). *The fight for our national intelligence*. London: P. S. King.

Cattell, R. B. (1965). *The scientific study of personality*. London: Penguin.

Cattell, R. B. (1972). *Beyondism*. New York: Praeger.

Cattell, R. B. (1987). *Beyondism* (2d ed.). New York: Praeger.

Cavalli-Sforza, L. L., & Bodmer, W. F. (1971). *The genetics of human populations*. San Francisco: Freeman.

Cherry, K. E., Matson, J. L., & Paclawskwj, T. R. (1997). Psychopathology in older adults with severe and profound mental retardation. *American Journal of Mental Retardation, 101*, 445–58.

Chorney, M. J., Chorney, K., Seese, N., Owen, M. J., Daniels, J., McGuffin, P., Thompson, L. A., Detterman, D. K., Benbow, C., Lubinski, D., Eley, T., & Plomin, R. (1998). A quantitative trait locus associated with cognitive ability in children. *Psychological Science, 9*, 159–66.

Clark, C. (1963). Eugenics and genetics. In G. Wolstenholme (Ed.), *Man and his future*. London: Churchill.

Clark, L. A., & Livesley, W. J. (1994). Two approaches to identifying the dimensions of personality disorder: Convergence on the five factor model. In P. T. Costa & T. A. Widiger (Eds.), *Personality disorders*. Washington, DC: American Psychological Association.

Clark, R. (1968). *J. B. S.: The life and work of J. B. S. Haldane*. New York: Coward-McCann.

Clarke, G. R., & Strauss, R. P. (1998). Children as income-producing assets: The case of teen illegitimacy and government transfers. *Southern Economic Journal, 64*, 827–56.

Clearie, A. F., Hollingsworth, L. A., Jamison, M. Q., & Vincent, M. L. (1985). International trends in teenage pregnancy: An overview of sixteen countries. *Biology and Society, 2*, 23–30.

Cleckley, H. (1941). *The mask of sanity*. St. Louis, MO: Mosby.

Clerici, M., Carta, I., & Cazzullo, C. (1989). Substance abuse and psychopathology: A diagnostic screening of Italian narcotic addicts. *Social Psychiatry and Psychiatric Epidemiology, 24*, 219–26.

Coghlan, A. (1995, November). Gene dream fades away. *New Scientist 25*, 14–15.

Cohen, C. (Ed.) (1996). *New ways of making babies: The case of egg donation.* Bloomington: Indiana University Press.

Coid, B, Lewis, S. W., & Reveley, A. M. (1993). A twin study of psychosis and criminality. *British Journal of Psychiatry, 162,* 87–92.

Coleman, D. (1995). International migration: Demographic and socioeconomic consequences in the United Kingdom and Europe. *International Migration Review, 29,* 155–206.

Compton, W. W., Helzer, J. E., & Hiou, H. G. (1991). New methods in cross-cultural psychiatry: Psychiatric illness in Taiwan and the United States. *American Journal of Psychiatry, 148,* 1697–1704.

Comptroller General of the United States. (1977). *Preventing mental retardation: More can be done.* Washington, DC: U.S. General Accounting Office.

Connor, J. M., & Ferguson-Smith, M. A. (1988). *Essential medical genetics.* Oxford: Blackwell.

Corder, E. H., Saunders, A. M., Strittmatter, W. J., Schmechel, D., Gaskell, P. C., & Small, G. W. (1993). Gene dose of apolipoprotein E4 allele and the risk of Alzheimer's disease in late onset families. *Science, 261,* 921–23.

Correctional Service of Canada. (1990). A mental health profile of federally sentenced offenders. *Forum on Corrections Research, 2,* 7–8.

Costa, P. T., & McCrae, R. R. (1990). Personality disorders and the five factor model of personality. *Journal of Personality Disorders, 4,* 362–71.

Costa, P. T., & McCrae, R. R. (1992a). The five factor model and its relevance to personality disorders. *Journal of Personality Disorders, 6,* 343–59.

Costa, P. T., & McCrae, R. R. (1992b). Four ways five factors are basic. *Personality and Individual Differences, 13,* 653–65.

Courtney, M. E. (1997). The politics and realities of transracial adoption. *Child Welfare, 76,* 749–79.

Cox, C. (1926). *The early mental traits of three hundred geniuses.* Stanford, CA: Stanford University Press.

Craig, R. J. (1988). A psychometric study of the prevalence of DSM-IV personality disorders among treated opiate addicts. *International Journal of Addictions, 23,* 115–24.

Crick, F. (1963). Eugenics and genetics. In G. Wolstenholme (Ed.), *Man and his future.* London: Churchill.

Crocker, A. G., & Hodgins, S. (1997). Criminality of noninstitutionalized mentally retarded persons. *Criminal Justice and Behavior, 24,* 432–54.

Crosier, A. (1996). Women's knowledge and awareness of emergency contraception. *British Journal of Family Planning, 22,* 87–90.

Crow, J. F. (1986). *Basic concepts in population, quantitative and evolutionary genetics.* San Francisco: Freeman.

Cutright, P. (1986). Child support and responsible male procreative behavior. *Sociological Focus, 19*, 27–45.

Cynkar, R. J. (1981). *Buck v. Bell*: Felt necessities v. fundamental values. *Columbia Law Review, 81*, 1418–461.

Daley, D., & Gold, R. B. (1995). Public funding for contraceptive, sterilization, and abortion services, fiscal year 1992. *Family Planning Perspectives, 25*, 244–51.

Darke, S., Hall, W., and Swift, W. (1994). Prevalence, symptoms, and correlates of anti–social personality disorder among methadone maintenance clients. *Drug and Alcohol Dependence, 34*, 149–54.

Darwin, C. (1871). *The descent of man and selection in relation to sex*. London: Macmillan.

Darwin, L. (1926). *The need for eugenic reform*. London: Murray.

Davenport, C. (1911). *Heredity in relation to eugenics*. New York: Henry Holt.

Davenport, C. B. (1916). Huntington's chorea in relation to heredity and eugenics. *American Journal of Insanity, 63*, 195–208.

Davenport, K. S., & Remmers, H. H. (1950). Factors in state characteristics related to average A-12 V-12 test scores. *Journal of Educational Psychology, 41*, 110–15.

Dawkins, R. (1976). *The selfish gene*. New York: Oxford University Press.

Dawson, D. (1986). Effects of sex education on adolescent behavior. *Family Planning Perspectives, 18*, 162–70.

Dearden, K., Hale, C., & Alvarez, J. (1992). The educational antecedents of teen fatherhood. *British Journal of Educational Psychology, 62*, 139–47.

De Jong, C. A., van den Brink, W., Harteveld, F. M., & van der Wielen, E. G. (1993). Personality disorders in alcoholics and drug addicts. *Comprehensive Psychiatry, 34*, 87–94.

Devlin, B., Fienberg, S. E., Resnick, D. P., & Roeder, K. (1997). *Intelligence, genes, and success: Scientists respond to the bell curve*. New York: Springer-Verlag.

Digman, J. M., & Takemoto-Chock, N. K. (1981). Factors in the natural language of personality: Reanalysis, comparison, and interpretation of six major studies. *Multivariate Behavioral Research, 16*, 149–70.

DiLalla, D. L., Carey, G., Gottesman, I. I., & Bouchard, J. J. (1996). Heritability of MMPI personality indicators of psychopathology in twins reared apart. *Journal of Abnormal Psychology, 105*, 491–99.

Dobzhansky, T. (1962). *Mankind evolving*. New Haven, CT: Yale University Press.

Donnelly, M., McGilloway, S., Mays, N., Perry, S., & Lavery, C. (1997). A three-to-six-year follow-up of former long-stay residents of mental handicap hospitals in Northern Ireland. *British Journal of Clinical Psychology, 36*, 585–600.

Donovan, P. (1995). The "family cap": A popular but unproven method of welfare reform. *Family Planning Perspectives*, *27*, 116–71.

Donovan, P. (1997). Can statutory rape laws be effective in preventing adolescent pregnancy? *Family Planning Perspectives*, *29*, 30–40.

Dryfoos, J. (1990). *Adolescents at risk*. New York: Oxford University Press.

Dugdale, R. L. (1877). *The Jukes: A study in crime, pauperism, disease, and heredity*. New York: Putnam.

Duijsens, I. J., & Dikstra, R. F. W. (1996). DSM-III-R and ICD-10 personality disorders and their relationship with the big five dimensions of personality. *Personality and Individual Differences*, *21*, 119–33.

Duncan, G., Harper, C., Ashwell, E., & Mant, D. (1990). Termination of pregnancy: Lessons for prevention. *British Journal of Family Planning*, *15*, 112–17.

Duncan, G. J., Hill, M. S., & Hoffman, S. D. (1988). Welfare dependence within and across generations. *Science*, *239*, 467–71.

Duster, T. (1990). *Backdoor to eugenics*. New York: Routledge.

Dykens, E. M., & Hodapp, R. M. (1997). Treatment issues in genetic mental retardation syndromes. *Professional Psychology*, *28*, 263–70.

Dykens, E. M., Hodapp, R. M., Ort, S., Finucane, B., Shapiro, L. R., & Leckman, J. F. (1989). The trajectory of cognitive development in males with fragile X syndrome. *Journal of the American Academy of Child and Adolescent Psychiatry*, *28*, 422–26.

Dykens, E. M., Ort, S., & Cohen, J. (1996). Trajectories and profiles of adaptive behavior in males with fragile X syndrome: Multicenter studies. *Journal of Autism and Developmental Disorders*, *26*, 287–301.

Eastabrooke, C. (1916). *The Jukes in 1913*. Cold Spring Harbor, NY: Eugenics Record Office.

Editorial. (1982, May 27). *New York Times*.

Elster, A. B., Lamb, M. E., & Tavare, J. (1987). Association between behavioral and school problems and fatherhood in a national sample of adolescent fathers. *Journal of Pediatrics*, *111*, 932–36.

Ermisch, J. F. (1991). *Lone parenthood: An economic analysis*. Cambridge: Cambridge University Press.

Essen, J., & Wedge, P. (1982). *Continuities in childhood disadvantage*. London: Heinemann.

European Parliament. (1990). *Ethical and legal problems of genetic engineering and human artificial insemination*. Strasbourg: Author.

European Parliament. (1997, March 11). Resolution on cloning. *Proceedings of the European Parliament*. Strasbourg.

Evans, G., Todd, S., Beyer, S., Felce, D., & Perry, J. (1994). Assessing the impact of the all-Wales mental handicap strategy: A survey of four districts. *Journal of Intellectual Disability Research*, 38, 109–33.

Eysenck, H. J. (1947). *Dimensions of personality*. London: Routledge and Kegan Paul.

Eysenck, H. J. (1973). *The inequality of man*. London: Routledge and Kegan Paul.

Eysenck, H. J. (1979). *Intelligence: Structure and measurement*. Berlin: Springer Verlag.

Eysenck, H. J. (1993). Creativity and personality: An attempt to bridge divergent traditions. *Psychological Inquiry, 4*, 238–46.

Eysenck, H. J. (1995). *Genius: The natural history of creativity*. Cambridge: Cambridge University Press.

Eysenck, H. J., & Gudjonsson, G. H. (1989). *The causes and cures of criminality*. New York: Plenum.

Eysenck, M. W. (1982). *Attention and arousal: Cognition and performance*. Berlin: Springer-Verlag.

Ezzell, C. (1987). First ever animal patent issues in United States. *Nature, 332*, 21.

Falconer, D. S. (1960). *Introduction to quantitative genetics*. London: Longman.

Falk, S., Paul, D. B., & Allen, G. (1998). Forward. *Science in Context, 11*, 329–30.

Farndale, N. (1994). The great strawberry hunt. *Country Life, 188*, 58–61.

Farrington, D. P. (1994). *Human development and criminal careers*. In M. Maguire, R. Morgan and R. Reiner (Eds.), *The Oxford handbook of criminology*. Oxford; Clarendon Press.

Feingold, W. (1976). *Artificial insemination*. New York: Academic Press.

Feldman, M. A. (1994). Parenting education for parents with intellectual disabilities: A review of outcome studies. *Research in Developmental Disabilities, 15*, 299–332.

Feldman, M. A. (1998). Preventing child neglect: Child-care training for parents with intellectual disabilities. *Infants and Young Children, 11*, 1–11.

Fischer, C. S., Hout, M., Jankowskei, M. S., Lucas, S. R., Swidler, A., & Voss, K. (1996). *Inequality by design: Cracking the bell curve myth*. Princeton, NJ: Princeton University Press.

Fisher, R. A. (1918). The correlations between relatives on the supposition of Mendelian inheritance. *Transactions of the Royal Society of Edinburgh, 52*, 399–433.

Fisher, R. A. (1924). The elimination of mental defect. *Eugenics Review, 16*, 114–16.

Fisher, R. A. (1928). Income tax rebates: The birth rate and our future policy. *Eugenics Review, 20*, 79–81.

Fisher, R. A. (1929). *The genetical theory of natural selection.* Oxford: Clarendon Press.

Fisher, R. A. (1932). Family allowances in the contemporary economic situation. *Eugenics Review, 24*, 87–95.

Fiske, D. W. (1949). Consistency of the factorial structure of personality ratings from different sources. *Journal of Abnormal and Social Psychology, 44*, 329–44.

Flanegan, J. C. (1939). A study of psychological factors related to fertility. *Proceedings of the American Philosophical Society, 80*, 513–23.

Fletcher, J. (1988). *The ethics of genetic control.* Buffalo, NY: Prometheus.

Flynn, J. R. (1984). The mean IQ of Americans: Massive gains 1932 to 1978. *Psychological Bulletin, 95*, 29–51.

Flynn, J. R. (1991). *Asian Americans: Achievement beyond IQ.* Hillsdale, NJ: Lawrence Erlbaum.

Folling, I. A. (1934). Uber Aussscheidung von Phenylbrenztranbesanre in den harn als Stoffwechselanomie in Verbindung mit Imbezilitat. *Zeitschrift fur Physiologisch Chemic, 227*, 169–76.

Francoeur, R. (1971). *Utopian motherhood new trends in human reproduction.* London: George Allen and Unwin.

Frick, P. J., O'Brien, B. S., Wooton, J. M., & McBurnett, K. (1994). Psychopathy and conduct disorders in children. *Journal of Abnormal Psychology, 103*, 700–707.

Fustenberg, F. F., Levine, J. A., & Brooks-Gunn, J. (1990). The children of teenage mothers: Patterns of early childbearing in two generations. *Family Planning Perspectives, 22*, 54–61.

Galton, D. J., & Galton, C. J. (1998). Francis Galton: And eugenics today. *Journal of Medical Ethics, 24*, 99–105.

Galton, F. (1865). Hereditary talent and character. *Macmillan's Magazine, 12*, 157–66; 318–27.

Galton, F. (1869). *Hereditary genius.* London: Macmillan.

Galton, F. (1874). *English men of science: Their nature and nurture.* London: Methuen.

Galton, F. (1883). *Inquiries into human faculty and its development.* London: Dent.

Galton, F. (1908a). *Memories of my life.* London: Methuen.

Galton, F. (1908b). Local association for promoting eugenics. *Nature, 78*, 645–47.

Galton, F. (1909). *Essays on eugenics.* London: Eugenics Society.

Galton, F., & Schuster, E. (1906). *Noteworthy families.* London: Murray.

Galton Institute (1999). Notes of the quarter. *Galton Institute Newsletter, 2*, 8.

Gauthier, A. H. (1991). *Family policies in comparative perspective*. Oxford, Nuffield College: Centre for European Studies.

Gay, P. (1988). *Freud: A life for our time*. New York: Norton.

Giami, A. (1998). Sterilization and sexuality in the mentally handicapped. *European Psychiatry, 13*, 113–19.

Gillon, R. (1998). Eugenics, contraception, abortion, and ethics. *Journal of Medical Ethics, 24*, 219–20.

Glaister, A., & Baird, D. (1998). The effects of self-administered emergency contraception. *New England Journal of Medicine, 339*, 1–4.

Glass, D. V. (1940). *Population policies and movements in Europe*. Oxford: Clarendon Press.

Glasse, A. (1998, April 29). U.S. and Norway used insane for Nazi-style tests. *The Times* (London), p. 22.

Goddard, E. (1993). *General Household Survey 1991*. London: Her Majesty's Stationery Office.

Goddard, H. H. (1912). *The Kallikak family: A study in the heredity of feeblemindedness*. New York: Macmillan.

Goerree, M. A., O'Brien, B. O., Goering, P., & Blackhouse, G. (1999). The economic burden of schizophrenia in Canada. *Canadian Journal of Psychiatry, 44*, 464–72.

Gold, M. A., Schein, A., & Coupey, S. M. (1997). Emergency contraception: A national survey of adolescent health experts. *Family Planning Perspectives, 29*, 15–24.

Goldman, D., Lappalainen, J., & Ozaki, N. (1996). Direct analysis of candidate genes in impulsive behaviors. In Ciba Foundation (Ed.), *Genetics of criminal and antisocial behavior*. Chichester, UK: Wiley.

Gooder, P. (1996). Knowledge of emergency contraception amongst men and women in the general population and women seeking an abortion. *British Journal of Family Planning, 22*, 81–84.

Goodson, P., Evans, A., & Edmundson, E. (1997). Female adolescents and the onset of sexual intercourse: A theory-based review of research 1984–1994. *Journal of Adolescent Health, 21*, 147–56.

Gordon, R. A. (1997). Everyday life as an intelligence test: Effects of intelligence and intelligence context. *Intelligence, 24*, 203–320.

Gordon, S. (1984). *Hitler, Germans, and the Jewish question*. Princeton, NJ: Princeton University Press.

Gottesman, J. J., & Goldsmith, H. H. (1995). Developmental psychopathology of anti-social behavior: Inserting genes into its ontogenesis and epigenesis. In C. Nelson (Ed.), *Threats to optimal development*. Hillsdale, NJ: Erlbaum.

Gottfredson, L. S. (1997). Mainstream science on intelligence (Editorial). *Intelligence*, *24*, 13–24.

Gotz, K. O., & Gotz, K. (1979). Personality characteristics of successful artists. *Perceptual and Motor Skills*, *49*, 919–24.

Gould, S. J. (1981). *The mismeasure of man*. New York: Norton.

Gould, S. J. (1985). *The flamingo's smile*. New York: Norton.

Graham, R. K. (1970). *The future of man*. North Quincy, MA: Christopher.

Graham, R. K. (1987). Combating dysgenic traits in contemporary society. *Mankind Quarterly*, *27*, 327–35.

Griffiths, M. (1990). Contraceptive practices and contraceptive failures among women requesting termination of pregnancy. *British Journal of Family Planning*, *16*, 16–18.

Guilford, J. P., & Zimmerman, W. S. (1949). *The Guilford-Zimmerman temperament survey*. Beverly Hills, CA: Sheridan Supply.

Gustavson, K. H., Hagberg, B., Hagberg, G., & Sars, K. (1977). Severe mental retardation in a Swedish county, etiological and pathogenetic aspects of children born 1959–1970. *Neuropadiatrie*, *8*, 293–304.

Haas–Wilson, D. (1997). Women's reproductive choices: The impact of Medicaid funding restrictions. *Family Planning Perspectives*, *29*, 228–33.

Hafner, H., & Boker, W. (1973). *Crimes of violence by mentally retarded offenders*. Cambridge: Cambridge University Press.

Hall, R. (1977). *Passionate crusader: The life of Marie Stopes*. New York: Harcourt Brace Jovanovich.

Hamerton, J. L., Evans, J. A., & Stranc, L. (1993). Prenatal diagnosis in Canada. In *Research Volumes of the Royal Commission on New Reproductive Technologies*. Ottawa: H. M. S. O.

Hammer, R. E., Palmiter, R. D., & Brinster, R. I. (1989). Partial correction of murine hereditary disorder by germ-line incorporation of a new gene. *Nature*, *311*, 65–67.

Hanauske-Abel, H. M. (1996). Not a slippery slope or sudden subversion: German medicine and national socialism in 1933. *British Medical Journal*, *313*, 1453–63.

Hansen, H., & Lykke-Olesen, L. (1997). Treatment of dangerous sexual offenders in Denmark. *Journal of Forensic Psychiatry*, *8*, 195–99.

Hare, R. D. (1983). Diagnosis of anti-social personality disorder in two prison populations. *American Journal of Psychiatry*, *140*, 887–90.

Hare, R. D., (1994). *Without conscience*. London: Warner.

Hare, R. D., Hart, S. D., & Harpur, T. J. (1991). Psychopathy and the DSM-IV criteria for antisocial personality disorder. *Journal of Abnormal Psychology*, *100*, 391–98.

Harlap, S., Kost, K., & Darroch, F. J. (1991). *Pregnancies occurring during contraceptive use.* New York: Alan Guttmacher Institute.

Harpur, T. J., Hare, R. D., & Hakstian, A. N. (1989). Two-factor conceptualization of psychopathy: Construct validity and assessment implications. *Psychological Assessment, 1,* 6–17.

Harris, J. (1992). *Wonderwoman and Superman.* Oxford: Oxford University Press.

Harris, J. (1998). *Clones, genes, and immortality.* Oxford: Oxford University Press.

Harris, P. (1993). The nature and extent of aggressive behavior among people with learning difficulties (mental handicap). in a single health district. *Journal of Intellectual Disability Research, 37,* 221–42.

Hart, B., & Spearman, C. (1912). General ability, its existence and nature. *British Journal of Psychology, 5,* 51–84.

Hayek, F. A. (1960). *The constitution of liberty.* London: Routledge and Kegan Paul.

Hayfron, J. E. (1998). The performance of immigrants in the Norwegian labor market. *Journal of Population Economics, 11,* 293–303.

Hayman, R. L. (1990). Presumptions of justice: Law, politics, and the mentally retarded parent. *Harvard Law Review, 103,* 1202–71.

Heaven, P. C. L. (1996). Personality and self–reported delinquency: Analysis of the big five personality dimensions. *Personality and Individual Differences, 20,* 47–54.

Hendriks, A. A. J., Hofstee, W. K. B., & De Raud, B. D. (1999). The five factor personality inventory FFPI. *Personality and Individual Differences, 27,* 307–25.

Henshaw, S. K., Koonin, L. M., & Smith, J. C. (1991). Characteristics of U.S. women having abortions, 1987. *Family Planning Perspectives, 23,* 75–79.

Henshaw, S. K., & Van Vost, J. (1994). Abortion services in the United States, 1991 and 1992. *Family Planning Perspectives, 26,* 100–114.

Herrman, L., & Hogben, L. (1932). The intellectual resemblance of twins. *Proceedings of the Royal Society, 53,* 105–29.

Herrnstein, R. J. (1971). *IQ in the meritocracy.* Boston: Atlantic–Little, Brown.

Herrnstein, R J. (1983). Some criminogenic traits of offenders. In J. Q. Wilson (Ed.), *Crime and public policy.* San Francisco: ICS Press.

Herrnstein, R. J., & Murray, C. (1994). *The bell curve.* New York: Free Press.

Hirschi, T., & Hindelang, M. J. (1977). Intelligence and delinquency: A revisionist review. *American Sociological Review, 42,* 571–87.

Hitler, A. (1943). *Mein Kampf* (R. Manheim, Trans.). Boston: Houghton Mifflin.

Hobbes, T. (1651). *Leviathan.* London: Murray.

Hoem, J. M. (1990). Social policy and recent fertility changes in Sweden. *Population and Development Review, 16,* 735–48.

Hoem, J. M. (1993). Public policy as the fuel of fertility: Effects of a policy reform on the pace of childbearing in Sweden in the 1980s. *Acta Sociologica, 36*, 19–31.

Hotz, R. L. (1997, February 18). Robert Graham, founder of exclusive sperm bank, dies. *Los Angeles Times*, p. 8.

Hudson, G., & Hawkins, R. (1995). Contraceptive practices of women attending for a termination of pregnancy—a study from South Australia. *British Journal of Family Planning, 21*, 61–64.

Hume, D. (1751). *Enquiry concerning the principles of morals.* Edinburgh: Hudson.

Humphreys, L. G. (1994). Intelligence from the standpoint of a pragmatic behaviorist. *Psychological Inquiry, 5*, 179–92.

Hutchings, B., & Mednick, S. R. (1977). Criminality in adoptees and their adoptive and biological parents: A pilot study. In S. A. Mednick and K. O. Christiansen (Eds.), *Biosocial bases of criminal behavior.* New York: Gardner Press.

Huxley, A. (1932). *Brave new world.* London: Chatto and Windus.

Huxley, J. S. (1936). Eugenics and society. *Eugenics Review, 28*, 11–31.

Huxley, J. S. (1941). *Man stands alone.* London: Harper and Row.

Huxley, J. S. (1942). *Evolution: The modern synthesis.* London: Allen and Unwin.

Huxley, J. S. (1962). Eugenics in evolutionary perspective. *Eugenics Review, 54*, 123–41.

In China, a sperm bank confines its donors to the scholarly set. (1999, June 28). *Chronicle of Higher Education.*

Inge, W. R. (1927). *Outspoken Essays.* London: Longmans, Green.

International dictionary of medicine and biology. (1986). New York: Wiley.

Isen, A. M., Daubman, K. A., & Nowicki, G. P. (1987). Positive affect facilitates creative problem solving. *Journal of Personality and Social Psychology, 52*, 1122–31.

Jackson, C., & Klerman, J. (1994). *Welfare, abortion and fertility.* Santa Monica, CA: Rand.

Jacobsberg, L., Frances, A., & Perry, S. (1995). Axis II diagnoses among volunteers for HIV testing and counseling. *American Journal of Psychiatry, 152*, 1222–24.

Jahoda, M. (1989). A helping hand to evolution? *Contemporary Psychology, 34*, 816–17.

James, M. E., Rubin, C. P., & Wilks, S. E. (1991). Drug abuse and psychiatric findings in HIV seropositive pregnant patients. *General Hospital Psychiatry, 13*, 4–8.

Jamison, K. R. (1993). *Touched with fire: Manic-depressive illness and the artistic temperament.* New York: Free Press.

Jang, K. L., Livesley, W. J., & Vernon, P. A. (1996). Heritability of the big five personality dimensions and their facets: a twin study. *Journal of Personality, 64,* 577–91.

Jencks, C. (1972). *Inequality.* London: Penguin.

Jensen, A. R. (1972). *Genetics and education.* London: Methuen.

Jensen, A. R. (1980). *Bias in mental testing.* London: Methuen.

Jensen, A. R. (1998). *The g factor.* Westport, CT: Praeger.

Jinks, J. L., & Fulker, D. W. (1970). Comparison of the biometrical, genetical, MAVA, and classical approaches to the analysis of human behavior. *Psychological Bulletin, 73,* 311–49.

Johnson, A. M., Wadsworth, J., Wellings, K., and Field, J. (1994). *Sexual attitudes and life styles.* Oxford: Blackwell.

Johnson, D. M. (1948). Applications of the standard score IQ to social statistics. *Journal of Social Psychology, 27,* 217–27.

Johnson, K. C., & Rouleau, J. (1997). Temporal trends in Canadian birth defects prevalences, 1979–1993. *Canadian Journal of Public Health, 88,* 169–76.

Jones, D. C., & Carr-Saunders, A. M. (1927). The relation between intelligence and social status among orphan children. *British Journal of Psychology, 17,* 343–64.

Jones, E. F., Forrest, J. D., Goldman, N., Henshaw, S., Lincoln, R., Rosoft, J. J., Westoff, C. F., & Wulf, D. (1986). *Teenage pregnancy in industrialized countries.* New Haven, CT: Yale University Press.

Jones, E. F., Forrest, J. D., Henshaw, S. K., Silverman, J., & Torres, A. (1989). *Pregnancy, contraception, and family planning services in industrialized countries.* New Haven, CT: Yale University Press.

Jones, P., & Cannon, M. (1998). The new epidemiology of schizophrenia. *Psychiatric Clinics of North America, 21,* 1–25.

Jones, S. (1993). *The language of the genes.* London: HarperCollins.

Joyce, T., & Kaestner, R. (1996a). State reproductive policies and adolescent pregnancy resolution: The case of parental involvement laws. *Journal of Health Economics, 15,* 579–607.

Joyce, T., & Kaestner, R. (1996b). The effect of expansions in Medicaid income eligibility on abortion. *Demography, 33,* 181–92.

Joyce, T., Kaestner, R., & Kwan, F. (1998). Is Medicaid pronatalist? *Family Planning Perspectives, 30,* 108–27.

Juda, A. (1949). The relationship between highest mental capacity and psychic abnormalities. *American Journal of Psychiatry, 106,* 296–304.

Judge, T. A., Martocchio, J. J., & Thoresen, C. J. (1997). Five factor model of personality and employee absence. *Journal of Applied Psychology, 82,* 745–55.

Kaku, M. (1998). *Visions.* Oxford: Oxford University Press.

Kamin, L. J. (1974). *The science and politics of IQ*. London: Penguin.

Karlsson, J. L. (1978). *Inheritance of creative intelligence*. Chicago: Nelson-Hall.

Kaye, H. L. (1987). *The social meaning of modern biology*. New Haven, CT: Yale University Press.

Kazue, S. (1995, January). Women rebuff the call for more babies. *Japan Quarterly*, 14–20.

Keltner, B. R. (1992). Caregiving by mothers with mental retardation. *Family and Community Health, 15*, 10–18.

Kelvin, R. P., Lucas, C. J., & Ojha, A. B. (1965). The relation between personality, mental health, and academic performance in university students. *British Journal of Social and Clinical Psychology, 4*, 224–53.

Kemp, T. (1957). Genetic-hygienic experiences in Denmark in recent years. *Eugenics Review, 49*, 11–18.

Kennedy, J. F. (1959). *A nation of immigrants*. New York: Random House.

Kessler, R. C., McGongale, K. A., Zhao, S., & Nelson, C. B. (1984). Lifetime and 12-month prevalence of DSM-III-R psychiatric disorders in United States. *Archives of General Psychiatry, 51*, 8–19.

Ketterlinus, R. D., Lamb, M. E., Nitz, K., & Elster, A. B. (1992). Adolescent nonsexual and sex related problem behaviors. *Journal of Adolescent Research, 7*, 431–56.

Kevles, D. J. (1985). *In the name of eugenics*. New York: Knopf.

Kevles, D. J. (1992). Out of eugenics: The historical politics of the human genome. In D. J. Kevles and L. Hood (Eds.), *The code of codes*. Cambridge, MA: Harvard University Press.

Kevles, D. J., & Hood, L. (1992). Reflections. In D. J. Kevles and L. Hood (Eds.), *The code of codes*. Cambridge, MA: Harvard University Press.

Keynes, M. (1993). Sir Francis Galton. In M. Keynes (Ed.), *Sir Francis Galton: The legacy of his ideas*. London: Macmillan.

Khantzian, E. J., & Treece, C. (1985). DSM-III psychiatric diagnosis of narcotic addicts. *Archives of General Psychiatry, 42*, 1067–71.

Kincheloe, J., Steinberg, S., & Gresson, A. (1996). *Measured lies*. New York: St. Martin's.

King, T. J., & Briggs, R. (1956). Serial transplantation of embryonic nuclei. *Cold Spring Harbor Symposium on Quantitative Biology, 21*, 271–90.

Kirby, D. (1984). *Sexuality education: An evaluation of programs and their effect*. Santa Cruz, CA: Network.

Kirby, D., Resnick, M. D., Downes, B., Kocher, T., Gunderson, S. P., Zelterman, D., & Blum, R. W. (1993). The effects of school-based health clinics in St. Paul on schoolwide birthrates. *Family Planning Perspectives, 25*, 12–16.

Kirby, D., & Waszak, C. (1992). School-based clinics. In B. C. Miller, J. J. Card,

R. L. Paickoff, & J. L. Peterson (Eds.), *Preventing Adolescent Pregnancy*. Santa Cruz, CA: Network.

Kirk, D., & Pillet, B. (1998). Fertility levels, trends, and differentials in sub–Saharan Africa in the 1980s and 1990s. *Studies in Family Planning, 29*, 1–22.

Kleinman, P. H., Millman, R. B., Robinson, H., Lesser, M., Hsu, C., Engelhart, P., & Finkelstein, J. (1994). Lifetime needle sharing: A predictive analysis. *Journal of Substance Abuse and Treatment, 11*, 449–55.

Koenig, R. (1997). Watson urges "Put Hitler behind us." *Science, 276*, 892.

Koestler, A. (1950). *The god that failed*. London: Methuen.

Kolata, G. (1997). *Clone*. London: Allen Lane.

Kopp, M. (1936). Legal and medical aspects of eugenic sterilization in Germany. *American Sociological Review, 1*, 766–70.

Koshland, D. E. (1988). The future of biological research: What is possible and what is ethical? *Science, 3*, 11–15.

Kosten, T. R., Rounsaville, B. J., & Kleber, H. D. (1982). DSM-III personality disorder in opiate addicts. *Comprehensive Psychiatry, 23*, 572–81.

Kraepelin, E. (1915). *Textbook of psychiatry*. Berlin: Springe.

LaFollette, H. (1980). Licensing parents. *Philosophy and Public Affairs, 9*, 182–97.

Landes, D. (1998). *The wealth and poverty of nations: Why some are so rich and some are so poor*. New York: Little Brown.

Lange, J. (1929). *Crime as destiny*. Leipzig: Thieme.

Langlors, B. (1980). Heritability of racing ability in thoroughbreds. *Livestock Production Science, 7*, 591–605.

LaPorta, R., Lopez-de-Silanes, F., Schleifer, A., & Vishny, R. (1999). *The quality of government*. Cambridge, MA: National Bureau of Economic Research.

Lasker, G. W. (1991). Introduction. In C. G. N. Mascie-Taylor and G. W. Lasker (Eds.), *Applications of biological anthropology to human affairs*. Cambridge: Cambridge University Press.

Lau, S. (1995). Report from Hong Kong. *Cambridge Quarterly of Healthcare Ethics, 4*, 364–66.

Laughlin, H. H. (1912, March). Eugenics. *Nature Study Review*, 110–11.

Lawrence, E. M. (1931). An investigation into the relation between intelligence and inheritance. [Monograph Supplement 16.] *British Journal of Psychology*, 1–80.

Laxova, R., Ridler, M., & Borven-Bravery, M. (1977). An etiological survey of the severely mentally retarded Hertfordshire children born between 1 January 1965 and 31 December 1967. *American Journal of Medical Genetics, 1*, 75–86.

Lederberg, J. (1963). Biological future of man. In G. Wolstenholme (Ed.), *Man and his future*. London: Churchill.

Lee, N. C., Rubin, G. L., & Borucki, R. (1988). The intrauterine device and pelvic inflammatory disease: New results from the women's health study. *Obstetrics and Gynaecology, 72,* 1–12.

Lehtinen, V., Lindholm, T., Veijola, J., & Vaisanen, E. (1990). The prevalence of PSE-CATEGO disorders in a Finnish adult population cohort. *Social Psychiatry and Psychiatric Epidemiology, 25,* 187–92.

Leibowitz, A., Eisen, M., & Chow, W. K. (1986). An economic model of teenage pregnancy decision making. *Demography, 23,* 67–77.

Le Pine, J. A., Hollenbeck, J. R., Ilgen, D. R., & Hedlund, J. (1997). Effects of individual differences on the performance of hierarchical decision-making teams: Much more than g. *Journal of Applied Psychology, 82,* 803–11.

Levin, M. (1997). *Why race matters.* Westport, CT: Praeger.

Levitt, S. D. (1996). The effect of prison population size on crime rates: evidence from prison overcrowding litigation. *Quarterly Journal of Economics, 64,* 319–52.

Levitt, S. D., & Donohue, J. (1999). *Legalized abortion and crime* (Working Paper). Stanford, CA: Stanford Law School.

Lewontin, R. C, Rose, S., & Kamin, L. (1989). *Not in our genes.* New York: Pantheon.

Lichter, D. T., McLaughlin, D. K., & Ribar, D. C. (1998). State abortion policy, geographic access to abortion providers and changing family formation. *Family Planning Perspectives, 30,* 281–87.

Lieberson, S., & Waters, M. C. (1988). *From many strands.* New York: Russell Sage Foundation.

Lindquist, P., & Allebeck, P. (1990). Schizophrenia and crime: A longitudinal follow-up of 644 schizophrenics in Stockholm. *British Journal of Psychiatry, 157,* 345–50.

Lippman, A. (1991). Prenatal genetic testing and screening: Constructing needs and reinforcing inequities. *American Journal of Law and Medicine, 17,* 15–50.

Loehlin, J. C. (1993). *Genetics and personality.* Thousand Oaks, CA: Sage.

Loehlin, J. C. (1997). Dysgenesis and IQ: What evidence is relevant? *American Psychologist, 52,* 1236–39.

Loehlin, J. C. (1998). Whither dysgenics? Comments on Preston and Lynn. In U. Neisser (Ed.), *The rising curve.* Washington, DC: American Psychological Association.

Loose, C. (1998, February 19). Birth control and bureaucracy. *Washington Post,* A10–12.

Ludmerer, K. (1972). *Genetics and American society.* Baltimore, MD: Johns Hopkins University Press.

Ludwig, A. M. (1992). Creative achievement and psychopathology: Comparison among professions. *American Journal of Psychotherapy, 46,* 330–56.

Ludwig, A. M. (1994). Mental illness and creative activity among female writers. *American Journal of Psychiatry, 151*, 165–66.

Lutton, W. (1999). Editorial. *The Social Contract, 9*, 271–72.

Lykken, D. T. (1995). *The antisocial personalities*. Hillsdale, NJ: Lawrence Erlbaum.

Lynn, R. (1979). The social ecology of intelligence in the British Isles. *British Journal of Social and Clinical Psychology, 18*, 1–12.

Lynn, R. (1980). The social ecology of intelligence in France. *British Journal of Social and Clinical Psychology, 19*, 325–31.

Lynn, R. (1988). *Educational achievement in Japan*. London: Macmillan.

Lynn, R. (1996). *Dysgenics: Genetic deterioration in modern populations*. Westport, CT: Praeger.

Lynn, R. (1997). Geographic variation in intelligence. In H. Nyborg & J. Gray (Eds.), *The scientific study of human nature*. Hillsdale, NJ: Lawrence Erlbaum.

Lynn, R. (1998). Dysgenics. *American Psychologist, 52*, 1431.

Lynn, R. (1999a). The attack on the bell curve. *Personality and Individual Differences, 26*, 761–765.

Lynn, R. (1999b). New evidence for dysgenic fertility for intelligence in the United States. *Social Biology, 46*,146–53.

Lynn, R. (in press). The geography of intelligence. In H. Nyborg (Ed.), *The g factor*. New York: Academic.

Lynn, R., Hampson, S., and Magee, M. (1984). Home background, intelligence, personality, and education as predictors of unemployment in young people. *Personality and Individual Differences, 5*, 549–57.

Lynn, R., and Lynn, A. (in press). On the relation between intelligence and happiness. *Journal of Social Psychology*.

Lynn, R., & Pagliari, C. (1994). The intelligence of American children is still rising. *Journal of Biosocial Science, 26*, 65–67.

Lyons, M. J., Tone, W. R., Eisen, S. A., & Goldberg, J. (1995). Differential heritability of adult and juvenile antisocial traits. *Archives of General Psychiatry, 52*, 906–15.

MacDonald, K. (1994). *A people that shall dwell alone*. Westport, CT: Praeger.

MacDonald, K. (1995). Evolution, the five factor model and levels of personality. *Journal of Personality, 63*, 525–67.

MacDonald, K. (1998). *Separation and its discontents*. Westport, CT: Praeger.

MacKenzie, D. (2000). A kinder, gentler killer. *New Scientist, 2245*, 34–36.

MacKinnon, D. W. (1978). *In search of human effectiveness*. New York: Creative Education Foundation.

Mackintosh, N. J. (1998). *IQ and human intelligence*. Oxford: Oxford University Press.

Mackintosh, N. J., & Mascie-Taylor, C. G. N. (1984). The IQ question. In *Education for all*. CMND Paper 4453. London: Her Majesty's Stationery Office.

Maller, J. B. (1933). Vital indices and their relation to psychological and social factors. *Human Biology, 5*, 94–121.

Manlove, J. (1997). Early motherhood in an intergenerational perspective: The experiences of a British cohort. *Journal of Marriage and the Family, 59*, 263–79.

Mann, D. M. A. (1993). Association between Alzheimer's disease and Down syndrome: Neuropathological observations. In J. M. Berg, H. Karlinsky, & A. J. Holland (Eds.), *Alzheimer's disease. Down syndrome and their relationship*. Oxford: Oxford University Press.

Marsiglio, W. (1987). Adolescent fathers in the United States: Their initial living arrangements, marital experience, and educational outcomes. *Family Planning Perspectives, 19*, 240–51.

Marsiglio, W., & Mott, F. L. (1986). The impact of sex education on sexual activity, contraceptive use, and premarital pregnancy among American teenagers. *Family Planning Perspectives, 18*, 151–62.

Martens, P. L. (1997). Immigrants, crime, and social justice in Sweden. In M. Tonry (Ed.), *Ethnicity, crime, and immigration*. Chicago: University of Chicago Press.

Mason, D. A., & Frick, P. J. (1994). The heritability of anti-social behavior. *Journal of Psychopathology and Behavioral Assessment, 16*, 301–23.

Masters, M. W. (1999). Republican or third party? *American Renaissance, 10*, 1–6.

Matthews, G., & Deary, I. J. (1998). *Personality traits*. Cambridge: Cambridge University Press.

Matthews, S., Riber, D., & Wilhelm, M. (1997). The effects of economic conditions and access to reproductive health services on state abortion rates and birth rates. *Family Planning Perspectives, 29*, 52–60.

McCabe, M. P., & Cummins, R. A. (1996, March). The sexual knowledge, experience, feelings, and needs of people with mild intellectual disability. *Education and Training in Mental Retardation and Developmental Disabilities*, 13–21.

McCormick, R. A. (1994). Blastomere separation: Some concerns. *Hastings Center Report, 24*, 14–16.

McDougall, W. (1921). *Is America safe for democracy?* New York: Macmillan.

McDougall, W. (1939). *Group mind*. Cambridge: Cambridge University Press.

McKenzie, J. (1989). Neuroticism and academic achievement: The Furneaux factor. *Personality and Individual Differences, 10*, 509–15.

McKusick, V. A. (1994). *Mendelian inheritance in man*. Baltimore, MD: Johns Hopkins University Press.

McLaren, J., & Bryson, S. E. (1987). Review of recent epidemiological studies of mental retardation: Prevalence, associated disorders, and etiology. *American Journal of Mental Retardation, 92,* 243–54.

Medawar, P. (1960). *The future of man.* London: Shenval.

Mednick, S. A., & Christiansen, K. O. (1977). *Biosocial bases of criminal behavior.* New York: Gardner.

Mednick, S. A., Gabrielli, W. F., & Hutchings, B. (1984). Genetic influences in criminal convictions: Evidence from an adoption cohort. *Science, 224,* 891–94.

Meezan, W., Katz, S., & Russo, E. M. (1978). *Adoptions without agencies: A study of independent adoptions.* New York: Child Welfare League of America.

Mercola, K. E., & Cline, M. J. (1980). The potentials of inserting new genetic information. *New England Journal of Medicine, 303,* 1297–1300.

Meier, K. J., & McFarlane, D. R. (1994). State family planning and abortion expenditures: Their effect on public health. *American Journal of Public Health, 84,* 1468–72.

Meredith, P., & Thomas, L. (1986). *Planned parenthood in Europe.* London: Croom Helm.

Merrick, J. C. (1995). Critically ill newborns and the law. *Journal of Legal Medicine, 16,* 189–209.

Meyer, C. E., Barthwick, S. A., & Eyman, R. K. (1985). Place of residence by age, ethnicity, and level of retardation of the mentally retarded and developmentally disabled population of California. *American Journal of Mental Deficiency, 90,* 266–70.

Mickelson, P. (1947). The feebleminded parent: A study of 90 family cases. *American Journal of Mental Deficiency, 51,* 644–45.

Mickelson, P. (1949). Can mentally deficient parents be helped to give their children better care? *American Journal of Mental Deficiency, 53,* 516–34.

Mill, J. S. (1859). *On liberty.* London: Macmillan.

Miller, P. W. (1999). Immigration policy and immigrant policy: The Australian points system. *American Economic Review, 89,* 192–97.

Model, S., Fisher, G., & Silberman, G. (1999). Black Caribbeans in comparative perspective. *Journal of Ethnic and Migration Studies, 25,* 187–212.

Moffitt, T. E. (1993). Adolescent limited and life course persistent antisocial behavior: A developmental taxonomy. *Psychological Review, 100,* 674–701.

Moffitt, T. E., Gabrielli, W. F., Mednick, S. A., & Schulsinger, F. (1981). Socioeconomic status, IQ, and delinquency. *Journal of Abnormal Psychology, 90,* 152–56.

Moore, K. A., & Erickson, P. I. (1985). Age, gender, and ethnic differences in sexual and contraceptive knowledge, attitudes, and behaviors. *Family and Community Health, 8,* 38–51.

Moore, K. A., Manlove, J., Glei, D. A., & Morrison, D. R. (1998). Non-marital school age motherhood: Family, individual, and school characteristics. *Journal of Adolescent Research, 13*, 433–57.

Moran, P. (1999). The epidemiology of antisocial personality disorder. *Social Psychiatry and Psychiatric Epidemiology, 34*, 231–42.

Morch, W. T., Skar, J., & Andersgard, A. B. (1997). Mentally retarded persons as parents: Prevalence and the situation of their children. *Scandinavian Journal of Psychology, 38*, 343–48.

Morel, B. A. (1857). *Traite des degenerescenses physiques, intellectuelles et morales de l'espece humain.* Paris: Larousse.

Morgan, P. (1999). *Adoption: The continuing debate.* London: Institute of Economic Affairs.

Morris, L., Warren, C. W., & Aral, S. O (1993). Measuring adolescent sexual behaviors and related health outcomes. *Public Health Reports, 108* (Suppl.), 31–36.

Mosher, S. W. (1998, February 27). In Peru, women lose the right to choose more children. *Wall Street Journal*, A19.

Moskowitz, E. H., Jennings, B., & Callahan, D. (1996). Long-acting contraceptives: Ethical guidance for policy makers and health care providers. In E. H. Moskowitz & B. Jennings (Eds.), *Coerced Contraception.* Washington, DC: Georgetown University Press.

Mughal, S., Walsh, J., & Wilding, J. (1996). Stress and work performance: The role of anxiety. *Personality and Individual Differences, 20*, 685–91.

Muller, H. J. (1935). *Out of the night.* New York: Vanguard.

Muller, H. J. (1939). The geneticists' manifesto. *Eugenical News, 24*, 63–64.

Muller, H. J. (1963). Genetic progress by voluntarily conducted germinal choice. In G. Wolstenholme (Ed.), *Man and his future.* London: Churchill.

Muller–Hill, B. (1988). *Murderous science: Elimination by scientific selection of Jews, gypsies and others in Germany.* Oxford: Oxford University Press.

Murray, C. (1984). *Losing ground.* New York: Basic.

Murray, C. (1990). *The emerging British underclass.* London: Institute of Economic Affairs, Health and Welfare Unit.

Murray, C. (1993, October 29). The coming white underclass. *Wall Street Journal*, p. 7.

Murray, I. (1999, February 3). Doctor calls for school-girl birth control implants. *The Times* (London), p. 9.

Mydans, S. (1993, June 27). Poll finds tide of immigration brings hostility. *New York Times*, Sec. 1, 16.

Myrdal, G. (1962). *Challenge to affluence.* New York: Pantheon.

National Foundation for Brain Research (1992). *The cost of schizophrenia.* Washington, DC: Author.

Neel, J. V. (1994). *Physician to the gene pool.* New York: Wiley.

Neisser, U. (1996). *Intelligence: Knowns and unknowns.* Washington, DC: American Psychological Association.

Nelson, J. R., Smith, D. J., & Dodd, J. (1990). The moral reasoning of juvenile delinquents: A meta-analysis. *Journal of Abnormal Child Psychology, 18,* 231–39.

Nesmith, J. D, Klerman, L. V., Kim, M., & Feinstein, R. A. (1997). Procreative experience and orientations toward paternity held by incarcerated adolescent males. *Journal of Adolescent Health, 20,* 198–203.

Neumann, P. J. (1994). The cost of a successful delivery with in vitro fertilization. *New England Journal of Medicine, 331,* 239–43, 270–71.

Newman, H. H., Freeman, F. N., & Holzinger, K. J. (1937). *Twins: A study of heredity and environment.* Chicago: University of Chicago Press.

Nigg, J. T., & Goldsmith, H. H. (1994). Genetics of personality disorders: perspectives from personality and psychopathology research. *Psychological Bulletin, 115,* 346–80.

Nock, S. L. (1998). The consequences of premarital fatherhood. *American Sociological Review, 63,* 250–63.

Oakley-Browne, M. A. Joyce, P. R., Wells, J. E., Bushnell, J. A., & Hornblow, A. R. (1989). Christchurch psychiatric epidemiology study, part 2: Six-month and other period prevalences of specific psychiatric disorders. *Australian and New Zealand Journal of Psychiatry, 23,* 327–40.

Ochse, R. (1991). The relation between creative genius and psychopathology: An historical perspective and a new explanation. *South African Journal of Psychology, 21,* 45–53.

Oden, M. H. (1968). The gifted child. Stanford, CA: Stanford University Press.

Office of Population, Censuses, and Surveys (1993). *Congenital malformation statistics.* London: Her Majesty's Stationery Office.

Ohsfeldt, R. L., & Gohmann, S. F. (1994). Do parental involvement laws reduce adolescent abortion rates? *Contemporary Policy Issues, 12,* 65–76.

O'Kane, A., Fawcett, D., & Blackburn, R. (1996). Psychopathy and moral reasoning: Comparison of two classifications. *Personality and Individual Differences, 20,* 505–14.

Ortmann, J. (1980). The treatment of sexual offenders: Castration and antihormone therapy. *International Journal of Law and Psychiatry, 3,* 443–51.

Osborn, F. H. (1953). *Preface to eugenics.* New York: Harper.

Osborn, F. H. (1974). History of the American Eugenics Society. *Social Biology, 21,* 115–26.

Palmiter, R. D., Brinster, R. L., Hanmer, M. E., & Trumbauer, M. G. (1982). Dramatic growth of mice that develop from eggs microinjected with metallutionein-growth hormone fusion genes. *Nature, 300*, 611–15.

Patton, J. R., Beirne-Smith, M., & Payne, J. S. (1990). *Mental retardation*. New York: Macmillan.

Patton, J. R., & Polloway, E. (1992). Learning disabilities: The challenge of adulthood. *Journal of Learning Disabilities, 25*, 410–15.

Paul, D. B. (1995). *Controlling human heredity*. Amherst, NY: Prometheus Books.

Pauling, L. (1959). Molecular disease. *American Journal of Orthopsychiatry, 29*, 682–92.

Pauling, L. (1968). Reflections on the new biology. *UCLA Law Review, 15*, 268–72.

Pearson, K. (1901). *National life from the standpoint of science*. London: Methuen.

Pearson, K. (1903). On the inheritance of the mental and moral characters in man. *Journal of the Anthropological Institute of Great Britain and Ireland, 33*, 179–237.

Pearson, K. (1912). *The groundwork of eugenics*. Cambridge, UK: Eugenics Laboratory.

Pearson, K. (1914). *The life, letters, and labours of Francis Galton*. Cambridge: Cambridge University Press.

Pearson, R. (1992). *Shockley on eugenics and race*. Washington, DC: Scott-Townsend.

Pearson, R. (1996). *Heredity and humanity: Race, eugenics, and modern science*. Washington, DC: Scott-Townsend.

Pebley, A. R., & Westoff, C. F. (1982). Women's sex preferences in the United States: 1970 to 1975. *Demography, 19*, 177–90.

Perkins, D. O., Davidson, J. D., Leserman, J., Liao, D., & Evans, D. (1993). Personality disorder in patients infected with HIV: A controlled study with implications for clinical care. *American Journal of Psychiatry, 150*, 309–15.

Petrill, S. A., Ploman, R., Berg, S., Johansson, B., Pedersen, N. L., Ahern, F., & McClearn, G. E. (1998). The genetic and environmental relationship between general and specific cognitive abilities in twins aged 8 and over. *Psychological Science, 9*, 183–88.

The pill off prescription? (1974). *Lancet, 2*, 933–34.

Pinker, S. (1997). *How the mind works*. New York: Norton.

Plato. (1956). *The republic*. London: Penguin. (Original work published about 380 B.C.)

Plomin, R., DeFries, J. C., McClearn, G. E., & Rutter, M. (1997). *Behavioral genetics*. New York: Freeman.

Plomin, R., & Petrill, S. A. (1997). Genetics and intelligence: What's new? *Intelligence, 24*, 53–77.

Polaneczky, M., Slap, G., & Forke, C. (1994). Use of Levonorgestrel implants (Norplant) for contraception in adolescent mothers. *New England Journal of Medicine, 331*, 1201–6.

Polloway, E. A., Smith, J. D., Patton, J. R., & Smith, T. E. C. (1996, March). Historic changes in mental retardation and developmental disabilities. *Education and Training in Mental Retardation and Developmental Disabilities*, 3–12.

Popenoe, P., & Johnson, R. H. (1918). *Applied eugenics*. New York: Macmillan.

Posner, R. A. (1989). The ethics and economics of enforcing contracts of surrogate motherhood. *Journal of Contemporary Health Law and Policy, 5*, 21–29.

Post, F. (1994). Creativity and psychopathology. *British Journal of Psychiatry, 165*, 22–34.

Prasher, V. P. (1996). Advances in Down's anomaly. *Current Opinion in Psychiatry, 9*, 312–16.

Preston, S. H., & Campbelll, C. (1993). Differential fertility and the distribution of traits: The case of IQ. *American Journal of Sociology, 98*, 997–1019.

Pritchard, J. C. (1837). *A treatise on insanity and other diseases affecting the mind.* Philadephia: Harwell, Barrington and Harwell.

Quay, H. C. (1987). Intelligence. In H. C. Quay (Ed.), *Handbook of juvenile delinquency*. New York: Wiley.

Rahier, J. M. (1998). Blackness, the racial/spatial order, migrations, and Miss Ecuador 1995–96. *American Anthropologist, 100*, 421–30.

Raine, A. (1993). *The psychopathology of crime*. New York: Academic.

Rani, A. S., Jyothi, A., Reddy, P. P., & Reddy, O. S. (1990). Reproduction in Down's syndrome. *International Journal of Gynaecology and Obstetrics, 31*, 81–86.

Ratzinger, J., & Bovone, A. (1987). *Instruction on respect for human life in its origin and on the dignity of procreation*. Vatican City, Italy: Vatican Polyglot Press.

Rawls, J. (1971). *A theory of justice*. Cambridge, MA: Harvard University Press.

Ree, M. J., & Earles, J. A. (1991). Predicting training success: Not much more than *g*. *Personnel Psychology, 44*, 321–32.

Reed, E. W., & Reed, S. C. (1965). Mental retardation: A family study. Philadelphia: Saunders.

Reed, S. C., & Anderson, V. E. (1973). Effects of changing sexuality on the gene pool. In F. F. La Cruz & G. D. LaVeck (Eds.), *Human sexuality and the mentally retarded*. New York: Brunner-Mazel.

Rees-Mogg, W. (1997, October 20). Promising start for cultural crusaders. *The Times* (London), p. 20.

Reilly, P. R. (1994). Eugenic sterilization in the United States. In A Milunsky & G. J. Annes (Eds.), *Genetics and the law*. New York: Plenum.

Reimann, R., Angleitner, A., & Strelau, J. (1997). Genetic and environmental influences on personality: A study of twins reared together. *Journal of Personality, 65*, 450–75.

Reschly, D. J., & Ward, S. M. (1991). Use of adaptive behavior measures and overrepresentation of black students in programs for students with mild mental retardation. *American Journal on Mental Retardation, 96*, 257–68.

Resnick, M. D., Chambliss, S. A., & Blum, R. W. (1993). Health and risk behaviors of urban adolescent males involved in pregnancy. *Families in Society, 74*, 366–74.

Richards, R. L. (1981). Relationships between creativity and psychopathology: An evaluation and interpretation of the evidence. *Genetic Psychology Monographs, 103*, 261–324.

Ridley, M. (1998, August). Eugenics and liberty. *Prospect*, 44–47.

Riordan, J., Rommens, J. M., Keren, B. S., & Alon, N. (1989). Identification of the cystic fibrosis gene: Cloning and characterization of complementary DNA. *Science, 245*, 1066–1071.

Robbins, C., Kaplan, H. B., & Martin, S. S. (1985). Antecedents of pregnancy among unmarried adolescents. *Journal of Marriage and the Family, 47*, 567–83.

Roberts, D. E. (1996). Race and the new reproduction. *Hastings Law Journal, 47*, 935–49.

Robertson, J. E. (1999). Cruel and unusual punishment in United States prisons: Sexual harassment among male inmates. *American Criminal Law Review, 36*, 1–52.

Robins, L. N. (1966). *Deviant children grown up*. Baltimore, MD: Williams and Wilkins.

Robins, L. N., & Regier, D. A. (1991). *Psychiatric disorders in America*. New York: Free Press.

Robins, L. N., West, P. A., & Herjanic, B. L. (1975). Arrests and delinquency in two generations. *Journal of Child Psychology and Psychiatry, 16*, 125–40.

Rogers, J. L. (1991). Impact of the Minnesota parental notification law on abortion and birth. *American Journal of Public Health, 81*, 294–300.

Rogers, L. (1999, July 4). Disabled children will be a "sin," says scientist. *Sunday Times* (London), p. 15.

Rounsaville, B. J., Weismann, M. M., Kleber, H., & Wilber, C. (1982). Heterogenicity of psychiatric diagnosis in treated opiate addicts. *Archives of General Psychiatry, 39*, 161–66.

Rowe, D. C. (1986). Genetic and environmental components of antisocial personality: A study of 265 twin pairs. *Criminology, 24*, 513–32.

Royal College of Obstetricians and Gynaecologists (1991). *Report of RCOG working party on unplanned pregnancy*. London: Author.

Royal College of Physicians (1989). *Prenatal diagnosis and genetic screening*. London: Author.

Rubenstein, D. S., Thomasma, D. C., Schon, E. A., & Zinaman, M. J. (1995). Germ-line therapy to cure mitrochondrial disease: Protocol and ethics of in vitro ovum nuclear transplantation. *Cambridge Quarterly of Healthcare Ethics*, *4*, 316–39.

Ruch-Ross, H. S., Jones, E. D., & Musick, J. S. (1992). Comparing outcomes in a statewide program for adolescent mothers with outcomes in a national sample. *Family Planning Perspectives*, *24*, 66–71.

Rudred, E., Ferrara, J., & Ziarnik, J. (1980). Living placement and absenteeism in community-based training programs. *American Journal of Mental Deficiency*, *84*, 401–4.

Rushton, J. P. (1995). *Race, evolution and behavior*. New Brunswick, NJ: Transaction.

Rushton, J. P. (1997). Impure genius—psychoticism, intelligence, and creativity. In H. Nyborg (Ed.), *The scientific study of human nature: Tribute to Hans J. Eysenck at eighty*. Oxford: Elsevier.

Russell, B. (1930). *The conquest of happiness*. London: Unwin.

Rutter, M. (1999, June 25). Do genes make a genius? *The Times (London) Higher Education Supplement*, 12.

Rutter, M., & Madge, N. (1976). *Cycles of disadvantage*. London: Heineman.

Saleeby, C. W. (1910). *Parenthood and race culture: An outline of eugenics*. New York: Morton.

Salgado, J. F. (1997). The five factor model in personality and job performance in the European community. *Journal of Applied Psychology*, *82*, 30–43.

Scally, B. (1973). Marriage and mental handicap: Some observations in Northern Ireland. In F. de la Cruz & G. La Veck (Eds.), *Human sexuality and the mentally retarded*. New York: Brunner-Mazel.

Scarr, S. (1984). *Race, social class, and individual differences in IQ*. London: Lawrence Erlbaum.

Scarr, S., & Weinberg, R. A. (1978). The influence of family background on intellectual attainments. *American Sociological Review*, *43*, 674–92.

Schilling, L. H. (1984). Awareness of the existence of post-coital contraception among students. *Journal of American College Health*, *32*, 244–46.

Schorr, A. (1970). Income maintenance and the birth rate. In T. Ford & G. De Jong (Eds.), *Social demography*. Englewood Cliffs, NJ: Prentice Hall.

Schweller, R. (1998). *Deadly imbalances: Tripolarity and Hitler's strategy of world conquest*. New York: Random House.

Scott, J. P., & Fuller, J. L. (1965). *Genetics and the social behavior of dogs*. Chicago: University of Chicago Press.

Scully, G. (1995). *Multiculturalism and economic growth*. Washington, DC: National Center for Policy Analysis.

Seabrook, J. (1994, March 28). Building a better human. *New Yorker*, 109–14.

Sharp, H. C. (1907). Vasectomy as a means of preventing procreation of defectives. *Journal of the American Medical Association, 51,* 1897–1902.

Shaw, M. W. (1984). To be or not to be? That is the question. *American Journal of Human Genetics, 36,* 1–9.

Sheridan, R., Llerena, J., Matkins, S., Debenham, P., Cawood, A., & Bobrow, M. (1989). Fertility in a male with trisomy 21. *Journal of Medical Genetics, 26,* 294–98.

Shockley, W. B. (1972). Dysgenics, geneticity, raceology. *Phi Delta Kappan*, (Suppl.), 297–312.

Short, S. E., & Fengying, Z. (1998). Looking locally at China's one child policy. *Studies in Family Planning, 29,* 373–87.

Sigafoos, J., Elkins, J., Kerr, M., & Attwood, T. A. (1994). A survey of aggressive behavior among a population of persons with intellectual disability in Queensland. *Journal of Intellectual Disability Research, 38,* 369–81.

Silberg, J., Meyer, J., Pickles, A., Simonoff, E., Eaves, L., Hewitt, J., Maes, H., and Rutter, M. (1996). Heterogeneity among juvenile antisocial behaviors: Findings from the Virginia twin study of adolescent behavioral development. In D. K. Ciba (Ed.), *Genetics of criminal and anti-social behavior*. Chichester, UK: Wiley.

Silver, L. M. (1996). *Remaking Eden*. New York: Avon.

Simms, M., & Smith C., (1986). *Teenage mothers and their partners*. London: Her Majesty's Stationery Office.

Simon, J. L. (1974). *The effects of income on fertility*. Chapel Hill: University of North Carolina Press.

Singapore Ministry of Health (1994). Personal communication.

Sinsheimer, R. (1969). The prospect of designed genetic change. *Engineering and Science, 32,* 8–13.

Sipe, C., Grossman, J., & Milliner, J. (1988). *Summer training and education program*. Philadelphia: Public/Private Ventures.

Sivin, I. (1988). International experience with Norplant and Norplant-2 contraceptives. *Studies in Family Planning, 19,* 81–94.

Smith, D. J. (1997). Ethnic origins, crime, and criminal justice in England and Wales. In M. Tonry (Ed.), *Crime, ethnicity and immigration*. Chicago: University of Chicago Press.

Smith, G. M. (1967). Usefulness of peer ratings of personality in educational research. *Educational and Psychological Measurement, 27*, 967–84.

Smith, J. D. (1994). Reflections on mental retardation and eugenics, old and new. *Mental Retardation, 32*, 234–38.

Smith, J. D. (1999). Thoughts on the changing meaning of disability. *Remedial and Special Education, 20*, 131–33.

Solivetti, L. M., & D'Onofrio, P. (1996). Some quantitative considerations on migration, crime, and justice in Italy. In *Migration and Crime, Proceedings of the International Conference on Migration and Crime*. New York: United Nations; ISPAC.

Soloway, R. A. (1990). *Demography and degeneration*. Chapel Hill: University of North Carolina Press.

Spearman, C. (1904). General intelligence, objectively determined and measured. *American Journal of Psychology, 15*, 201–93.

Spencer, H. (1868). *Social statistics*. London: Williams and Norgate.

Spencer, H. (1874). *Study of sociology*. London: Macmillan.

Stafford, R. E. (1961). Sex differences in spatial visualization as evidence of sex-linked inheritance. *Perceptual and Motor Skills, 13*, 428.

Stampfer, M. J. (1988). A prospective study of past uses of oral contraceptive agents and risk of cardiovascular diseases. *New England Journal of Medicine, 319*, 1313.

Stattin, H., & Magnusson, D. (1991). Stability and change in criminal behavior up to age 30. *British Journal of Criminology, 31*, 327–46.

Steinberg, D. L. (1997). A most selective practice: The eugenic logic of IVF. *Women's Studies International Forum, 20*, 33–48.

Steinbock, B. (1996). The concept of coercion and long-term contraception. In C. H. Moskowitz and B. Jennings (Eds.), *Coerced contraception*. Washington, DC: Georgetown University Press.

Sterilized in Alberta. *The Economist* (1996, November 9), p. 38.

Stern, C. (1973). *Principles of human genetics*. San Francisco: Freeman.

Stevens-Simon, C., Kelly, L., & Singer, D. (1996). Absence of negative attitudes towards childbearing among pregnant teenagers: a risk factor of repeat pregnancy. *Archives of Pediatric and Adolescent Medicine, 150*, 1037–43.

Stout, J., & Rivara, F. (1989). Schools and sex education, "Does it work"? *Pediatrics, 83*, 375–79.

Suzuki, D., & Knudtson, P. (1990). *Genetics: The clash between the new genetics and human values*. Cambridge, MA: Harvard University Press.

Taylor, J., & Whitney, G. (1999). U.S. racial profiling in the prevention of crime:

Is there an empirical basis? *Journal of Social, Political, and Economic Studies, 24,* 485–510.

Tellegen, A., Lykken, D. T., Bouchard, T. J., Wilcox, K. J., Segal, N. L., & Rich, S. (1988). Personality similarity in twins reared apart and together. *Journal of Personality and Social Psychology, 54,* 1031–1039.

Te Nijenhuis, J. (1997). *Comparability of test scores for immigrants and majority group members in The Netherlands.* Enschede, Netherlands: Print Partners.

Terman, L. M. (1917a). The intelligence quotient of Francis Galton in childhood. *American Journal of Psychology, 28,* 209–18.

Terman, L. M. (1917b). *The Stanford-Binet intelligence test.* Stanford, CA: Stanford University Press.

Terman, L. M. (1922). Were we born that way? *World's Work, 44,* 660–82.

Terman, L. M. (1925). *Genetic studies of genius: Vol. 1.* Stanford, CA: Stanford University Press.

Terman, L. M., & Oden, M. H. (1959). *The gifted group in mid-life.* Stanford, CA: Stanford University Press.

Testart, J. (1995). The new eugenics and medicalized reproduction. *Cambridge Quarterly of Health Care Ethics, 4,* 304–12.

Thiessen, D. (1990). Hormonal correlates of sexual aggression. In L. Ellis & H. Hoffman (Eds.), *Crime in biological, social, and moral contexts.* New York: Praeger.

Thomas, C. (1997, September 30). Indian barbers help to trim births. *The Times,* (London), p. 10.

Thomas, H. (1981). *An unfinished history of the world.* London: Pan.

Thomas, H., & Krail, R. (1991). Sex differences in speed of mental rotation and the X-limited hypothesis. *Intelligence, 15,* 17–32.

Thompson, D., Clare, I., & Brown, H. (1997). Not such an ordinary relationship. *Disability and Society, 12,* 573–92.

Thompson, W. R. (1954). The inheritance and development of intelligence. *Proceedings of the Association for Research in Nervous and Mental Disease, 33,* 209–31.

Thomson, G. (1946). The trend of national intelligence. *Eugenics Review, 38,* 9–18.

Thornberry, T. P., Smith, C. A., & Howard, G. J. (1997). Risk factors for teenage fatherhood. *Journal of Marriage and the Family, 59,* 505–22.

Thorndike, E. L. (1913). Eugenics: with special reference to intellect and character. *Scientific Monthly, 83,* 130–34.

Thorndike, E. L., & Woodyard, E. (1942). Differences within and between communities in the intelligence of children. *Journal of Educational Psychology, 33,* 641–56.

Thorne, J. D., & Collocott, T. C. (1984). *Chambers biographical dictionary*. Edinburgh: Chambers.

Tobin, J. (1998, February 4). Court: Retarded can be sterilized. *Detroit News*, p. D1.

Tonry, M. (1997). *Ethnicity, crime, and immigration*. Chicago: University of Chicago Press.

Toppen, J. T. (1971). Unemployment: Economic or psychological? *Psychological Reports*, 28, 111–22.

Tournier, P. (1997). Nationality, crime, and criminal justice in France. In M. Tonry (Ed.), *Ethnicity, crime and immigration*. Chicago: University of Chicago Press.

Tracy, P. E., Wolfgang, M. H., & Figlio, R. M. (1990). *Delinquency in two birth cohorts*. New York: Plenum.

Trussell, J. (1988). Teenage pregnancy in the United States. *Family Planning Perspectives*, 20, 262–72.

Trussell, J., & Stewart, F. (1992). The effectiveness of postcoital hormonal contraception. *Family Planning Perspectives*, 24, 262–64.

Tryon, R. C. (1940). Genetic differences in maze learning in rats. In *39th Yearbook of the National Society for the Study of Education*. Bloomington, IN: Public School Publishing.

Tucker, W. H. (1994). *The science and politics of racial research*. Chicago: University of Chicago Press.

Turner, D. (1996). Interview with H. J. Eysenck. *Right Now*, 11, 4–6.

Tyler, P. E. (1993, December 22). China weighs using sterilization and abortion to stop abnormal births. *New York Times*, p. 15.

United States Office of Technology Assessment (1987). *Artificial insemination*. Washington, DC: Author.

United States Supreme Court Reports (1927). *Judgment in the Case of* Buck v. Bell. Washington, DC: Author.

Valle Silva, N. (1988). Updating the cost of not being white in Brazil. In P. M. Fontaine (Ed.), *Race, class, and power in Brazil*. Los Angeles: University of California.

Verkerk, A. J., Piretti, M., & Sutcliffe, I. S. (1991). Identification of a gene (FMR-1) containing a CGG repeat coincident with a breakpoint cluster region exhibiting length variation in fragile X syndrome. *Cell*, 65, 904–14.

Verlinsky, Y., Pergament, E., & Strom, C. (1990). The preimplantation diagnosis of genetic diseases. *Journal of In Vitro Fertilization and Embryo Transfer*, 7, 1–5.

Vines, G. (1995, October 28). Every child a perfect child? *New Scientist*, 14–15.

Vining, D. R. (1982). On the possibility of the re-emergence of a dysgenic trend with respect to intelligence in American fertility differentials. *Intelligence, 6*, 241–64.

Vining, D. R. (1986). Social versus reproductive success: The central theoretical problem of human sociobiology. *Behavioural and Brain Sciences, 9*, 167–216.

Vining, D. R. (1995). On the possibility of the re-emergence of a dysgenic trend: An update. *Personality and Individual Differences, 19*, 259–65.

Visser, F. E., Aldenkamp, A. P., van Huffelen, A. C., Knilman, M., Overweg, J., & van Wijk, J. (1997). Prospective study of the prevalence of Alzheimer-type dementia in institutionalized individuals with Down syndrome. *American Journal on Mental Retardation, 101*, 400–412.

Von Verschuer, O. F. (1957). Uber den genetischen ursprung der begabung. *Akad Wissen Literatus, Jahrb*, 230–46.

Vukov, M., Baba-Milkic, N., Lecic, D., Mijalkovic, S., & Marinkovic, J. (1995). Personality dimensions of opiate addicts. *Acta Psychiatrica Scandinavia, 91*, 103–7.

Wagner, M. M., & Blackorby, J. (1996). Transition from high school to work or college: How special education students fare. *The Future of Children, 6*, 103–20.

Wagner, T., & Hopper, P. (1981). Microinjection of a rabbit beta-globin gene into zygotes and its subsequent expression in adult mice and their offspring. *Proceedings of the National Academy of Sciences, 78*, 6376–78.

Wahlsten, J. (1990). Gene map of mental retardation. *Journal of Mental Deficiency Research, 34*, 11–27.

Wakayama, T., Perry, A. C., Zuccotti, M., Johnson, K. R., & Yanagimachi, R. (1998). Full–term development of mice from enucleated oocytes injected with cumulus cell nuclei. *Nature, 394*, 369–74.

Wallace, A. R. (1890). Human selection. *Popular Science Monthly, 38*, 90–102.

Warnock, M. (1987). Do human cells have rights? *Bioethics, 1*, 8.

Weatherall, D. J. (1991). *The new genetics and clinical practice.* Oxford: Oxford University Press.

Weaver, T. R. (1946). The incidence of maladjustment among mental defectives in military environments. *American Journal of Mental Deficiency, 51*, 238–46.

Webb, A. M. C., Russell, J., & Elstein, M. (1992). Comparison of Yuzpe regimen, Danazol, and Mifepristone (RU 486) in oral postcoital contraception. *British Medical Journal, 305*, 927–30.

Webb, B. (1948). *Our partnership.* London: Methuen.

Webb, S. (1896). *The difficulties of individualism.* London: Fabian Society.

Wehman, P., Kregel, J., & Seyforth, J. (1985). Transition from school to work for individuals with severe handicaps. *Journal of the Association for Persons with Severe Handicaps, 10,* 132–36.

Weinberg, R. A., Scarr, S., & Waldman, I. D. (1992). The Minnesota transracial adoption study: A follow-up of IQ test performance in adolescence. *Intelligence, 16,* 117–35.

Weiss, V. (1992). Major genes for general intelligence. *Personality and Individual Differences, 13,* 1115–34.

Wells, E. S., Hutchings, J., & Gardner, J. S. (1998). Using pharmacies in Washington State to expand access to emergency contraception. *Family Planning Perspectives, 30,* 288–90.

Wells, H. G. (1905). *A modern utopia.* London: Dent.

Wertz, D. C. (1998). Eugenics is alive and well: A survey of genetic professionals around the world. *Science in Context, 11,* 493–510.

West, D., & Farrington, D. P. (1977). *The delinquent way of life.* London: Heineman.

Westman, J. C. (1994). *Licensing parents: Can we prevent child abuse and neglect?* New York: Plenum.

Westman, J. C. (1996). The rationale and feasibility of licensing parents. *Society, 34,* 46–52.

Westoff, C. F. (1988). Contraceptive paths toward the reduction of unintended pregnancy and abortion. *Family Planning Perspectives, 20,* 4–13.

Westoff, C. F., Calot, G., & Foster, A. D. (1983). Teenage fertility in developed nations: 1971–1980. *Family Planning Perspectives, 15,* 105–10.

Weyl, N. (1989). *The geography of American achievement.* Washington, DC: Scott-Townsend.

White, M., & Gribbin, J. (1993). *Einstein.* New York: Simon and Schuster.

Widiger, I. A., & Trull, T. J. (1992). Personality and psychopathology: an application of the five factor model. *Journal of Personality, 60,* 363–93.

Wiggins, J. S., & Pinus, A. L. (1989). Conceptions of personality disorders and dimensions of personality. *Psychological Assessment, 1,* 305–16.

Wille, R., & Beier, K. M. (1989). Castration in Germany. *Annals of Sex Research, 2,* 103–34.

Wilmut, I., Schnieke, A., McWhis, J., Kind, A., & Campbell, K. H. (1997). Viable offspring derived from fetal and adult mammation cells. *Nature, 390,* 27.

Wilson, J. Q., & Herrnstein, R. J. (1985). *Crime and human nature.* New York: Simon and Schuster.

Winston, A. S. (1997). Genocide as a scientific project. *American Psychologist, 52,* 182–83.

Wolf, C. (1995). *Long-term economic and military trends, 1994–2015.* Santa Monica, CA: Rand.

Wolpoff, M., & Caspari, R. (1997). *Race and human evolution.* New York: Simon and Schuster.

Wood, P. G. (1999). To what extent can the law control human cloning? *Medicine, Science and Law, 39,* 5–10.

Woods, F. A. (1913). Heredity and the hall of fame. *Popular Science Monthly, 82,* 445–52.

Woodward, L. J., & Fergusson, D. M. (1999). Early conduct problems and later risk of teenage pregnancy in girls. *Development and Psychopathology, 11,* 127–42.

World Health Organization. (1997, March 11). Press release on cloning, 20.

Yeung, A. S., Lyons, M. J., Waternaux, C. M., Faraune, S. V., & Tsuang, M. T. (1993). The relationship between DSM-III personality disorders and the five factor model of personality. *Comprehensive Psychiatry, 34,* 227–34.

Yule, W., Gold, R. D., & Busch, C. (1982). Long-term predictive validity of the WPPSI: An 11-year follow-up study. *Personality and Individual Differences, 3,* 65–71.

Zabin, L. S. (1992). School-linked reproductive health services: The Johns Hopkins Program. In B. C. Miller, J. J. Card, R. L. Paikoff, & J. L. Peterson (Eds.), *Preventing adolescent pregnancy.* London: Sage.

Zabin, L. S., & Hayward, S. C. (1993). *Adolescent sexual behavior and childbearing.* London: Sage.

Zabin, L. S., Smith, E. A., Hirsch, M. B., & Hardy, J. B. (1986). Ages of physical maturation and first intercourse in black teenage males and females. *Demography, 23,* 595–605.

Zeanah, C. H., Boris, N. W., & Larrien, J. A. (1997). Infant development and developmental risks: A review of the past 10 years. *Journal of the Academy of Child and Adolescent Psychiatry, 36,* 165–78.

Zelnick, M., & Shah, F. K. (1983). First intercourse among young Americans. *Family Planning Perspectives, 15,* 64–70.

Ziebland, S., & Scobie, S. (1995). Could a publicity campaign for emergency contraception reduce the incidence of unwanted pregnancy? *British Journal of Family Planning, 21,* 68–71.

Zuckerman, H. (1977). *Scientific elite: Nobel laureates in the United States.* New York: Simon and Schuster.

Zuckerman, M. (1991). *Psychobiology of personality.* Cambridge: Cambridge University Press.

Zuckerman, M. (1992). What is a basic factor and which factors are basic? Turtles all the way down. *Personality and Individual Differences, 13,* 675–81.

Zuckerman, M. (1994). *Behavioral expressions and biosocial bases of sensation seeking.* Cambridge: Cambridge University Press.

Index

Abortion, 38, 41, 66, 68; eugenic effect, 182ff, 251; Europe, 183; legalization of, 38, 183, 250; United States, 183, 250
Abrahamse, A. E., 192
Accardo, P. J., 105
Achondroplasia, 61, 262
Adoption, 6, 31; effects of, 6
Agreeableness, 111
Alleles, 139
Allowance, child, 21
Alpert's syndrome, 61
Alzheimer's disease, 64, 253
Amaurotic family idiocy, 99
American Association on Mental Retardation, 97, 98
American Breeders' Association, 24
American Civil Liberties Union (ACLU), 200
American Genetics Association, 24, 37; Committee on Eugenics, 24
American National Academy of Sciences, 40
American National Opinion Research Center (NORC), 91
American Psychological Association (APA) 79
American Task Force, 79
Amniocentesis, 248
Anderson, E., 195
Anderson, V. E., 101
Anderson, W. F., 254, 272

Andreason, N. C., 75
Anencephaly, 251
Animal breeding, 9
Anti-Semitism, 29–30
Antisocial personality disorder, 117
Anxiety, 109
Arabs, 90
Artificial insemination by donor (AID), 246, 278–81; elite sperm banks, 260, 278–79; and ethical issues, 259–61
Auletta, Ken, 128
Auschwitz, 28
Authoritarian states, future of eugenics in, 292ff; future power of, 304
Ayer, Sir Alfred, 95

Baird, P. A., 60
Bakewell, Robert, 9, 153
Balfour, Arthur, 293
Baroody A. J., 103
Barrick, M. R., 109–11
Bauer, Gary, 170
Beethoven, Ludwig van, 66, 81
Bell Curve, The (Richard Herrnstein & Charles Murray), 85–87, 275
Benbow, Camilla, 304
Beveridge, Sir William, 23
Beyondism (Raymond Cattell), 40
Billig, Michael, 138
Binet, Alfred, 26, 81

Biotechnology, 243; development in, 245ff; ethical issues, 258ff; future in authoritarian states, 298ff; future in democracies, 278ff
Birch, H. G., 100
Birth control, 23, 27, 32–33, 42, 92
Blackaby, D., 312
Blacker, Carlos, 49, 216
Blacks, 37, 100–101, 193, 196, 211, 238, 265, 277
Bland, R. C., 119
Blank, Robert, 70, 184
Boaz, Franz, 40
Bodmer, Sir Walter, 51–53, 138, 264
Bojorquez, J., 260
Bok, G., 28
Borjas, G. J., 223
Bouchard, Thomas J., 7, 23, 142, 148, 154
Brebner, John, 113
Breeding techniques, 9, 150ff; animals and plants, 150; thoroughbreds, 150
Brennan, P. A., 158, 246
Brewer, M., 279
Brickley, M., 103
Brigham, C. C., 36
Brinchmann, Berit S., 67–68
Britain, acceptance of eugenics in, 19
British Eugenics Society, 21, 37, 49
Broadhurst, Peter, 152
Broberg, G., 27, 276
Brock, J., 248
Brody, Nathan, 80
Broman, S. J., 100
Bromham, D. R., 176
Brown, B. S., 104
Bruce, E. J., 69
Bruiniks, R. H., 104
Brunner, H. G., 158
B-thalassemia, 254
Buck, Carrie, 34, 232
Buck v. Bell, 34, 231
Budiansky, S., 152

Burack, J. A., 102
Burleigh, M., 29
Burt, Sir Cyril, 22–23, 87, 109
Burton, R., 177
Buss, David M., 267

Caesar, Julius, 90
Caldwell, J. H., 246
Campanella, Tommaso, 3
Cancer, 154, 252–54
Cantor, C., 251
Card, J. J., 167
Cardiovascular disorders, 251
Carrell, Alexis, 27, 37, 42
Cartwright, A., 217
Castration, 201
Cattell, J. M., 82
Cattell, Raymond B., 22–23, 39–40, 55, 109–11, 216
Cavalli-Sforza, Luigi, 51–53, 159
Celibacy, 40
Cerebral palsy, 98, 100, 106
Cherry, K. E., 104
Children: allowances for, 21; rearing of, 13; restrictions on numbers of, 14;
China, 9, 14, 16, 41, 43; acceptance of eugenics in, 41, 294–96; future global supremacy, 314; high IQ, 222, 305
Chorion villus sampling, 249
Chornley, M. J., 156
Chromosome disorders, 65, 97–98, 248ff
Churchill, Sir Winston, 23, 293
Civic worthiness, 11
Civitas Solis (Tommaso Campanella), 3
Clark, L. A., 114
Clarke, Colin, 239
Clarke, G. R., 193
Clarke, R., 21
Clearie, A. F., 168
Cleckley, Hervey, 117

Cleft lip, 154
Cleft palate, 154
Clerici, M., 121
Clinton, Bill, 170, 268
Cloning, 254; ethical issues, 267; in
    future authoritarian states, 302;
    future developments, 282
Cohen, Cynthia, 247
Cold Spring Harbor, 24
Coleman, David, 311
Color blindness, 63
Committee on Eugenics, of the
    American Genetics Association,
    24
Compton, W. W., 119
Comptroller General, 107
Concentration camps, 28, 29
Condoms, 32
Connor, J. M., 154
Contraception, 23, 27, 31, 33, 42, 92,
    174; emergency, 176
Costa, Paul T., 109–12, 156
Courtney, M. E., 211
Cox, Catherine, 82, 153
Craig, R. J., 122
Crick, Francis, 20, 22, 42, 304
Crime, 6, 31, 85, 87, 89, 95; reduced
    by abortion, 183
Criminality, 6, 31, 85, 104
Criminals, 34, 39, 42–43, 83, 85,
    104; habitual, 10, 31; and social
    class, 31; sterilization of, 34, 42
Critically ill newborns, 69
Crocker, A. G., 104
Crosier, A., 123
Crow, J. F. D., 152
Cutwright, P., 196
Cycle of deprivation and disadvan-
    tage, 129, 198
Cynkar, R. J., 232
Cystic fibrosis, 62, 99, 248, 253

Daedalus (J. B. S. Haldane), 21
D'Alembert, Jean, 6

Daley, D., 175
Darke, S., 122
Darwin, Charles, 5, 7, 19
Darwin, Leonard, 24
Davenport, Charles, 24, 30, 146
Davenport, K. S., 88
Dawson, D., 171
Deafness, 66
Dearden, K., 124
Deary, Ian J., 193
De Jong, C. A., 122
Depression, 73–75, 77, 154
Depressive psychosis, 72
Descent of Man (Charles Darwin), 19
Devlin, B., 86
Diabetes, 64, 69, 154
Digman, J. M., 111, 113
DiLalla, D. L., 73
Disease, genetic causes of, 30
Dobzhansky, Theodosius, 66
Dolly, 255
Dominant genes, 24, 60, 144
Donnelly, M., 103
Donovan, P., 194
Dostoevsky, Fyodor, 66
Down's syndrome, 65, 70, 98–101,
    248, 250–52
Dryden, John, 74
Dryfoos, J., 170–72, 174
Duchenne's muscular dystrophy, 63,
    252
Dugdale, Richard, 129
Duijsens, I. J., 113–14
Duncan, G., 177, 182
Duster, Troy, 256
Dwarfism, 61, 262
Dykens, E. M., 99–100
Dysgenic fertility, 21, 24, 30, 33, 39,
    42, 48
Dysgenics, 7, 8, 18–24, 26, 30, 40–
    41, 57–58; in developing nations,
    308; East Asia, 312; Europe, 311;
    United States, 309
Dysgenics (R. Lynn), 7, 37, 41, 57, 86

Easterbrooke, A. H., 129
Edison, Thomas, 94
Egg donation, 249, 261
Einstein, Albert, 81, 93, 95, 304
Elite, genetic, 21
Elite sperm banks, 260
Elster, A. B., 123–24
Embryo selection, 252; ethical issues, 263; future developments, 283
Emergency contraception, 176
Emigration, 12, 13
Epilepsy, 64, 66, 154
Erles, J. A., 80
Ermish, J. F., 166, 194
Essen, J., 131
Ethical principles of eugenics, 225ff; 258ff
Eugenic Education Society, 19
Eugenics: as an academic discipline, 10; acceptance of, 18; aims of, 47; associations, 11; in authoritarian states, 292ff; beauty, 52; in Britain, 19; classical eugenics, 135ff; criticisms of, 137–38; decline of, 37; ethical principles, 225ff, 258ff; future developments in democratic societies, 265ff; genetic foundations, 137–49; in Germany, 27; immigration, 35–37, 222; licenses for parenthood, 205ff; military effectiveness, 53, 222, 292; national strength, 53, 292; negative eugenics, 10, 15, 165ff, 187ff; the new eugenics, 243ff; objectives, 45ff; positive eugenics, 10, 215ff; sterilization, 187ff; in the United States, 24; world eugenic state, 307ff
Eugenics: Galton and After (Carlos Blacker), 49
Eugenic Society, 16, 19
Eugenics Quarterly, 37
Eugenics Record Office, 24
Eugenics Review, 37

Eugenic Sterilization Law, 28
Eugenic utopia, 14
Euphenics, 13
Europe, future of, 311
Eutelegenesis, 279
Euthanasia, 28
Evans, G., 103
Evolution, the Modern Synthesis (Julian Huxley), 21
Exomphalos, 253
Extroversion, 110
Eysenck, Hans J., 75, 76, 80, 92, 93, 109, 126, 156, 158, 162
Eysenck, Michael W., 109
Ezzell, C., 253

Falconer, D. S., 163
Falk, S., 225
Family planning clinics, 23, 32–33
Faraday, Michael, 94
Farndale, Nicholas, 9, 151
Farrington, David, 121, 132
Federation for American Immigration Reform, 236
Feingold, W., 246
Feldman, M. A., 105, 106
Fertility, 8, 10, 19–21, 26, 31; dysgenic fertility, 21; and social class, 31
Fetal biopsy, 248
Finland, sterilization in, 34
Fischer, C. S., 86
Fischer, Eugen, 28
Fisher, Sir Ronald A., 8, 20, 42, 148, 216
Fiske, D. W., 111
Flanagan, J. C., 228
Fletcher, J., 71
Flynn, Jim R., 93, 222
Folling, Ivar, 31
Fonda, Jane, 170
Fragile X syndrome, 63, 99
Francoer, R., 246
Frick, P. J., 114

Fustenberg, F. F., 130
Future of Man (Peter Medawar), 21

Galactosemia, 99, 101
Galileo, 76
Gallstones, 64, 154
Galton, C. J., 29
Galton, D. J., 29
Galton, Sir Francis, 4–18, 30, 41–42, 66, 135; his IQ, 153; *Hereditary Genius* (1869), 4–9, 12, 19, 48; *Inquiries into Human Faculty* (1883), 4; *Memories* (1908) 8, 17; objectives of eugenics, 47–49; positive eugenics, 215; ultimate triumph of eugenics, 320
Galton Institute, 37
Gauthier, A. H., 217
Gay, P., 29
Gene processes, 137ff; selection, 150ff
Genes: additive, 137ff; dominant, 24, 144; recessive, 24, 146; X-linked, 147
Gene therapy, 254; ethical issues, 270; germ-line and somatic-line, 270–73
Genetical Theory of Natural Selection, The (Ronald Fisher), 20
Genetic deterioration, 19, 22
Genetic diseases and disorders, 43, 59–71; future treatment, 280
Genetic elite, 21
Genetic engineering, 25, 253; future developments, 283
"Geneticists' Manifesto, The" (Hermann Muller), 25, 48, 49
Genetic mutations, 24
Genetics, Mendelian, 20, 137–49
Genius, 83, 87, 95
German measles, 98
Germany, 28–30, 33, 73; eugenic programs, 216; Eugenics Society, 28

Giami, A., 182
Gillon, R., 225
Glaister, A., 178
Glass, David V., 217
Glasse, A., 197
Goddard, H. H., 139
Goerree, M. A., 73
Gold, M. A., 177
Goldman, D., 158
Gooder, P., 177
Goodson, P., 123
Gordon, Robert A., 84
Gordon, S., 239
Gottesman, J. J., 158
Gottfredson, Linda S., 79
Gould, Stephen J., 36, 40–41, 232
Graham, Robert, 39, 260
Greece, ancient, 20
Gribbin, John, 95, 99
Griffiths, M., 182
Gudjonsson, Gisli, 84, 158
Guilford, J. P., 111

Haas-Wilson, D., 184
Habitual criminals, 10, 12, 115
Hafner, H., 73
Haldane, John B. S., 20–21, 25, 49
Hall, R., 33
Hammer, R. E., 254
Hammerton, J. L., 250
Hanauske-Abel, H. M., 28
Hansen, H., 202
Happiness, 54, 91, 113
Hare, Robert, 114, 117, 119–20
Harland, Sydney C., 49
Harlap, S., 123
Harpur, T. J., 114, 119
Harris, John, 66, 213
Harris, P., 104
Hayek, Friedrich, 230, 259
Hayfron, J. E., 223
Hayman, R. L., 239
Health, 4–5, 30, 41–42, 49–50
Heart disease, 64, 154

Heaven, Patrick C., 112
Hemophilia, 63, 248
Hendriks, A. A., 113
Henshaw, S. K., 184
Hereditary Genius (Francis Galton), 4,
    5, 6, 7, 8, 12, 19, 48
Hereditary Health Courts, 28
Heredity, 22, 30, 42
*Heredity and Humanity: Race, Eugen-
    ics and Modern Science* (Roger
    Pearson), 40
Heredity in Relation to Eugenics
    (Charles Davenport), 24
Heritability: intelligence, 155;
    multifactorial diseases, 153;
    personality, 156; polygenetic
    disorders, 153; psychopathic
    personality, 157
Herrman, L., 31
Herrnstein, Richard J., 32, 36, 80,
    85–87, 120, 156, 162, 222–23,
    235, 238, 275
Hispanics, 37
Hitler, Adolf, 29–30, 54, 90, 126,
    239, 314
HIV, 38
Hobbes, Thomas, 230
Hoem, J. M., 218
Hogben, Lancelot, 31, 49
Holmes, Oliver Wendell, 232
Holocaust, 30, 239
Hotz, R. L., 39
Hudson, G., 182
Huguenots, 12
Human biotechnology. *See* Biotech-
    nology
Human Fertilization and Embryology
    Act (Britain), 261
Hume, David, 273
Humphries, Lloyd G., 79
Huntington's disease (chorea), 30,
    60–61, 146, 248
Hutchins, B., 6
Huxley, Aldous, 303
Huxley, Sir Julian, 20, 21, 25, 49;

Evolution, the Modern Synthesis
    (1942), 21; *Man Stands Alone*
    (1941), 21
Hydrocephalus, 251
Hypertelorism, 99
Hypertension, 64, 154
Hyperthyroidism, 154
Hypomania, 73

Immigration, 12–13, 35–37, 42;
    Chinese and Japanese, 222; costs,
    237; dysgenic, 57, 307ff; ethical
    issues, 234; Jews, 222; in United
    States, 309
Immigration Act of 1924, 235
Immigration Act of 1965, 37
Inge, William, 23–24
Inheritance, 6
Inquiries into Human Faculty, (Francis
    Galton), 4
Insane, sterilization of, 34
Institute for the Study of Man, 40
Intelligence, 78–96, 98; deterioration
    of, 23; East Asian nations, 222,
    305; and ethnic background, 36;
    future eugenic developments, 282;
    genes for, 26, 31, 42; and happi-
    ness, 91; intergenerational trans-
    mission, 140ff; low, 83–85; and
    social class, 31
Introversion, 110
Is America Safe for Democracy?
    (William McDougall), 26
Isen, A. M., 75
IUD, 178

Jackson, C., 193
Jacobsberg, L., 122
Jahoda, Marie, 40
James, M. E., 122
Jamison, K. R., 72, 74, 75
Jang, K. L., 157
Japan, 27, 34, 90; sterilization laws
    in, 34; unplanned births, 168
Jencks, Christopher, 80

Jensen, Arthur R., 32, 37, 79, 80, 146, 156
Jews, 13, 29, 222; Nazi killing of, 29, 239
Jinks, J. L., 156, 159
Job proficiency, 109–13
Johnson, A. M., 166
Johnson, K. C., 60
Jones, E. F., 171, 178
Jones, P., 73
Jones, Steve, 137
Joseph, Sir Keith, 294
Journal of Biological Science, 37
Joyce, T., 184, 217
Juda, A., 74
Judge, T. A., 113
"Just Say No" campaigns, 169–70

Kaiser Wilhelm Institute for Anthropology, Human Heredity, and Eugenics, 28
Kaku, M., 254
Kallikak family, 129
Kamin, Leon J., 22, 36, 40, 155
"Kantsaywhere," 13–16
Karlsson, J. L., 74, 75
Kaye, H. L., 239
Kazue, S., 217
Keens, Michael, 9
Keens' Imperial, 9
Keltner, B. R, 106
Kelvin, R. P., 109
Kemp, T., 197
Kennedy, John F., 235
Kessler, R. C., 119
Ketterlinus, R. D., 124
Kevles, D. J., 29, 36, 40, 197, 232, 264
Keynes, Maynard, 2
Keynes, Milo, 137
Kincheloe, J., 86
King, T. J., 255
Kirby, D., 171, 173
Kirk, D., 308
Kleinman, P. H., 122
Klinefelter's syndrome, 65

Koenig, R., 138
Koestler, Arthur, 296
Kolata, G., 268
Kopp, Marie, 28, 216
Koshland, Daniel, 272
Kosten, T. R., 122
Kraepelin, Emile, 117

LaFollette, Hugh, 205–7
Landes, David, 305
Lange, Johannes, 31
Langlois, B., 152
LaPorta, R., 237
Lasker, G. W., 67
Lau, S., 266
Laughlin, Harry, 24, 38
Lawrence, E. M., 31
Lech-Nyhan syndrome, 63
Lederberg, Joshua, 24, 25, 42
Lee, N. C., 179
Lee Kuan Yew, 219
Lehtinen, V., 119
Leibowitz, A., 103, 166
Lenz, Fritz, 28
Leopard's syndrome, 61
Le Pine, J. A., 113
Leprosy, 154
Levin, Michael, 7
Levitt, S. D., 124, 183
Lewontin, Richard, 40
L'Homme, cet inconnu (Alexis Carrell), 27
Licenses for parenthood, 14, 205ff
Lichter, D. T., 184
Lieberson, S., 36
Life, Letters, and Labours of Francis Galton (Karl Pearson), 14
Lindquist, P., 73
Lippman, Abby, 256
Loehlin, John, 57, 157, 163, 309
Loose, C., 181–82
Lowe's syndrome, 64
Ludmerer, Kenneth, 34
Ludwig, A. M., 74, 75
Lutton, Wayne, 237

Lykken, David, 84, 117, 121, 158, 209ff, 297
Lynn, Angharad, 91
Lynn, Richard, 7, 32, 37, 57, 83, 88, 91, 93, 222, 293, 304–5, 309, 312
Lyons, M. J., 57

MacDonald, K., 36, 109, 112, 222, 239
MacKenzie, D., 151
MacKinnon, D. W., 125
Mackintosh, Nicholas J., 79, 80, 156, 238, 312
Maller, J. B., 87
Man and Superman (George Bernard Shaw), 23
Manic depression, 74, 75, 154
Manic depressive psychosis, 64, 72, 73, 77
Mann, D. M. A., 65
Man Stands Alone (Sir Julian Huxley), 21
Marfan's syndrome, 61
Marsiglio, W., 123, 171, 195
Martel, Charles, 90
Martens, P. L., 312
Mason, D. A., 157
Masters, M. W., 310
Maternal serum screening, 248
Matthews, G., 193
McCabe, M. P. 106
McCormick, R. A., 267
McDougall, William, 25, 26
McKenzie, J., 110
McKie, Robert, 52, 264
McLaren, J., 100
Measured Lies (J. Kincheloe, S. Steinberg & A. Gresson), 86
Medawar, Sir Peter, 20, 21, 22, 42
Mednick, Sarnoff, 121, 158
Meezan, W., 211
Meier, K. J., 184
Mein Kampf (Adolf Hitler), 29
Memories (Sir Francis Galton), 8, 17

Mendel, Gregor, 30, 138
Mendelian genetics, 20, 97, 99, 138ff
Mengele, Joseph, 2
Mental Deficiency Act, 35
Mental illness, 72–77
Mental retardation, 10, 34, 39, 42–43, 78, 97–107; children, 106; fertility, 106; heritability, 154; sterilization, 34, 42
Mercola, K. E., 254
Meredith, P., 182
Merrick, J. C, 70
Mickelson, P., 105
Microcephaly, 99
Military strength, 89–96, 222
Mill, John Stuart, 82, 230, 259
Miller, P. W., 223
Mini-Mental State Examination, 102
Minnesota Multiphasic Personality Inventory (MMPI), 73
Model, Suzanne, 312
Modern Utopia, A (H. G. Wells), 24
Moffit, T. E., 118–19
Moore, K. A., 166, 196
Moral character, 4–5, 17, 41–42, 48–50
Moral imbecility, 117
Moran, P., 118
Morch, W. T., 106
Morel, Benedict, 57
Morgan, Patricia, 167
"Morning-after pill," 176
Morris, L., 167
Mosher, S. W., 192
Moskowitz, E. H., 175
Mughal, S., 109
Muller, Hermann, 21, 24, 25, 39, 42, 48, 112, 279; "The Geneticists' Manifesto," 25, 48–49; objectives of eugenics, 48
Muller-Hill, B., 28, 197
Multifactorial diseases, 64
Multiple sclerosis, 64, 154
Multiracial societies, 237

Murray, Charles, 32, 36, 57, 80, 85–88, 156, 192ff, 223, 235, 238, 275
Muscular dystrophy, 255
Myrdal, Gunnar, 128

Napoleon, 90
National Foundation for Brain Research, 73
National Life (Karl Pearson), 19
Natural selection, failure of, 19
Nazis, 28, 34
Needham, Joseph, 49
Negative eugenics, 10, 165ff, 187ff; licenses for parenthood, 205ff
Neill, James, 66
Neisser, Ulrich, 79
Nelson, J. R., 118
Nesmith, J. D., 123
Neuroticism, 109
Newman, H. H., 31, 233
Newton, Sir Isaac, 76, 81, 304
New York Times, 279
Nigg, J. T., 157
Nock, S. L., 166, 195
Norplant, 179
Norway, sterilization in, 34, 106

Oakley-Browne, M. A., 119
Objectives of eugenics, 4, 5
Obscenity laws, 32, 33
Ochse, R., 75
Oden, M. H., 161
Ohsfeldt, R. L., 184
O'Kane, A., 118
Openness to experience, 111
Origin of Species (Charles Darwin), 7
Ortmann, J., 202
Osborn, Frederick, 25, 26, 37
Our Ostriches (Marie Stopes), 92
Outspoken Essays (William Inge), 24

Pagliari, Claudia, 93
Parenthood, license for, 14
Parenting, 105, 106

Pascal, Blaise, 82, 153
Patton, J. R., 66, 103
Paul, Diane, 37, 38, 225
Pauling, Linus, 24, 25, 42
Pearson, Karl, 14, 19, 20
Pearson, Roger, 39, 40, 58
Pebley, A. R., 266
Perkins, D. O., 122
Personality, 82, 108–15; future eugenic developments, 282
Personality disorder, 117
Petrill, S. A., 95, 99
Phenylketonuria (PKU), 31, 62, 99
Pinker, Steven, 299
Pioneer Fund, 228
Planned Parenthood Federation of America, 190
Plato, 3, 9, 54, 151
Plomin, R., 72, 73, 99, 141, 144
Polaneczky, M., 175
Polloway, J., 103
Polygenetic inheritance, 148
Poor, undeserving, 128
Popenoe, P., 38
Popper, Karl, 94
Population quality, 13
Pornography, 38
Positive eugenics, 10, 15, 215ff; ethical obligations, 220; financial incentives, 215; immigration, 222; in Singapore, 219
Posner, R. A., 211
Post, F., 51, 66, 74–75, 126
Prasher, V. P., 65
Preface to Eugenics (Frederick Osborn), 26
Pregnant women, 13, 41, 70
Prenatal diagnosis, 248; ethical issues, 261
Prenatal tests, 41
Preston, Samuel, 162
Pritchard, John, 117
Procreation, licensing of, 22, 203ff
Psoriasis, 154

Psychiatric illnesses, 73
Psychopathic personality, 10, 113,
    116–33; and AIDS, 122; costs,
    124; creativity, 125; crime, 119;
    definition, 117–18; drug abuse,
    121; intelligence, 127; prevalence,
    118; teenage fathers, 123; teenage
    mothers, 123; underclass, 128;
    unemployment, 121
Pyloric stenosis, 69, 154

Race, 4, 40, 277; and crime, 310; and
    IQ, 308
Race hygiene, 28
Raine, Adrian, 84, 121, 158, 202,
    231
Rani, A. S., 65
Ratzinger, J., 259
Rawls, John, 276
Recessive genes, 24, 146; disorders,
    62
Ree, Malcolm J., 80
Reed, E. W., 101, 161
Regression to the mean, 160ff
Reproduction, selective, 16, 30
Republic, The (Plato), 3, 9
Resnick, M. D., 124
Retinoblastoma, 69
Richards, R. L., 7
Ridley, Matt, 230
Rights, social and individual, 38, 41,
    43, 48, 276
Robbins, C., 124
Roberts, Dorothy, 265
Robertson, J. E., 122
Robins, L. N., 130
Roe v. Wade, 183
Rogers, J. H., 184
Rogers, Lois, 265
Roman Catholic Church, 40, 259,
    262
Rome, ancient, 20
Roosevelt, Theodore, 27, 293
Rose, Stephen, 40

Rounsaville, B. J., 122
Rowe, David C., 117
RU 486 (Mifepistone), 176
Rubenstein, D. S., 272
Ruch-Ross, H. S., 167
Rudred, E., 103
Rush, Benjamin, 117
Rushton, J. Philippe, 125–27, 234,
    310, 318
Russell, Bertrand, 23
Rutter, Sir Michael, 81, 130, 141,
    143

Saleeby, Caleb, 19
Salgado, J. F., 113
Sanger, Margaret, 33
Sarcoidosis, 154
Scally, B., 106
Scarr, Sandra, 6, 161, 162
Schilling, L. H., 177
Schizophrenia, 64, 72–73, 154
School-based clinics, 171
Schorr, A., 217
Schumann, Robert, 142
Schuster, Edgar, 11
Schweller, R., 314
Scotland, decline of intelligence in,
    100–101
Scully, G., 237
Seabrook, J., 295
Selection, 7–8, 16–20, 41, 48
Selective breeding, 8–10; bees, 151;
    corn, 152; dogs, 151; effectiveness,
    159; horses, 151; rats, 152; roses,
    151; sheep, 151; strawberries, 165;
    thoroughbred racehorses, 152
Selective reproduction, 16, 30
Semen banks, 278
Senile dementia, 154
Sex education, 170
Sex selection, 252, 265
Shakespeare, William, 81
Sharp, H. C., 197
Shaw, George Bernard, 23

Shockley, William, 39; payments for
    sterilization, 188, 228
Short, S. E., 315
Sickle cell anemia, 25, 62, 248, 253,
    270
Silberg, J., 157
Silver, Lee M., 229, 235, 260
Simms, M., 184
Simon, J. L., 191
Singapore, 40, 43, 219, 286
Single mothers, 85, 87, 95, 162
Sipe, C., 171
Sivin, I., 179
Sjorgen-Larssen syndrome, 62
Smith, G. M., 113
Smith, J. D, 66, 275
Social Biology, 37
Social class, and intelligence, 31
Society for the Study of Social
    Biology, 37
Sociobiology, 26
Sociopathic personality, 117
Socrates, 9
Solivetti, L. M., 224
Soloway, Richard, 33, 66, 138
Soto's syndrome, 61
Soviet Union, 16, 18
Spain, dysgenic processes and
    decline, 8, 48
Spearman, Sir Charles, 22, 79
Spearman's g, 48, 79, 95
Spencer, Herbert, 7, 19, 230
Sperm bank, 21, 25, 39; elite, 25, 39
Spina bifida, 64, 154, 248, 251
Steinbock, B., 190, 199
Steiner, George, 66
Sterilization, 27–28, 34, 37, 41–42,
    181, 187ff; of criminals, 199;
    effectiveness, 143, 146; ethics of,
    229; in Europe, 34, 197; financial
    incentives for, 188; in India, 191;
    in Japan, 34; laws, 12, 24, 28, 196;
    of the mentally retarded, 27, 37,
    42, 196; in Peru, 191; in

Scandinavia, 34, 276; Shockley
    plan, 188; in the United States,
    34, 196
Stern, C., 147
Stevens-Simon, C., 123
Stopes, Marie, 23, 33, 92
Stout, J., 171
Strength, national, 18
Suzuki, David, 270–72
Sweden, 28, 34, 73, 98–99; steriliza-
    tion in, 34
Switzerland, sterilization in, 34

Talipes equinovarus, 154
Taylor, Jared, 310
Tay-Sachs disease, 62, 248, 251
Teenage mothers, 130–31, 166
Tellegen, A., 7
Te Nijenhuis, J., 314
Terman, Lewis, 25, 26; studies on
    intelligent California children, 26,
    153, 161, 233
Termination of pregnancy, 38, 41, 66,
    70, 71
Testart, Jacques, 138, 229, 264
Thiessen, D., 202
Thomas, C., 176
Thomas, Hoben, 148
Thomas, Hugh, 90, 151
Thompson, D., 104, 105
Thompson, W. R., 152
Thomson, Sir Godfrey, 22–23
Thornberry, T. P., 124
Thorndike, E. L., 88
Thorndike, Edward, 25, 26
Thorne, J. D., 93
Tobin, J., 198
Tonry, Michael, 224, 238
Toulouse-Lautrec, Henri, 66
Tourette's syndrome, 61
Tournier, P., 312
Tours, battle of, 90
Tracy, P. E., 120–21
Transgenic animals, 253

Trussell, J., 171
Tryon, R. C., 152
Tuberculosis, 154
Tucker, W. H, 29
Turner, D., 92
Turner's syndrome, 65
Twin studies, 6, 7, 22–23
Tyler, P. E., 295

Ultrasound scan, 248
Underclass, 11, 128, 207; future
    genetic, 289
Unemployment, 83, 85, 87, 89, 95,
    103
United States: eugenics movement,
    24; future dysgenic forces, 309;
    single teenage mothers, 168;
    sterilization laws, 34
Unmarried mothers, 166ff

Veiss, Volkmar, 140
Verkerk, A. J., 99
Verlinsky, Y., 254
Vernon, Phillip A., 156
Vining, Daniel R., 15, 106, 166
Visser, F. E., 65
Von Verschuer, Otmar, 28, 142
Vukov, M., 122

Wadsworth, Michael, 166
Wagner, M. M., 103
Wagner, Richard, 76
Wagner, T., 253
Wahlsten, J., 99
Wakayama, T., 255
Wallace, Alfred R., 19
Warnock, Mary, 267
Wasserman, Jacob, 29
Watson, James, 22, 275, 304
Weatherall, D. J., 250, 254
Weaver, T. R., 103
Webb, A. M. C., 176
Webb, Beatrice, 23
Webb, Sydney, 19

Wehman, P., 103
Weinberg, Richard A., 6, 233
Weiss, Volkmar, 92, 93
Welfare dependency, 85, 87, 95
Welfare fathers, 195
Welfare mothers, 189ff
Wellington, Duke of, 90
Wells, E. S., 177
Wells, H. G., 24, 132
Wertz, D. C., 256, 294, 313
West, D., 132
Westman, John, 124–25, 207ff, 221
Westoff, C. F., 168–69, 178
Weyl, Nathaniel, 222
White, M., 95
Whitney, Glayde, 310
Widiger, I. A., 109, 113
Wiggins, J. S., 113
Wille, R., 202
Wilmut, Ian, 255, 268
Wilson, Charles, 27
Wilson, Edward, 26
Winston, A. S, 29
Wittgenstein, Ludwig, 93
Wolpoff, Milford, 234
Wood, P. G., 279
Woods, F. A., 142
Woodward, L. J., 162

X-linked disorders, 63
X rays, effects on genetic mutations,
    24
XYY syndrome, 65

Yerkes-Dodson law, 109
Yeung, A. S., 114
Yule, William, 81

Zabin, L. S., 170, 172, 174, 196
Zelnik, M., 196
Ziebland, S., 182
Zuckerman, H., 222
Zuckerman, Marvin, 109, 112, 125,
    127, 156

About the Author

RICHARD LYNN is Professor Emeritus of Psychology, University of Ulster, Northern Ireland. Professor Lynn has held positions at the University of Exeter and The Dublin Economic and Social Research Institute. Among his earlier publications are *Educational Achievement in Japan* and *Dysgenics: Genetic Deterioration in Modern Populations* (Praeger, 1996).